Springer Oceanography

The Springer Oceanography series seeks to publish a broad portfolio of scientific books, aiming at researchers, students, and everyone interested in marine sciences. The series includes peer-reviewed monographs, edited volumes, textbooks, and conference proceedings. It covers the entire area of oceanography including, but not limited to, Coastal Sciences, Biological/Chemical/Geological/Physical Oceanography, Paleoceanography, and related subjects.

Emil Stanev

Trajectories in Oceanography

Springer

Emil Stanev
Institute of Coastal Systems—Analysis
and Modeling
Helmholtz-Zentrum Hereon
Geesthacht, Schleswig-Holstein, Germany

ISSN 2365-7677 ISSN 2365-7685 (electronic)
Springer Oceanography
ISBN 978-3-031-33719-2 ISBN 978-3-031-33720-8 (eBook)
https://doi.org/10.1007/978-3-031-33720-8

This Springer imprint is published by the registered company Springer Nature Switzerland AG
The registered company address is: Gewerbestrasse 11, 6330 Cham, Switzerland

To all from whom I have learned and with whom I have worked, with gratitude.

Preface

Dear reader,
In this book, I will tell you about the atmospheric sciences, in which I graduated, and the ocean sciences, with which I have been predominantly concerned. In both cases, physical processes are my main interest. The challenge is to give an assessment of the dynamics of the development of these sciences over the last fifty years. Without this, any "snapshot" of today's state of affairs would be incomplete or subjective.

Albert Einstein said, "You do not really understand something unless you can *explain* it to your *grandmother*". I want to present in an "easily digestible form" some scientific problems I have worked on without much using the language scientists speak. Hence, I will present the material so that its read does not need special knowledge to understand it. I want to explain not only what we know, but also how we know it. In other words, I will not only be giving a collection of facts, but also I will explain most of them. I want to answer the questions how to learn from nature and how to learn about the ocean. Unlike many other meteorological and oceanographic books, in which the scientific issues are structured into several scientifically distinct sections, the structure of this book roughly follows my first encounter with individual scientific issues. In this sense, the narrative is streamlined by my trajectory in science. Along this trajectory, I will present short essays on different scientific issues. More weight will be given to issues I have worked on. Therefore, the book cannot be regarded as a balanced presentation of individual oceanographic subjects. I hope that the diversity of my research activities allows for a wide range of oceanographic issues to be covered. My choice of examples affects, of course, the references to studies used to support the individual stories.

The personal choice of examples is something that has to do not only with my personal preference, but also with the specificity of ocean sciences in general. There are books about the Global Ocean, which is often named "Blue Ocean", regional seas known as the "Green Ocean", and the coastal ocean, sometimes referred to as the "Brown Ocean". The basic issues of the regional and coastal oceanography are illustrated with the examples of European seas and several European coastal and estuarine environments. Not only similarities and differences between physical oceanography of these basins will be discussed, but also the interactions between physical and biogeochemical processes.

The relationship between discoveries and the specific individuals behind them is often invisible. In some textbooks, you do not find the name of the person who created or contributed to the creation of a theory or a scientific field. In scientific articles, the names are there, but the touch of the person is missing. Not to mention that the public does not read scientific articles. You might wonder whether you remember who discovered the laser, without which today's world would be very different. Moreover, what about smaller discoveries? Many people not looking for such connections, I attribute to the fact that today we increasingly live only with the present, abstracting ourselves from the past. I try to show my students a picture of the development of science and not only describe its present state. Hence, I will sketch some portraits of my colleagues.

Science provides a particular case of continuity. Old theory is refuted, updated, or developed further; new theories incorporate successful earlier theories as approximations.[1] Sometimes I wonder how much we care about yesterday's state of science. If we are blind to what was, would we know where we are now and where we want to get to tomorrow?[2] There is a fundamental and unresolved problem here. As information increases, the ability to grasp "everything important" decreases. Is it important today to have read the writings of Galileo and Kepler? Isaac Newton (1642–1727) wrote in 1675, "If I have seen further it is by standing on the shoulders of Giants". In his days, giants were few in number and "tall in stature". Are there "giants" today? Perhaps, they are not as "tall in stature" and are difficult to identify. What I cannot reconcile myself with is that many materials look like rediscoveries often repackaged in a new way, sometimes without reference to the original sources.[3] Partially, this is because many similar scientific issues in Earth sciences can be presented as new discoveries for the individual regional or coastal cases. Therefore, sometimes it is hard to find the principal novelty.

I hope that my narrative, which follows the development of oceanography in recent decades, will help you understand the relationship of ocean sciences to other sciences and other areas of human activity. I want to give a more complex view of the world in its totality, various interconnections and relationships, and interdisciplinary thinking.

Again, using the allegory with trajectories, one can use different trajectories to describe the development of science. The more trajectories are used, the more representative the "ensemble" is. Trajectories could represent the development of science for one regional basin, the development of the understanding of physical processes

[1] Newton's law of gravity can be regarded as an approximation of the theory of Einstein, which is valid for objects moving with velocities much less than that of light.

[2] Steven Weinberg (1933–2021) was an American theoretical physicist and Nobel Laureate in physics (1979) for his contributions to the unification of the weak force and electromagnetic interaction between elementary particles. He is one of the founders of the Standard Model of particle physics. In his 2015 book *To Explain the World: The Discovery of Modern Science*, he says that to know the history is not less important than to know the present state of science.

[3] The credit for scientific discovery normally goes to the first to publish.

over certain scales, or the development of technology, or even personal scientific development. I try to intertwine several such trajectories in my story to give you a complete view of two important developments: firstly, the continuity from large-scale to small-scale process (such as global-regional-coastal) and secondly, the evolution of basic science to operational oceanography and the provision of marine services.

Here, I would like to recall the compelling 1993 article of the American scientist Carl Wunsch in the Policy Forum of the journal Science. He says: "Looking back, it is clear that most of the advances have taken place in the context of pure basic science driven by curiosity". Perhaps, this period can also be called the romantic period of oceanography, as this is usually associated with the flavor and emotion of the work at sea and the excitement of encounters with unknown processes and their theoretical explanation. However, there comes a time when oceanographers, according to C. Wunsch, reached the point where they could "...describe the real and potential application of their work to a multitude of societally important problems. An abbreviated listing of those applications would include weather forecasting; climate change; sea-level rise and related shorefront processes including water supply, fisheries management, and pollution trajectories; and the ever present military issues". Clearly, oceanographic knowledge finds wide application to a host of problems of serious public interest, in particular at the present time, when society depends on the ocean more than ever. Now, in line with the perception of the ocean as the driver of economic and social well-being, new trends are emerging from individual-oriented science, or "small science", to research and development which align with national and international priorities. Most of my time in science has been spent in the period of transformation of oceanography toward a new stage dominated by the fusion between curiosity-driven ocean research and active actions supporting socio-economic development. In my narrative, I will describe some important steps of this conversion, mostly at European level, that preceded the UN Decade of Ocean Science for Sustainable Development, one major step toward a global consolidation of efforts to protect the oceans and the environment.

Geesthacht, Germany Emil Stanev

About This Book

This book presents the author's personal view of the development of oceanography over the last 50 years. It describes the evolution of scientific practice alongside the technological development of measurements and research relying on computer modeling. Observations from satellites and in situ platforms such as autonomous robots are presented, along with applications of artificial intelligence in oceanography. The author explains not only what we know, but also how we know it. Therefore, he not only presents facts but also explains most of them. The presentation of the material does not exceed the level of complexity of undergraduate education in science, i.e., the reader does not need much special knowledge to understand it. One emphasis is on transforming basic research into applied science and services. This reflects the growing public awareness of ocean issues and the response of different actors and their commitment to addressing sustainable development and the various problems associated with global climate change and sea level rise. Through personal recollections, the author relates the current state of science to past developments. One of the aims of this book is to highlight the consolidation of European and regional scientific communities and their cooperation in the scientific development of operational oceanography, especially at regional and coastal scales. The book could be of interest to scientists and students in various fields of oceanography, to practitioners in the field, and to the general readership interested in environmental sciences.

I am grateful to the University of Sofia in Bulgaria, the University of Oldenburg, and the Hereon Research Center in Germany, who supported my research. Special thanks go to my daughter, Emilia Staneva, for her help with language editing and many helpful comments on the text.

Contents

Acronyms

ADCP	Acoustic Doppler Current Profiler
AISBL	Association Internationale Sans But Lucratif
ASSEMBLAGE	ASSEssMent of the BLAck Sea sedimentary system since the last Glacial Extreme
AUV	Autonomous Underwater Vehicle
AWI	Alfred-Wegener-Institut
BAW	Bundesanstalt für Wasserbau
BGC	BioGeoChemical
BMBF	Bundesministerium für Bildung und Forschung
BONUS	Baltic Organisations Network for Funding Science
BOOS	Baltic Operational Oceanographic System
BSEC	Black Sea Economic Cooperation
BSH	Bundesamts für Seeschifffahrt und Hydrographie
BSHcmod	The operational circulation model of BSH
CERN	Conseil Européen pour la Recherche Nucléaire
CFC	Chlorofluorocarbon
CMEMS	Copernicus Marine Environment Monitoring Service
COD	Coastal Ocean Darkening
CODAR	Coastal Ocean Dynamics Applications Radar
CoMSBlack	Cooperative Marine Science Program for the Black Sea
CoSiNE	Carbon, Silicon and Nitrogen Ecosystem model
COSS-TT	Coastal Ocean and Shelf Seas Task Team
COSYNA	Coastal Observing System for Northern and Arctic Seas
DEM	Digital Elevation Map
DFG	Deutsche Forschungsgemeinschaft
DMI	Danish Meteorological Institute

E-AIMS	Euro-Argo Improvements for the GMES Marine Service
ECOOP	European COastal-shelf sea OPerational observing and forecasting system
EEA	European Environment Agency
EGS	European Geophysical Society
EGU	European Geosciences Union
EMB	European Marine Board
EMSA	European Maritime Safety Agency
ENEA	Italian National Agency for New Technologies, Energy and Sustainable Economic Development
ENIAC	Electronic Numerical Integrator and Computer
Envisat	Environmental satellite
EOOS	European ocean observing system
EOV	Essential Ocean Variables
ERIC	European Research Infrastructure Consortium
EROS	European River Ocean System
ERSEM	European Regional Seas Ecosystem Model
ESA	European Space Agency
ESFRI	European Strategy Forum on Research Infrastructures
ESKP	Earth System Knowledge Platform
EUG	European Union of Geosciences
EUMETSAT	European Organisation for the Exploitation of Meteorological Satellites
EuroGOOS	European Global Ocean Observing System
FerryBox	Automated measurement systems used to continuously determine physical and BGC parameters in surface waters
FIELD_AC	Fluxes, Interactions and Environment at the Land–Ocean boundary. Downscaling, Assimilation and Coupling
FINO	Forschungsplattform In Nord- und Ostsee
FORTRAN	FORmula TRANslation
GCOS	Global Climate Observing System
GEO	Group on Earth Observations
GEOSS	Global Earth Observing System of Systems
GETM	General Estuarine Transport Model
GFDL	Geophysical Fluid Dynamics Laboratory
GFZ	GeoForschungsZentrum
GHER	GeoHydrodynamics and Environment Research
GKSS	Gesellschaft für Kernenergieverwertung in Schiffbau und Schifffahrt

GMES	Global Monitoring for Environment and Security
GODAE	Global Ocean Data Assimilation Experiment
GOOS	Global Ocean Observing System
GOV	GODAE OceanView
GPS	Global Positioning System
GRACE	Gravity Recovery and Climate Experiment
GTOS	Global Terrestrial Observing System
HBM (HIROMB-BOOS Model)	High-Resolution Model for the Baltic Sea—Baltic Operational Oceanography System
HF	High Frequency
HPA	Hamburg Port Authority
HWK	Hanse-WissenschaftsKolleg
HYCOM	HYbrid Coordinate Ocean Model
HYDROBLACK '91	International observing mission in the Black Sea accomplished with the ships from three Black Sea riparian countries
HYPOX	In situ monitoring of oxygen depletion in hypoxic ecosystems of coastal and open seas, and land-locked water bodies
HZG	Helmholtz-Zentrum Geesthacht
IABO	International Association for Biological Oceanography
IAMAS	International Association of Meteorology and Atmospheric Sciences
IAPSO	International Association for the Physical Sciences of the Oceans
IBI	Iberia-Biscay-Irish
ICBM	Institut für Chemie und Biologie des Meeres
ICBSS	International Centre for Black Sea Studies
IFREMER	Institut français de recherche pour l'exploitation de la mer
IMGA	Istituto per lo studio delle Metodologie Geofisiche Ambientali—CNR
INGV	Istituto nazionale di geofisica e vulcanologia
IOC	Intergovernmental Oceanographic Commission of UNESCO
IPCC	Intergovernmental Panel on Climate Change
IT	Information Technology
IUGG	International Union of Geodesy and Geophysics
JERICO	Toward a Joint European Research Infrastructure network for Coastal Observatories
JONSWAP	Joint North Sea Wave Observation Project

JPI	Joint Programming Initiative
KSD	"KS" KüstenSysteme + "D" Daten
MASRI	Infrastructure for Sustainable Development of Marine Research, Including the Participation of Bulgaria in the European Infrastructure Euro-Argo
MERIS	MEdium Resolution Imaging Spectrometer
MERSEA	Marine Environment and Security for the European Area
MFC	Monitoring and Forecasting Centres
MHI	Marine Hydrophysical Institute of the Ukrainian National academy of Sciences
MIT	Massachusetts Institute of Technology
MODE	Mid-Ocean Dynamic Experiment
MOM	Modular Ocean Model
MONGOOS	Mediterranean Operational Network for the Global Ocean Observing System
MyOcean	Development and pre-operational validation of upgraded GMES Marine Core Services and capabilities
MyWave	A pan-European concerted and integrated approach to operational wave modeling and forecasting—a complement to GMES MyOcean services
NASA	National Aeronautics and Space Administration
NCAR	National Center for Atmospheric Research
NEMO	Nucleus for European Modeling of the Ocean
NERSC	Nansen Environmental and Remote Sensing Center
NOAA	National Oceanic and Atmospheric Administration
NOOS	North West Shelf Operational Oceanographic System
NRL	Naval Research Laboratory
NWO	Netherlands Organization for Scientific Research
ONR	Office of Naval Research
OSE	Observing System Experiments
OSSE	Observing System Simulation Experiment
PdE	Puertos del Estado
POEM	Physical Oceanography of the Eastern Mediterranean
PREVIMER	Système pré-opérationnel de prévision côtière à court-terme
PROVOR	Profiling Float

RANS	Reynolds Averaged Navier–Stokes
ROMS	Regional Ocean Modeling System
ROOS	Regional Ocean Observing System
SAC	Science Advisory Committee
SCHISM	Semi-implicit Cross-scale Hydroscience Integrated System Model
SCOR	Scientific Committee on Oceanic Research
SDG	Sustainable Development Goals **743**
SELFE	Semi-implicit Eulerian–Lagrangian Finite-Element model for cross-scale ocean circulation
SESAME	Southern European Seas: Assessing and Modeling Ecosystem changes
SfS	Science for Stability
SIDERI	Strengthening International Dimension of Euro-Argo Research Infrastructure
SIO	Second Institute of Oceanography
SSC	Stennis Space Center
STAC	Scientific and Technical Advisory Committee
STOI	Spatio-Temporal Optimal Interpolation
STREMP	Spatial and temporal resolution limits for regional mass transport and mass distribution in Mediterranean and Black Sea catchments
SWAMP	Sea Wave Modeling Project
TAC	Thematic Assembly Center
TU-BLACK SEA	Ecosystem modeling as a management tool for the Black Sea: a regional program of multi-institutional cooperation
UKMO	UK Met Office
UN	United Nations
US	United States
USA	United States of America
USSR	Union of Soviet Socialist Republics
UV	Ultraviolet
WAM	Wave model
WCRP	World Climate Research Programme
WERA	WavE RAdar
WIMO	Wissenschaftliche Monitoringkonzepte für die Deutsche Bucht
WISE	Waves in Shallow water Environment
WMO	World Meteorological Organization
WSR	Wadden Sea Region
WWW	World Wide Web

In Search of a Destination

In this chapter, I describe my path to science, starting from my first encounters with physics and ending with the moment I took my first steps in physical oceanography. One of the purposes of this chapter is to present some personal motivations for pursuing science. To address this, I need to make a general introduction to science as a driver for achieving personal excellence and freedom. I will evoke the similarities and differences between science and art and the role of curiosity and creativity in these two areas of human activity. Reflections on the earth sciences, meteorology, physical oceanography, and geophysics are given by way of comparison between earth sciences and physics. I recall the statement of the Norwegian scientist V. Bjerknes that meteorology would become an exact science, a true physics of the atmosphere, only after it had developed in a way that would make it possible to predict accurately the future states of the atmosphere, knowing its present state. This is one of the main streamlines of the book, describing the progress through using observations and numerical models to provide reliable forecasts of the atmosphere and oceans. A significant part of this chapter is devoted to fluid mechanics and numerical modeling as a powerful approach for predicting geophysical fluid motions. Along this path, I briefly describe my first experience with numerical modeling of the Black Sea.

1 Early Challenges

I was born in 1950 in Sofia, Bulgaria. Around the same time, my parents, Vassil and Dafina, had completed their university studies in engineering and French philology, respectively.[1] It didn't take long for me to find my vocation; it happened relatively early, in middle school. Mrs. Vassileva, the physics teacher, was one of the "guilty parties" to my career orientation. I was particularly attracted by the experiments

[1] The first years of my life and the story of my family are described in my book "Sea of People and Numbers" (Stanev 2022).

E. Stanev, *Trajectories in Oceanography*, Springer Oceanography,
https://doi.org/10.1007/978-3-031-33720-8_1

she did with Heron of Alexandria's steam engine, for instance. She demonstrated with enviable skill and conviction the conversion of thermal or electrical energy into mechanical energy and the emergence of forces and motion. Although her experiments were not as spectacular as the illusionistic performances of the many magicians of the circus or television, there was something more attractive to them. Knowing and explaining the peculiarities of natural processes was, for me, more exciting than illusion, in other words, to discover was, much more than to deceive.

The second "culprit" in my early orientation turned out to be my mother. She was working in the foreign trade system, involved in the import of products, which were used in nuclear engineering, radioisotopes, special chemicals, and research instruments. When shipments of radioisotopes were arriving at the airport, she sometimes took me with her. They had to be released quickly from customs because of their short decay time and delivered to a hospital, or to the Physical Institutes of the Bulgarian Academy of Sciences (BAS). This was my first encounter with the world of physicists. The impressive apparatus and the sometimes-strange connections and cables between the various instruments and measuring systems, which often seemed confused, provoked my curiosity. I felt that these people were different from those I knew, and something drew me to them.

My First Laboratory Works

The electric arc was very impressive. To build it, graphite electrodes had to be found, and this was not easy. The fortunate circumstance was that such electrodes were discarded in the garbage at the printing house next door to our home. At that time, the safety of experiments was not respected; we used neither goggles nor gloves. I had to make stands (holders) for the electrodes and provide a proper electrical supply. A high current density electrical discharge is created between the electrodes, the current through a non-conductive medium such as air produces a plasma, which is accompanied by light emission, the temperature reaches thousands of degrees, and you hear a slight hiss.

No less exciting is the electrolysis, a process of releasing a metal onto the electrodes when current flows through the electrolyte. For a child, to galvanize (coat with metal) the electrode plate was fascinating. A constant current was used to do this. I remember well my last experiment as a young child; I had run an alternating current by mistake. The vessel I was experimenting in boiled over, and my whole laboratory setup fell apart.

Taking photographs and producing the first pictures was a lot of fun. Before my father brought me a photocopier on one of his business trips from Moscow, I used to do this at the Pioneer Palace. It is unforgettable for a boy to see how, in the semidark room, illuminated with a faint reddish light, the image you have taken gradually emerges on the white photographic paper and becomes clearer with time.

For my early orientation, the opportunities offered by the Pioneer Palace[2] must have been decisive. There were groups on astronomy, electrical engineering, photography, etc. I shall never forget how happy I was when I made my own electric bell and electric motor, or when my first transistor radio began to receive the Radio Sofia program.

Einstein said, "Education is what is left when we forget everything we learned in school". I share this opinion; however, there are unforgettable events from my school years, and they may have been decisive for my later professional path.

2 Science and Art, Curiosity, and Creativity

Years later, I realized my dream of becoming a student at the Faculty of Physics of the University of Sofia "St. Kliment Ohridski". I will skip here this period of my life recalling only one important memory, which I will use as an introduction to the topic of curiosity and creativity, the major drivers of scientists. Our lecturer in philosophy, who taught "Dialectical and Historical Materialism[3]", was Ivanka Apostolova, who dealt with the philosophical problems of physics. Her classes were the only useful of all the ideological ones. I do not know what kind of teaching program she had to follow and to what extent she stuck to it, but I remember that most of her classes were on modern Western philosophical theories, and they were very interesting. Maybe she was supposed to criticize them, but she was rather providing information. Ivanka Apostolova liked to speak about the foundations of existentialism, based on the philosophical theories of Søren Kierkegaard[4] and Friedrich Nietzsche.[5] Søren Kierkegaard studied the question how man uses his personal freedom. He believes that each individual is responsible for finding and pursuing his or her own purpose in life. According to Jean-Paul Sartre,[6] "Man is condemned to be free; because once thrown into the world, he is responsible for everything he does" In addition, "Life has no meaning a priori… It is up to you to give it a meaning, and value is nothing but the meaning that you choose".

[2] The pioneer organization in Bulgaria during the communist regime, in addition to political activities among children, was also engaged in different activities in various scientific and technical fields. In the period 1951–1990, the Sofia Theological Seminary was transformed into the "Pioneer Palace".

[3] In the universities of communist countries, it was compulsory to study ideological disciplines.

[4] Søren Aabye Kierkegaard (1813–1855) was a Danish philosopher, essayist, and religious writer. He is often regarded as a pioneer of existential philosophy.

[5] Friedrich Wilhelm Nietzsche (1844–1900) was a German classical philologist and philosopher.

[6] Jean-Paul Charles Aymard Sartre (1905–1980) was a French novelist, playwright, philosopher, critic of religion, and publicist. He is considered a mastermind and chief exponent of existentialism and a paragon of twentieth-century French intellectuals. For his complete works, the Nobel Committee awarded him in 1964 the Nobel Prize for Literature, which he refused to accept. For this action, he cited his principle of not accepting official honors, since "a writer should not allow himself to be turned into an institution".

Before I met I. Apostolova, I had not thought about the similarities and differences between science and the arts. In one of her seminars, she brought up this topic, giving us some definitions and, to my surprise, arguing that the distance between science and art was not great. The dominating opinion is that people of science are boring and analytical, while people of art are free, high-spirited, emotional, and irrational. In the following, I will try to synthesize some of my knowledge, conclusions, observations, and judgments that I was stimulated to make by discussions with I. Apostolova. These understandings also explain to a certain extent why I am doing science. At many places, I will cite thoughts of great scientists and artists because they express some basic principles about science and art the best.

Art combines creative and technical skills. They are a product of the human individual and engage the esthetic sensibility of the audience. What matters is the individuality contained in artistic images, the personal and emotional view of the world. Unlike art, science is dominated by generalizations; that is, it reflects the world in terms of the general laws that govern it. In his research work, the scientist seeks to know the unknown, guided by his curiosity. Information about natural phenomena is gathered, measurable data are worked with, and what is measured is analyzed and used as a basis for building theoretical models. The models are tested under controlled conditions. Scientists publish the results of their research, driven by the desire to be first.[7] With that, they risk that double-checking their theory can refute it. Without risk, there is neither science nor art.

To put it more clearly, it is inherent in science to self-correct. Incorrect or incomplete scientific concepts do not live long. Consistent tests and comparisons with nature kill such concepts eventually, if they are not correct. Einstein, when he learned that the book *One Hundred Authors Against Einstein* was published, countered: "If I were wrong, then one [author] would have been enough!".[8]

Art is unique. It would hardly come to mind to any gifted artist to repaint the Mona Lisa by correcting what he did not like. On the contrary, science is constantly increasing human knowledge and understanding through new and deeper research. Scientific theories endure not only rejection, but also further development, because science is cumulative. It builds on continued experience and accumulated knowledge.[9] However, at its most fundamental level, science is not undertaken for any practical reason (Weinberg 2015); in this sense, it has something in common with art.

[7] The calculus controversy is perhaps the biggest "priority dispute" in science over who had first invented calculus with major players Isaac Newton and Gottfried Wilhelm Leibniz. Leibniz had published his work first in 1684 (he began working on his variant of calculus in 1674), but Newton had begun working on a "geometric" form of calculus (the method of fluxions and fluents) in 1666. Today, the consensus is that G. W. Leibniz and I. Newton independently invented and described the calculus. More frequently used notations follow the Leibniz formalism.

[8] Citation is from Hawking and Mlodinow (2005).

[9] Steven Weinberg says the following about science: "With all its imperfections, modern science is a technique that is sufficiently well tuned to nature, so that it works – it is a practice that allows us to learn reliable things about the world. In this sense, it is a technique that was waiting for people to discover it" (Weinberg 2015).

Great intellectuals who have worked in the sciences or humanities have provided an insight into the interaction of art and science. The French mathematician, physicist, philosopher, and theorist of science, one of the most important mathematicians of the nineteenth century, Henri Poincaré (1854–1912), says: "The scientist does not study nature because it is useful to do so. He studies it because he takes pleasure in it, and he takes pleasure in it because it is beautiful. If nature were not beautiful it would not be worth knowing, and life would not be worth living". That sounds nice. It seems that the beautiful, which humans appreciate, motivates scientists. H. Poincaré continues: "I am not speaking, of course, of the beauty which strikes the senses, of the beauty of qualities and appearances. I am far from despising this, but it has nothing to do with science. What I mean is that more intimate beauty which comes from the harmonious order of its parts, and which a pure intelligence can grasp". Now we understand what H. Poincaré means. He speaks of beauty in science, and he is allowed to do so because he is one of its genial representatives. He is above all a mathematician and sees nature as a harmony that can be painted with mathematical grace.[10, 11]

I am reminded of another renowned Frenchman, Antoine de Saint-Exupéry: "It is only with the *heart* that one *can see* rightly; what is essential is invisible to the eye", he says in *The Little Prince*, jumping to the other extreme, the spiritual one. In his book "Land of Men", A. de Saint-Exupéry defined a very difficult criterion of perfection: "Perfection is achieved, not when there is nothing more to add, but when there is nothing left to take away". This is in line with what I. Newton says, "Truth is ever to be found in the simplicity and not in the multiplicity and confusion of things". As far as diversity is concerned, here arises a "small problem" with the earth sciences where diversity is the main issue and brings out much of the beauty of the world.[12]

Between *intelligence* (H. Poincaré) and *heart* (A. de Saint-Exupéry), there is an infinite space for experimentation, an ocean in which harmony and perfection manifest themselves in various forms. The German poet Johan Wolfgang von Goete

---

[10] The reason Einstein considered general relativity must be right is that it was too beautiful a theory to be wrong. When it came to the question what if observations had not agreed with the theory, Einstein said: "I would have been sorry for the dear Lord, for the theory is correct".

[11] A possible explanation of the esthetic issue is formulated by Steven Weinberg: "…But we seek beauty in our theories, and use aesthetic judgments as a guide in our research. Some of us think that this works because we have been trained by centuries of success and failure in physics research to anticipate certain aspects of the law in nature, and through this experience we have come to feel that these features of the nature's laws are beautiful" (Weinberg 2015).

[12] Alexander von Humboldt (1769–1859) claims: "In this great chain of causes and effects, no single fact can be considered in isolation". To this he adds: "What gave me the main impetus was the endeavour to understand the phenomena of physical things in their general context, nature as a whole moved and animated by inner forces". With that, he can be considered as a forerunner to the James Lovelock's idea of the Earth as a living organism (the Gaia-Hypothesis).

was one of the most active actors playing on this vast scene. Today, he is famed for his literary work, but he was a passionate scientist, too.[13]

I find it appropriate to introduce these many thoughts in the section, in which I talk about science and art and personal freedom, two favorite topics of I. Apostolova. The quest for perfection is perhaps simply the individual quest for freedom. How it is to be attained is indicated by Aesop[14]: "Because it is a state of mind, freedom cannot be given". Also in science, the fruits do not hang in the low branches. It takes a great struggle and sacrifice of oneself. Yane Sandanski (1872–1915)[15] seems to find the right words: "To live is to struggle—the slave for freedom, the free for perfection". Scientists, whom I consider to be at the top of the pyramid of perfection, take great pleasure when something is successfully explained. This can be well understood from what Claudius Ptolemy wrote in *Almagest*[16]: "I know that I am mortal and the creature of a day; but when I search out the massed wheeling circles of the stars, my feet no longer touch the Earth, but, side by side with Zeus himself, I take my fill of ambrosia, the food of Gods" (Quotation borrowed from Weinberg 2015).

What is common in science and in art is creativity. Strong scientists are those in whom the two fundamentals—analytical skills and intuition—are unified. I suppose this is also true of people of art. Mae Jamieson, an American physician and the first woman astronaut of color, states: "Sciences provide an understanding of a universal experience, Arts are a universal understanding of a personal experience….. arts and sciences are avatars of human creativity[17]". Both, scientists and artists, are searching for the *atoms of the beautiful*, the simplicity and beauty of natural laws (according to I. Newton and H. Poincaré), or perfection (as defined by A. de Saint-Exupéry). Obviously, the search for beauty is at the apex of the human intellect in science and in art.

Measurements and data of natural phenomena constitute the basis of concepts and theories. Can it be the other way around? There are few but impressive cases in the history of science where theories outpace practice. Predicting things that never happened is something no less great than art. Such predictions become especially relevant today. How will the planet's climate evolve? What will happen and when?

---

[13] In March 1794, Goete visited the two brothers Alexander and Wilhelm von Humboldt in Jena. Their friend Friedrich Schiller also joined. The meeting of these four intellectual giants gave rise to the "Little Academy", a lively exchange of ideas, especially on scientific issues. Alexander von Humboldt, the younger of the two brothers, an explorer, and scientist, was claiming that nature had to be experienced through a feeling. The British romantic poet Robert Southey claimed that A. von Humboldt was "among travelers what Wordsworth is among poets" (Wulf 2015). Charles Darwin (1809–1882) talks about Humboldt's writing as a "rare union of poetry with science".

[14] Aesop (probably lived in the sixth century B.C.) was an ancient Greek poet of fables and parables.

[15] Bulgarian revolutionary and leader in the liberation struggles of the Macedonian Bulgarians.

[16] Claudius Ptolemy (approximately from 80 to 160 of our era), a mathematician, astronomer, astrologer, geographer, and music theorist, who lived in or around the city of Alexandria, in the Roman province of Egypt, wrote in the second century AD *the Almagest*. This is one of the most influential scientific texts in history, which canonized a geocentric model of the universe.

[17] https://nwcreations.com/ted-talk-thursday-mae-jemison-teach-arts-sciences-together/, last accessed on 03.02.2023.

3 From the Ionosphere to the Ocean

The Department of Meteorology and Geophysics was established in 1946 within the Faculty of Physics and Mathematics of the University of Sofia. After the separation of Physics and Mathematics into two independent faculties in 1963, the Department of Meteorology and Geophysics became part of the Faculty of Physics. In 1959, the head of the Bulgarian Hydro Meteorological Service in the period 1950–1959, Lyubomir Krastanov (Fig. 1), was elected a full professor at the University of Sofia.[18] He headed the department in the period 1958–1974. From 1962 to 1968, he was the president of the BAS. In the next year, 1969, he lectured me on "Introduction to Meteorology".

As the end of university life approached, the question of professional profiling and the preparation of a thesis arose. The level of teaching was steadily increasing. It was highest under Stoycho Panchev, who recently published the world's first book titled "Random Functions and Turbulence" (Panchev 1971). He offered in the study program a special course on this subject. From the distance of the half-century that has passed since then, I guess that such courses were possible at that time in only few leading world universities. In 1970, at the age of only 36, S. Panchev, who had previously specialized in meteorology in the USSR, the USA, and Japan, became the youngest professor at the University of Sofia.

Stoycho Panchev

Since I consider S. Panchev (Fig. 2) as my first scientific supervisor, I would like to say a few more words about him. There was a distinct element of sincerity about him, which could be read in his intense blue eyes. He was not an orator, his speech was very broken; it improved a little over the years, but it never matched his intellectual potential. He had an athletic poise; he was an avid tennis player. In 1974, my first year as a graduate student, S. Panchev became the head of the Department of Meteorology and Geophysics, an activity he carried out until 1994. From 1979 to 1983, he was the dean of the Faculty of Physics at the University of Sofia. In 1985, he was elected a corresponding member, and in 1997, he was elected as an academician of the BAS. In the period 1988–1991, he became the vice president of the BAS.

At the very beginning of my meteorological studies, I was attracted by global problems. At the heart of atmospheric dynamics are the so-called Rossby waves

[18] The research of L. Krastanov focused on the phase transitions of water in the atmosphere and cloud formation processes. Together with Ivan Stransky, one of the most famous Bulgarian scientists, he proposed a theory about the mechanism of crystal growth at a crystal surface or interface (Stranski and Krastanow 1938). On December 4, 2022, their paper has been cited 967 times in international journals (not counting citations to Stranski-Krastanov's growth model without citing the original article), which is a very good achievement for Bulgarian scientists.

Fig. 1 Lyubomir Krastanov (1908–1977). Photo from Rachev (2015), courtesy of the Annual of the University of Sofia St. Kliment Ohridski

Fig. 2 Stoycho Panchev (1933–2014). Photo courtesy of the Annual of the University of Sofia St. Kliment Ohridski, Faculty of Physics, vol. 108, p. 155

(named after a Swedish scientist[19]) also known as planetary waves, which circle the planet Earth from east to west. They are most pronounced at mid-latitudes, and we more often observe the effects associated with their existence, namely the origin and propagation of atmospheric fluctuations, such as cyclones and anticyclones.[20] Rossby waves are due to the rotation of the planet. Rotation causes special forces (see Sect. 2.1 in chapter "Further into the Ocean") that deflect the moving fluid particle to the right of its direction of motion in the Northern Hemisphere and to the left in the Southern Hemisphere. This force is called the Coriolis[21] force. It arises when the motion is described in a rotating co-ordinate system (i.e., a rotating Earth), and its direction is perpendicular to the direction of motion. It is a fictitious, or pseudo-force, because it only appears in non-inertial (moving, as the rotating Earth is) reference frames and disappears in inertial ones. The difference in the mode of rotation at the equator and the poles of a sphere is the cause of Rossby waves. If the Earth were a cylinder, there would be no such waves, because at every point on the surface of the cylinder, the rotation is the same.

In the ionosphere,[22] so-called magneto hydrodynamic forces are added to the forces causing motion. They owe their origin to the fact that the gas in the ionosphere is ionized. Thus, the motion is influenced by the Earth's magnetic field, unlike the motion in the troposphere[23] where the gases are electrically neutral. Because of that, other forces are added, and as a result, the propagation of Rossby waves changes. I was interested in ionospheric dynamics and decided to do my master's thesis[24] under the supervision of S. Panchev on this topic. One of the goals of my thesis was to develop a theoretical model for the propagation of the Rossby waves in the

[19] Carl-Gustaf Arvid Rossby (1898–1957) was a Swedish meteorologist who contributed significantly to the description of large-scale motions of the atmosphere using fluid dynamics. He was the first to describe the concept of potential vorticity (see Sect. 2 in chapter "Further into the Ocean") and also made contributions to oceanography.

[20] A cyclone is a system of winds that rotates in an anticlockwise direction around a center of low atmospheric pressure. In anticyclones, winds rotate in a clockwise direction around a center of high atmospheric pressure. Cyclones are generally associated with cloudy and rainy weather. Anticyclones are predictors of nice weather.

[21] The French mathematician and mechanical engineer Gaspard-Gustave de Coriolis (1792–1843) introduced the Coriolis force in 1835.

[22] The ionosphere is that part of the atmosphere which contains large quantities of ions and free electrons. It makes up the bulk of the high atmosphere. Ionization of the gas molecules is due to the solar radiation.

[23] The troposphere (from ancient Greek τροπή tropé "turn, change" and σφαίρα sphaira "sphere") is the lowest layer of the Earth's atmosphere. The troposphere extends from the ground to the beginning of the stratosphere. Its thickness ranges from ~ 8 km at the poles to ~ 18 km at the equator. The troposphere contains about 90% of all the air in the Earth's atmosphere. The stratosphere is a layer of the Earth's atmosphere that lies between ~ 10 and 50 km above the Earth. It is characterized by an increase in temperature with increasing height.

[24] At that time, it was called diploma work in Bulgaria.

ionosphere. I think I did it well, but my attempt to get a job or a Ph.D. position at the Institute of Space Studies in Sofia was not a success.

In the summer or fall of 1973, I learned that S. Panchev started to be interested in the theory of ocean turbulence. It was not difficult for him to transfer his familiar atmospheric theories to the ocean and develop them further. Because of his good reputation and reliability (I stress the latter because it was a major quality of his), he established links with leading oceanographers. The Shirshov Institute of Oceanology of the Academy of Sciences of the USSR had invited him to take part in 1974 in an expedition on the research ship Mendeleev in the South Pacific. "You know", he said, "marine research is very interesting. We do not have specialists in this field. I cannot devote myself entirely to it, because my main occupations will remain in meteorology. Don't you want to fill that niche? We can open a Ph.D. position for you". I had no other offers, and this new one was intriguing, as S. Panchev was far superior to his Bulgarian peers. In the field of physics of atmosphere and ocean, no better supervisor could be found in Bulgaria, and I accepted his offer to become his Ph.D. student.

4 About the Waters in Which I Started "Sailing"

4.1 Meteorology, Oceanography, Physics

I remember Academician L. Krastanov jocking whether meteorology was physics. This question is not superfluous, having in mind that meteorology uses the classical methods of physics, unlike astrophysics or particle physics, for example, which use and develop the most modern scientific fields. There is, however, something much more fundamental. Norwegian physicist and meteorologist Vilhelm Bjerknes, one of the fathers of modern meteorology, said in his inaugural address in Leipzig (Bjerknes 1914), "Just as the earth is a part of the entire cosmos, so geophysics can be conceived as a part of cosmical physics. It is divided into three branches: Physics of the atmosphere, Physics of the hydrosphere and Physics of the rigid earth; and has thus an extraordinarily extensive scope". Further he continued: "The physics of the atmosphere treats of the same subject as meteorology. For this very reason, however, these two sciences must not be confounded with each other. The distinction is marked for the reason that physics ranks among the so called exact sciences, while one may be tempted to cite meteorology as an example of a radically inexact science. Meteorology becomes exact to the extent that it develops into a physics of the atmosphere". As far as concepts about exact and inexact sciences in the past were concerned, Bjerknes stresses on the capabilities of different sciences to predict future events: "The essential difference, however, between meteorology and astronomy was that atmospheric motions could only be studied qualitatively, whereas stellar motions might also be studied quantitatively and indeed so accurately that, from the known position of a star at one time, its position could be successfully predicted for a remote date".[25] He ends his inaugural address, stressing on the daily prediction of the weather: "If only the calculation shall agree with the facts, the scientific victory will be won. Meteorology would then have become an exact science, a true physics of the atmosphere".

The issue of similarities and differences between meteorology and the rest of physical sciences has multiple aspects. While the basic approach in physics is to isolate an individual process and study it in as much depth as possible, meteorology brings together hydrodynamic processes with different spatial and temporal scales that interact. Hence, an integrated consideration of multiple processes is needed in order to fully address the motions in the atmosphere. The goal is to achieve a complete and accurate study (including in regional-geographical terms) of the motions and other processes in the atmosphere, as well as their prediction. Here, meteorology is clearly approaching geography. When it comes to climate change, the problem is complicated by the processes of atmospheric chemistry and biological processes in the atmosphere, ocean and land.

Essential is the scientific method that is used and the depth of research work in the various sciences. Geography provides rather a description of what is being observed, answering the "what" question and does not answer the "how" and "why" questions. The latter two questions are addressed by physics. This is why climatology, for example, is more a part of geography, while physical oceanography ("graphy" has a historical character) is more physics than a descriptive science.

On the occasion of the award of the 2021 Nobel Prize in Physics, I was very pleased to hear the speech of Tors Hans Hansson, a member of the Nobel Committee.[26] He stressed that physics is not only the science of simple and orderly phenomena, but is much more. Earth's climate change is a striking representative of a class of complex phenomena requiring deep intuition and perfect handling of the mathematical techniques. After so many years of basic research in the physical sciences of the atmosphere and ocean, the time had come for this prestigious committee to appreciate the importance of studying complex physical processes.

[25] Jacques Laskar showed that the motion of the planets in the Solar System is chaotic, which prohibits any accurate prediction of their trajectories beyond a few tens of millions of years (Laskar 2012, arXiv:1209.5996).

[26] See https://www.nobelprize.org/prizes/physics/2021/summary/, last accessed on 03.02.2023 and https://www.nobelprize.org/prizes/physics/2021/prize-announcement/, last accessed on 03.02.2023.

Nobel Laureate Klaus Hasselmann[27]

On October 5, 2021, the Nobel Prize in Physics was awarded for groundbreaking contributions to our understanding of complex physical systems with one half jointly to Syukuro Manabe and Klaus Hasselmann for the physical modeling of Earth's climate, quantifying variability and reliably predicting global warming and the other half to Giorgio Parisi for the discovery of the interplay of disorder and fluctuations in physical systems from atomic to planetary scales.

I had the chance to interact with Klaus Hasselmann many times and formed a very positive opinion of him as a great personality and scientist. Below I will only refer to what Heather M. Hill and Andrew Grant write in the issue of October 5, 2021, of Physics Today about Klaus Hasselmann's Nobel Prize:

"Inspired by Brownian motion, in which the observed motion of macroscopic particles is the result of continuous random movement at the microscopic level, Hasselmann developed a stochastic climate model (Hasselmann 1976) that treated changes to the climate as the integrated effects of continuous random weather disturbances.

Hasselmann's insights into deciphering the causes of observable effects at long time scales also led him to consider how to separate an anthropogenic climate signal from natural climate fluctuations. The models by Manabe and others had allowed researchers to explore the potential for global warming due to increased CO_2, but it was unclear how much of that warming was the result of external radiative forcing (the signal) versus natural fluctuations (the noise). In a 1979 paper (Hasselmann 1979), Hasselmann provided a statistical blueprint for climate scientists to examine their model results and distinguish signals from noise. He also identified telltale data signatures that cannot be explained by natural fluctuations.

A 2019 review in *Nature Climate Change* (Santer et al. 2019) calls Hasselmann's work 'the first serious effort to provide a sound statistical framework for identifying a human-caused warming signal'. In the decades since, climate scientists have been able to demonstrate with increasing confidence that rising global temperatures and other observed climatic effects are the direct result of human activities".

4.2 *Numerical Modeling*

Fluid mechanics, developed over the past few centuries, provide the equations that ideally describe hydrodynamic processes. The basis is given by the second law of

[27] https://www.nobelprize.org/uploads/2021/10/advanced-physicsprize2021.pdf, last accessed on 03.02.2023.

motion of I. Newton, according to whom the acceleration of a particle is the quotient between the force acting on it and its mass.[28] It seems very simple. However, the problem is that the particles that make up a fluid are infinitely many. The second problem is that, unlike solids, fluids change shape and volume and can contract and expand, not to mention the complexities caused by the warming and cooling of the atmosphere and ocean and the change of density. All of these processes can be described by mathematical equations, which are, however, valid for true fluids, i.e., those in which all movements, from the smallest to the largest, are resolved with equal completeness. These equations have been known since the late eighteenth and early nineteenth centuries and bear the names of Henri Navier[29] and George Stokes.[30] Because of their generality, their solutions can be used to understand many and sometimes very different processes.[31]

Vilhelm Bjerknes, who pioneered numerical weather forecasting (Bjerknes 1904) by applying physical laws to the atmosphere, set out the problem of computing the future state of the atmosphere given its present state: "We must apply the equations of theoretical physics not to ideal cases only, but to the actual existing atmospheric conditions as they are revealed by modern observations. These equations contain the laws according to which subsequent atmospheric conditions develop from those that precede them. It is for us to discover a method of practical utilizing the knowledge contained in the equations. From the conditions revealed by the observations we must learn to compute those that will follow" (Bjerknes 1914). Unfortunately, it was too early and technologically impossible at the beginning of the twentieth century to shorten the distance between theory and practice, and Bjerknes had understood this as early as 1904: "There can be no question of a strictly analytical integration of the system of equations. As is well known, calculating the motion of three points that influence each other according to a law as simple as Newton's already far exceeds the means of today's mathematical analysis. There is evidently no hope of knowing the movements of all points of the atmosphere which are influenced by much more complicated interactions".

Years after the first insights of Bjerknes, it was understood that the equations governing motion in the atmosphere describe something much more complete and complex than we can calculate even with the most powerful computers. It is unlikely that there will ever be computational tools that can fully resolve everything in a fluid. For these reasons, the equations of fluid dynamics are solved on a discreet grid of points, which makes it possible to describe the real processes only approximately (i.e., to describe not real fluids, but model fluids).

---

[28] The famous equation $F = m.a$ does not appear in the fundamental work by I. Newton *"Principia" (Mathematical Principles of Natural Philosophy)*, where one finds few equations (Smith 2008).

[29] Claude Louis Marie Henri Navier (1785–1836) was a French mathematician and physicist.

[30] Sir George Gabriel Stokes (1819–1903) was an Irish-British mathematician and physicist.

[31] Fox-Kemper (2018) presented a comprehensive review of almost all important aspects of the motion and typical types of variability at different scales of the ocean, as well as the unresolved processes and their parameterizations. Reading this review requires a thorough knowledge of mathematics and physics.

A major task of numerical modeling of the atmosphere and ocean is to develop methods that reduce ideal equations governing fluids to solvable equations, which are usually of an algebraic nature. The next task is to transform these equations to computer-understandable codes.[32] As simple as this may sound at a first glance, it was a long path to achieve some practically positive results. Defining numerical grids (Fig. 3) and representing equations by finite differences, finite volumes, finite elements, and spectral methods is a whole science in itself. It turns out that solutions are accurate if fields vary smoothly on the mesh, and hence, meshes must resolve the smallest features to guarantee convergence. The finer the grid is (the better the resolution is), the more details of the real motion are captured by the numerical model. However, here comes the problem of the reconciliation between the available computational power and the quality of numerical solutions. According to the methods of computational fluid dynamics (CFD), to resolve processes in a fluid where the Reynolds number $\mathrm{Re} = UL/\nu$ is ten times bigger would need a thousand-fold increase in the computational cost. Here, U and L are the typical velocity and length scales and ν the kinematic viscosity. The above applies to so-called direct numerical simulations (DNS), which are used to solve Navier–Stokes equations. However attractive, DNS seem to be inapplicable to numerical modeling of climate and weather.

Smoothed versions of the Navier–Stokes equations, such as Reynolds averaged Navier–Stokes equations, allow coarser grids while sacrificing accuracy. These equations, which are widely used today in atmospheric and ocean numerical models, bring another problem, which is how adequately the integral effect of motions, which are filtered in the process of the averaging of equations, is accounted for. Here, we speak of subgrid-scale processes, i.e., processes that are not resolved by filtering. These processes can be very important for motions in the atmosphere and ocean. To take them into account, specific parameterizations are developed, based either on theoretical concepts or on empirical relationships.

Many applications of numerical modeling are not global, but have been developed for specific parts of the atmosphere or the ocean. For such applications, the state of the fluid at the model boundary or the fluxes of heat, salt, momentum, or other mater through boundaries are needed to be known. Such data, named boundary conditions data, largely determine the interaction between the model domain and the fluid outside. Unfortunately, available observations do not provide all the needed data at the model boundaries. Therefore, data from global models, or models covering larger model domains, are used as boundary conditions. In such cases, the terms *nested models* or *child and parent models* are used. Three-dimensional model domains have also upper and lower boundaries. In the ocean models, the sea surface and bottom are such boundaries. At these boundaries, reliable data have to be prescribed. At the bottom of the ocean, one uses the so-called *no-slip condition*[33] and assumes *zero flux of* heat and salt (insulation), which is well applicable to most ocean floors. As far

[32] The reviews of Griffies et al. (2010) and Fox-Kemper et al. (2019) describe the present-day situation with ocean modeling, as well as the challenges and prospects for ocean circulation models.

[33] In dynamics of viscous fluids, it is assumed that velocity at the bottom is zero.

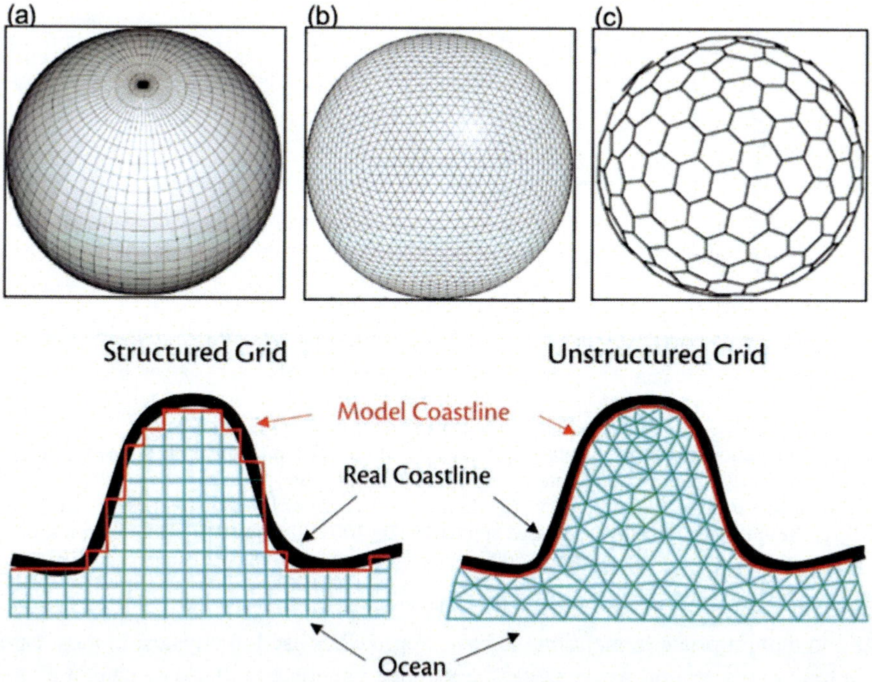

Fig. 3 Top: three examples of a global rectangular or latitude–longitude grid (**a**), triangular grid (**b**), and hexagonal grid (**c**). From Collins et al. (2013). This is an open access article distributed under the terms of the Creative Commons Attribution License (http://creativecommons.org/licenses/by/3.0). Bottom: an example of fitting a structured grid (left) and an unstructured grid (right) to a simple coastal embayment. The true coastline is shown in black and the model coastline in red. Note how the unstructured triangular grid can be adjusted so that the model coastline follows the true coastline, while the structured grid coastline is jagged, which can result in unrealistic flow disturbance close to the coast. The figures are from Chen et al. (2006) and are made available under the terms of the Creative Commons Attribution 4.0 International License (https://creativecommons.org/licenses/by/4.0/)

as the upper boundary conditions are concerned, the data we use have to accurately represent the fluxes of momentum, heat, and vapor between atmosphere and ocean. In the coupled atmosphere–ocean models,[34] these fluxes are computed in the models.

Numerical ocean modeling, which today is no less important than oceanographic measurements, data processing, and interpretation, was "just getting started" in the 1960s and followed some 20 years after the successes in numerical modeling of atmospheric processes.[35] Numerical modeling being an important research branch,

[34] By coupled atmosphere–ocean models, we would most commonly be referring to a numerical model of the atmosphere, usually with an associated land-surface model, which is "two-way coupled" (at least daily, if not more frequently, e.g., hourly) to a numerical model of the ocean, often with an associated sea-ice model (Harris 2018).

[35] A review on the historical development of concepts and practices of weather and ocean predictions has recently been provided by Pinardi et al. (2017).

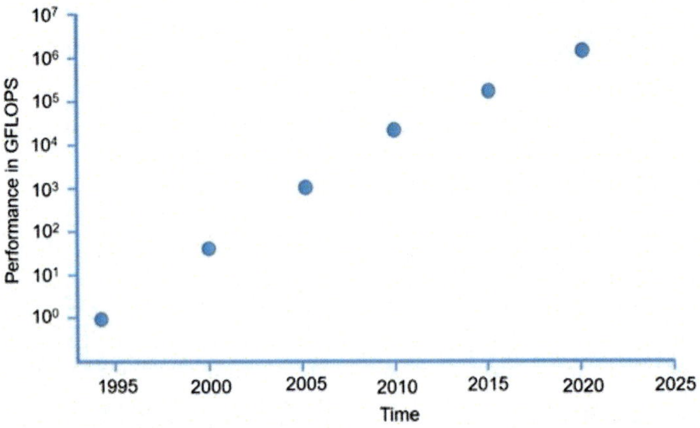

Fig. 4 Evolution of supercomputers performance. The logarithmic *y*-axis shows performance in GFLOPS. In 1965, Gordon E. Moore, the cofounder of Intel, found out that *the number of transistors on a microchip would double every year or two. This is known as* Moore's Law. From 1980 to 2010s the speed of computers increased approximately 100,000 times. The symbols correspond to the performance of supercomputer in 500th place. Adapted from Alvarez Fanjul et al. (2022)

I find it appropriate to mention the basic approaches used in physics. In most cases, in their experimental work, researchers notice phenomena that they cannot explain using existing theories. They then develop concepts or mathematical theories to explain these phenomena. There are also researchers, who use primarily a theoretical-mathematical approach. Their results are more due to precise mathematical work.[36] One issue on which mathematicians and physicists do not fully agree is the rigor. While for the mathematicians, definitions and assumptions must be precise and the deductions exact, physicists are more opportunistic demanding only enough precision and certainty (Weinberg 2015).

Numerical modeling is a mixture of mathematics, physical theory, experimentation with a model, as well as the interpretation of the results that leads to new discoveries. Numerical modeling, which provides a powerful tool for conducting controlled experiments, became in the second half of the twentieth century the main tool for ocean experimentation. It thus made it possible to test the hypotheses put forward as well as to repeat earlier experiments by other scientists in order to double-check theories.

[36] Eugene Paul Wigner (1902–1995) was a Hungarian-American physicist. He received the Nobel Prize in Physics in 1963 "for his contributions to the theory of the atomic nucleus and the elementary particles, particularly through the discovery and application of fundamental symmetry principles". In his famous article "*The unreasonable effectiveness of mathematics in the natural sciences*", he writes*:* "The miracle of the appropriateness of the language of mathematics for the formulation of the laws of physics is a wonderful gift which we neither understand nor deserve. We should be grateful for it and hope that it will remain valid in future research and that it will extend, for better or for worse, to our pleasure, even though perhaps also to our bafflement, to wide branches of learning" (Wigner 1960).

Fig. 5 John Vincent Atanasoff (1903–1995). *Credit* American Institute of Physics (AIP) Emilio Segrè Visual Archives, Physics Today Collection

Fig. 6 Blagovest Sendov (1932–2020) was Rector of Sofia University (1973–1979) and President of BAS (1988–1991) Photo courtesy of the Museum of Mathematics and Informatics in Bulgaria, Institute of Mathematics and Informatics of BAS, 2017 (http://mmib.math.bas.bg), last accessed on 02.02.2023

The First Computers and Numerical Modeling

Numerical modeling of atmospheric and ocean dynamics requires enormous computing resources. One of the first applications of the Electronic Numerical Integrator and Computer (ENIAC) in the fifties was in the field of numerical prediction of atmospheric dynamics. This fundamental step in science was possible due to the work of John von Neumann and several other leading meteorologists in the USA. Curious is that in the meteorological and oceanographic literature, ENIAC is said to be the world's first computer. There is a dispute about primacy. In 1973, John Vincent Atanasoff (Fig. 5) was identified by a court ruling in the patent case *Honeywell Inc. v. Sperry Rand* as the inventor of the first automatic electronic digital computer, a non-programmable special-purpose machine that later became known as the Atanasoff-Berry computer. His father, Ivan Atanasov, was born in 1876 in the village of Boyadzhik (Yambol region, Bulgaria; at that time, it still belonged to the Ottoman Empire), shortly before John's grandfather was killed during the April Uprising against the Turks. In 1889, Ivan Atanasov immigrated to the USA. The well-known Bulgarian mathematician Blagovest Sendov (Fig. 6) was the first to trace John Atanasoff's address and establish correspondence with him, the latter visiting Bulgaria for the first time in 1970 at the invitation of the BAS.

4.3 *About My Work*

The amalgamation of different scientific fields and activities in numerical modeling was for me a big challenge at the time when I started my Ph.D. in 1974 and is still until now. One of the fellows of the Geophysical Institute of the BAS, Prof. Nikola Godev, who lectured me on synoptic meteorology, had some conceptual views on the influence of friction in the lower atmosphere on the dynamics of cyclones and anticyclones. Stoycho Panchev suggested that I apply these concepts to the ocean.

In Bulgaria, the Soviet factor in science was manifested in the framework of the Council for Mutual Economic Assistance (COMECON) of Eastern European countries. Since S. Panchev was well known in the Soviet Union, it was not difficult for him to attract Artem Sarkisyan, at that time a leading scientist in the field of numerical modeling of ocean currents, to co-operate on this subject. Artem Sarkisyan gained recognition in the oceanographic community with the so-called diagnostic model for calculating ocean currents. This was essentially a computer analysis of currents from predefined data of temperature and salinity. He had developed a theory that evaluated how Joint Effect of Baroclinicity[37] and Bottom Relief (JEBAR) shapes

[37] In fluid dynamics, the baroclinicity of a stratified fluid is a measure of deviation of the gradient of pressure from the gradient of density. A simpler case is the barotropic flow, for which density surfaces coincide with the pressure ones.

the ocean currents (Sarkisyan 2006). My Ph.D. work boiled down to adding to the numerical model of ocean dynamics the effect of bottom friction, which I had already studied theoretically. The application of the numerical model enhanced by the new physical processes I introduced had to be tested in the Black Sea.

I completed my Ph.D. in 1977. The procedure at this time was such that the Ph.D. student had to ask independent scientist to send comments on his work. I was much encouraged by positive feedback from the world's leading experts: Kirk Bryan, Robert Haney, Norman Heaps, Robert Molinari, A. Sarkisyan, David Leichtman, etc. Instead of talking about my work, I will give short extracts of two reviews, which were posted to S. Panchev.

Robert Haney stated: "…I have often thought about this problem because the effect of bottom friction is not included in the work of Holland (1973) and others, and because only small deviations of flow due to friction are needed to produce significant vertical velocities at the top of the BBL.[38] It seems that your wonderful student Mr. Stanev has answered my question…".

Kirk Bryan claimed: "…I believe that the peculiarities of topographic drag in both the ocean and the atmosphere are a major unsolved problem that requires both experimental work and theoretical studies such as that of Dr. Stanev". Unfortunately, these early works remained unknown to most of the world's leading scientists because I did not publish them in the West, led by the fact that I had already published them in Soviet journals.

References

Alvarez Fanjul E, Ciliberti S, Bahurel P (eds) (2022) Implementing operational ocean monitoring and forecasting systems. IOC-UNESCO, Paris, France, p 376 and Annexes (GOOS-275). http://doi.org/10.25607/OBP-1774

Bjerknes V (1904) Das Problem der Wettervorhersage, betrachtet vom Standpunkte der Mechanik und der Physik (The problem of weather prediction, considered from the viewpoints of mechanics and physics). Meteorol Z 21:1–7 (translated and edited by Volken E, Bronnimann S (2009) Meteorol Z 18(6):663–667)

Bjerknes V (1914) Meteorology as an exact science. Mon Weather Rev 42(1):11–14

Chen C, Beardsley R, Cowles G (2006) An unstructured grid, finite-volume coastal ocean model (FVCOM) system. Oceanography 19(1):78–89

Collins SN, James RS, Ray P, Chen K, Lassman A, Brownlee J (2013) Grids in numerical weather and climate models. In: Climate change and regional/local responses, vol 256, IntechOpen. http://doi.org/10.5772/55922

Fox-Kemper B (2018) Notions for the motions of the oceans. In: Chassignet EP, Pascual A, Tintoré J, Verron J (eds) New frontiers in operational oceanography (Exeter: GODAE OceanView), 811. http://doi.org/10.17125/gov2018.ch02

Fox-Kemper B, Adcroft A, Böning CW, Chassignet EP, Curchitser E, Danabasoglu G, Eden C, England MH, Gerdes R, Greatbatch RJ, Griffies SM, Hallberg RW, Hanert E, Heimbach P, Hewitt ET, Hill CN, Komuro Y, Legg S, Le Sommer J, Masina S, Marsland SJ, Penny SG, Qiao

[38] Bottom Boundary Layer.

F, Ringler TD, Treguier AM, Tsujino H, Uotila P, Yeager SG (2019) Challenges and prospects. Front Mar Sci 6:65. https://doi.org/10.3389/fmars.2019.00065

Griffies SM, Adcroft AJ, Banks H, Boning CW, Chassignet EP, Danabasoglu G et al (2010) Problems and prospects in large-scale ocean circulation models. In: Hall J, Harrison DE, Stammer D (eds) Proceedings of the OceanObs'09 conference: sustained ocean observations and information for society, 21–25 Sept 2009, vol 2. ESA Publication WPP-306, Venice

Harris C (2018) Coupled atmosphere-ocean modelling. In: New frontiers in operational oceanography. Retrieved from http://purl.flvc.org/fsu/fd/FSU_libsubv1_scholarship_submission_1536243251_5afbe0aa

Hasselmann K (1976) Stochastic climate models part I. Theor Tellus 28(6):473–485

Hasselmann K (1979) On the signal-to-noise problem in atmospheric response studies. In: Shaw DB (ed) Meteorology of tropical oceans. Royal Meteorological Society, London, pp 251–259

Hawking SW, Mlodinow L (2005) A briefer history of time. Bantam Books, 162 p

Holland WR (1973) Baroclinic and topographic influences on the transport in western boundary currents. Geophys Fluid Dyn 4(3):187–210. https://doi.org/10.1080/03091927208236095

Panchev S (1971) Random functions and turbulence. Pergamon, Oxford, p 444

Pinardi N, Cavaleri L, Coppini G, De Mey P, Fratianni C, Huthnance J, Lermusiaux PFJ, Navarra A, Preller R, Tibaldi S (2017) From weather to ocean predictions: an historical viewpoint. J Mar Res 75:103–159

Rachev NH (2015) Meteorology and geophysics in the 125-history of Sofia University St. Kliment Ohridski teaching in physics. Annual of the Sofia University St. Kliment Ohridski, Anniversary edition 50 years faculty of physics, 5–25 (In Bulgarian: Метеорологията и геофизиката в 125 годишната история на преподаване на физика в Софийския университет "Св. Климент Охридски", Годишник на СУ, Юбилейно издание 50 години Физически факултет)

Santer BD, Bonfils CJW, Fu Q, Fyfe JC, Hegerl GC, Mears C, Painter JF, Po-Chedley S, Wentz FJ, Zelinka MD, Zou C-Z (2019) Celebrating the anniversary of three key events in climate change science. Nat Clim Change 9:180–182. https://doi.org/10.1038/s41558-019-0424-x

Sarkisyan AS (2006) Forty years of JEBAR—the finding of the joint effect of baroclinicity and bottom relief for the modeling of ocean climatic characteristics. Izv Atmos Ocean Phys 42:534–554

Smith G (2008) Newton's Philosophiae Naturalis Principia Mathematica. In: Zalta EN (ed) The Stanford encyclopedia of philosophy, Winter 2008 edn. https://plato.stanford.edu/archives/win2008/entries/newton-principia/

Stanev E (2022) A sea of people and numbers. St. Kliment Ohridski University Press, 744 p (In Bulgarian: Станев, Е. „Море от хора и числа", Издателство на Софийския университет Св. Кл. Охридски)

Stranski IN, Krastanow L (1938) Zur Theorie der orientierten Ausscheidung von Ionenkristallen aufeinander. Abhandlungen der Mathematisch-Naturwissenschaftlichen Klasse IIb. Akademie Der Wissenschaften Wien 146:797–810

Weinberg S (2015) To explain the world: the discovery of modern science. Allen Lane (Penguin Books Ltd), London; Harper, New York. ISBN 0241196620

Wigner EP (1960) The unreasonable effectiveness of mathematics in the natural sciences. Richard Courant lecture in mathematical sciences delivered at New York University, May 11, 1959. Commun Pure Appl Math 13:1–14

Wulf A (2015) The Invention of Nature: The Adventures of Alexander von Humboldt, the Lost Hero of Science, London: John Murray (Hodder & Stoughton), 2015, 472 pp.

Further into the Ocean

Regional oceanography is presented in this part by focusing on the European semi-enclosed seas, among them the Black Sea, the Marmara Sea, the Mediterranean Sea, and the Baltic Sea. The concepts of dynamics of estuarine and lagoon basins are introduced with a focus on the role of the strait exchange for the water mass formation. The propagation of the Mediterranean Sea water in the Atlantic Ocean is also presented. To facilitate the understanding of results, the physical principles of dynamics are explained, along with a general knowledge about the acting forces and balances between them. The POLIMODE program during the 1970s is used as an introduction to the issue of eddies in the ocean, basic principles of dynamics of rotating fluids, and different types of motion depending on their characteristic scales. Biogeochemical specificities of the marine environment are highlighted taking as a typical example the Black Sea with its anoxic deep water.

Some early steps in the direction of providing scientific services are also presented. The application of tracer techniques in the study of processes and pollution in the Black Sea is illustrated through examining changes in the Black Sea radioactivity resulting from the Chernobyl disaster. It is demonstrated that the experimental study of the intrusion in the deep layers of the Black Sea, together with numerical modeling of this process, represents a powerful mean of improving the quality of numerical models, making it possible to recover the spatial and temporal variation of concentrations of caesium and strontium.

The continuous progress of Mediterranean and Black Sea numerical modeling over the last 30 years and some important findings in these basins are also described. Recent developments are put in the context of several fundamental research projects. Physical Oceanography of the Eastern Mediterranean (POEM) developed patterns for international cooperation applicable to other areas of the World Ocean. The Cooperative Marine Science Program for the Black Sea (CoMSBlack), which adopted the concepts of POEM, was followed by several NATO and EU-funded projects oriented toward using Ecosystem Modeling as a Management Tool and understanding the

E. Stanev, *Trajectories in Oceanography*, Springer Oceanography,
https://doi.org/10.1007/978-3-031-33720-8_2

complex interactions between human activities and the marine environment. Scientific actions, which supported the Convention for the Protection of the Black Sea against Pollution (Bucharest Convention, 1992), are also presented.

This chapter provides also individual essays on the satellite oceanography and data assimilation in oceanography, as well as a number of applications of numerical modeling to study the dynamics of the European regional seas, Arctic and part of the Atlantic Ocean.

1 Bulgarian Oceanography and the Black Sea

1.1 Introduction

Historically, Bulgarian oceanography started in 1932, when the Marine Biological Station with the Aquarium (today the Institute of Fish Resources at the Agricultural Academy) was established in Varna (Bekov and Petrova-Klisarova 2013). Less than a year before I took my first steps in oceanography, the Institute of Marine Research and Oceanology (IMIO) of the BAS was established on July 1, 1973, with its headquarters in Varna. The first director of IMIO, Eng. Zdravko Belberov, headed also the section "Dynamics of the coastal zone". In 1984, IMIO was renamed the Institute of Oceanology (IO-BAS). In 1976, the motor vessel "Izsledovatel 2" entered into service. It was a small ship with a limited area of navigation, mainly in Bulgarian waters. In 1984, the vessel "Akademik", with a displacement of 1225 tons and a length of 55.5 m, was purchased and converted for scientific research. The crew and scientific staff were about 40 people in total.

Stoycho Panchev's plans for me were to take charge of a new unit in the Department of Meteorology and Geophysics. With my appointment as research assistant and that of my colleague Georgi Miloshev, who had graduated in Odessa in hydrometeorology, as a physics laboratory assistant, the beginning of physical oceanography at the University of Sofia was institutionalized. The Laboratory of Physics and Dynamics of the Sea was established in 1977 within the Department of Meteorology and Geophysics. A little later, Vassil Roussenov, who had completed his ocean studies in Japan, joined us.

Bulgarian physical oceanography is primarily regional.[1] It deals mainly with the Black Sea (Fig. 1), which can be seen as a small-scale laboratory for the study of processes that are common to different areas of the World Ocean. In particular, studies of the ventilation of this sea, i.e., the exchange between surface and deep waters, can facilitate the understanding of similar processes that took place in the paleo-ocean[2]

[1] Regional oceanography addresses the specific processes and states of individual oceanic basins and seas.

[2] From Greek prefix "palaios" old, ancient.

when its vertical stratification was stronger, similarly to the present-day stratification of the Black Sea. The small size of the sea makes it a useful test region for developing numerical models, which can then be applied to larger scales. These models can be used to simulate different scenarios for the evolution of the marine environment. This scientific method gives researchers a powerful tool to assess fundamental ocean characteristics, which cannot be directly measured. Numerical models also represent an important complement to scarce observations and provide scientific guidance for observational practice.[3]

The Black Sea is an almost enclosed basin, connected to the Marmara Sea and the Azov Sea by the narrow and shallow Bosporus and Kerch Straits, respectively. Its catchment area covers large parts of Europe and Asia, about five times the size of the sea itself (Fig. 1). The inflow from rivers amounts to approximately 3×10^2 km^3 water per year (Ünlüata et al. 1990). The Danube River provides over 60% of the entire runoff reaching the Black Sea. If it were not flowing out of the Bosporus, the sea level would be rising by about 75 cm per year. The difference between precipitation and evaporation does not change this estimate much because they roughly compensate each other.

The Black Sea is a typical estuarine-type[4] ocean basin due to the large freshwater flow. River flow, precipitation and bottom current in the strait of Bosphorus bring in as much water as the sum of the water evaporated into the atmosphere plus the water transported by the surface current in the Sea of Marmara (see Fig. 2[5]). The total volume of water flowing out of the sea is approximately twice the volume of water flowing into the sea with the bottom current in the Bosporus Strait. Therefore, the ratio between salinity of the outflowing and inflowing waters scales as 1:2 (Özsoy and Ünlüata 1997). Unlike other similar estuarine basins (e.g., the shallow Baltic Sea), the Black Sea is a deep basin with a maximum depth of ~ 2200 m and a wide shelf in its northwestern part, starting in the Crimea region and ending around the northern Bulgarian border (Fig. 1).

The specific sources of fresh water from rivers and saltier waters from the Mediterranean, as well as the control of the Bosporus Strait on the exchange of waters between the Black Sea and Mediterranean contributes to the creation of a strong vertical stratification between the waters in the upper 100 m and the deep waters throughout the basin. This stratification limits vertical exchange because mixing of waters in highly stratified fluids is weak (similar to what we observe between a layer of oil and water). This creates conditions for unique chemical and biological processes to occur.

---

[3] The various issues related to observational practices in oceanography will be introduced gradually at different places in the book. For this reason, I will only mention here that by marine/ocean observations we will mean the collection of data using classical and robotic equipments, such as profiling floats, surface buoys, seafloor landers, ships, satellites, etc., measuring the physical and biogeochemical properties of seawater.

[4] Inland seas can be considered as two types: the estuarine type, where precipitation and river discharge are greater than evaporation and hence salinity is reduced, and the lagoonal type (e.g., Mediterranean Sea), where evaporation is dominant.

[5] Notice that salinity in this figure is given in ‰, *per mile*. In some older studies, parts per thousand (ppt) is used, which is kg salt per kg water. The accepted nowadays unit is the practical salinity unit (psu), which is non-dimensional.

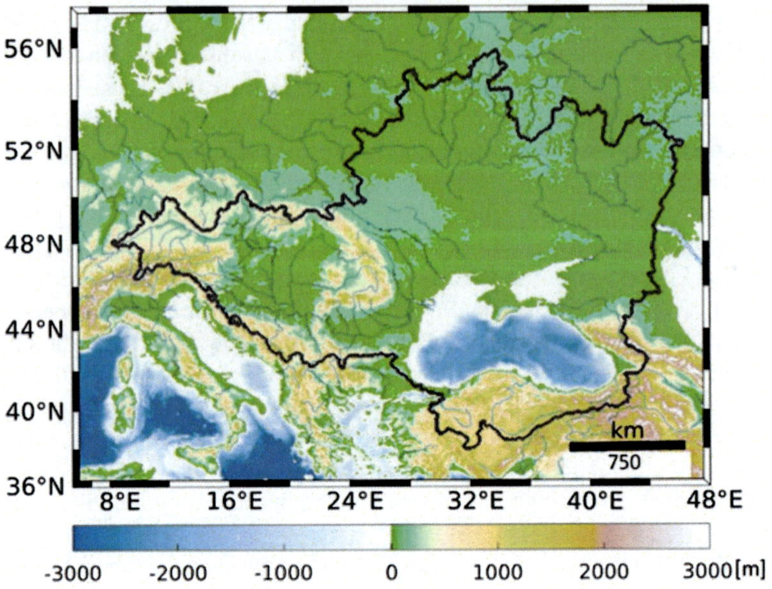

Fig. 1 Black Sea and its catchment area (the black line). The colors show part of the topography of Europe and the bathymetry of the sea. The boundary between white and light blue marks approximately the position of the continental slope. Courtesy of M. Ricker

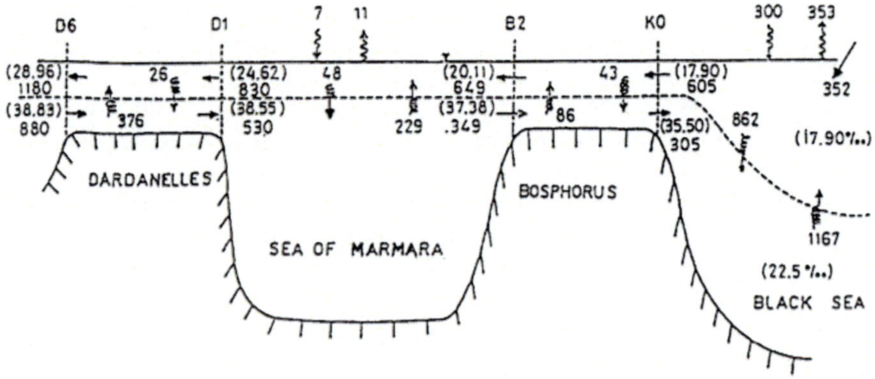

Fig. 2 Annual mean volume fluxes in the Turkish Straits System. The fluxes are given in units of $km^3\ yr^{-1}$ ($1\ km^3\ yr^{-1} = 31.7\ m^3\ s^{-1}$). The numbers in the brackets are average salinity values. The interface between less salty surface water and saltier deep water is schematically shown by the dashed line. Water fluxes (in two directions) across this interface are also shown. From Özsoy and Ünlüata (1997), with permission from Elsevier

1.2 *The Circulation of the Black Sea*

1.2.1 Wind-Driven Circulation. General Concepts

The major driving force for ocean circulation is provided by mechanical factors (wind, atmospheric pressure and tides) and thermohaline[6] factors associated with spatial variations of water density. Spatial variations in water density may be due to direct heating by solar radiation or cooling from long-wave radiation emitted from the ocean, cooling or heating due to contact between the atmosphere and the ocean, evaporation, precipitation, and fresh water sources due to river runoff and ice melting. In the following, brief explanations of how motion is created in the ocean due to mechanical driving by wind will be provided before addressing Black Sea circulation.

1.2.1.1 Ekman Drift

The frictional force between the atmosphere and the ocean is directed in the direction of the wind. Assuming that no forces other than the Coriolis force and the frictional force are acting, there will be a balance between these two forces, i.e., they will be equal and with opposite directions. Alternatively, the ocean will continuously accelerate, which we rule out as a possibility in our simplified model. Knowing that the Coriolis force is directed perpendicular to the direction of motion (to the right in the Northern Hemisphere) leaves, strange as this may sound, the motion for the entire water column[7] to be directed 90° to the right of the wind direction. This transport is called drift transport or Ekman drift, after the Norwegian oceanographer who first explained it. According to Ekman's theory, at the surface of the ocean, current is directed 45° to the right of the wind direction. The deviation to the right increases with increasing depth, and the current velocity vector describes the so-called Ekman spiral (Fig. 3).

1.2.1.2 Simplified Representation of the General Circulation in the Ocean

On a planetary scale, the most characteristic features of wind field are the westerlies (wind from the west) in the midlatitudes and trade winds (wind from the east) in the tropics. These winds are depicted in the diagram in Fig. 4 with bold arrows. The Ekman model states that the westerlies create a drift to the south; the drift created by the trade winds is directed to the north (see the block arrows). These two drift

[6] The term "thermohaline" (from the ancient Greek θερμός thermós, "warm" and ἅλς háls, "salty") describes the property of seawater related to temperature or salinity.

[7] In many places in the text we will use and understand "water column" as a column of water from the surface of the sea to the bottom. This is known as the pelagic zone (from Ancient Greek πέλαγος *(pélagos)* "open sea".

Fig. 3 Ekman spiral

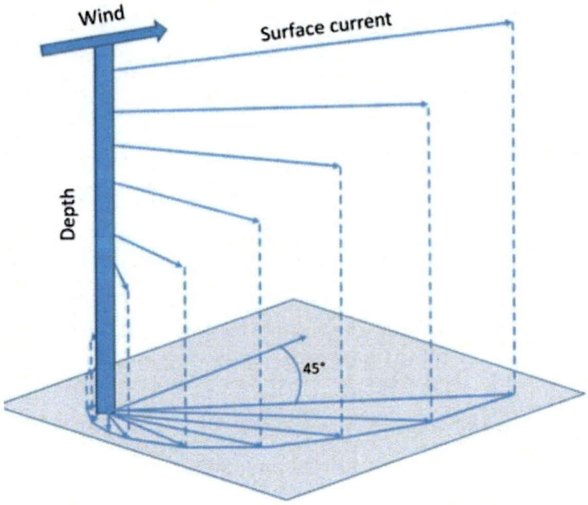

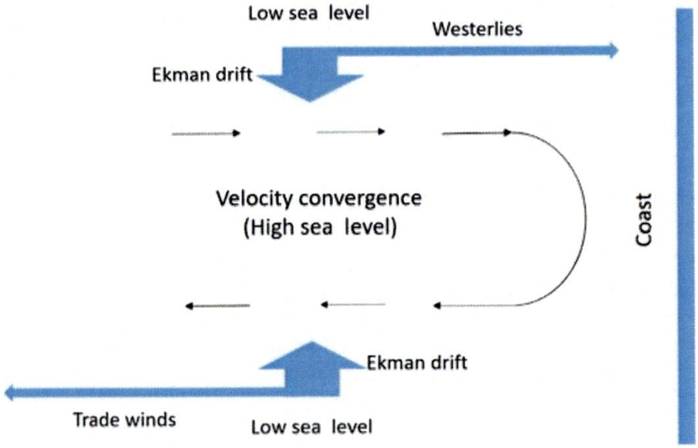

Fig. 4 Schematic representation of currents in a homogeneous ocean of uniform depth, bounded by meridional coastal wall on the east. The wind field is represented by the thin-shaded arrows. The Ekman layer transports (perpendicular and to the right of the wind) are indicated by the block arrows. The thin lines with arrows on them illustrate the geostrophic horizontal flow. See Stommel (1957)

currents (convergent) create accumulation of water in the interior of the ocean and sea level rise there. If we consider the diagram as representing part of the Atlantic, the convergence zone corresponds to the Sargasso Sea.

The elevated sea level in the interior of the ocean results in the creation of a pressure gradient force directed from high to low pressure (high to low sea level). In the absence of any other forces, the Coriolis force balances the pressure gradient. Similarly, to the case of drift currents, the flow is perpendicular to the direction of two forces. It is called geostrophic flow, and is such that, in the northern hemisphere, the elevated sea level is to the right looking in the flow direction. In the diagram of Fig. 4, the geostrophic current is directed eastward in the midlatitudes and westward in the tropics (see thin arrows). The circulation in Fig. 4 is anticyclonic, i.e., the rotation is clockwise. This simplified model explains the anticyclonic gyre in the middle of Atlantic Ocean (Sargasso Sea) as caused by the specific wind torque.

1.2.2 On the Mechanisms of the Black Sea Circulation

Numerical experiments of Stanev (1990) made it possible for the first time to quantify the separate effects of the mechanical and thermal factors setting the Black Sea in motion. Of fundamental importance was the discovery of the role of wind vorticity as a major driving force in creating the cyclonic (counterclockwise) general circulation of the Black Sea. It turns out that the spatial variations of the wind over the Black Sea are the opposite of the known ocean case, which is presented in Fig. 4. Therefore, the sea level in the basin interior is lower compared to that in coastal areas and the circulation is cyclonic.

The recently deceased doyen of oceanography Walter Munk (Fig. 5), largely contributed to the development of the theory of the wind-driven ocean circulation (Munk 1950). I met him for the first time in person after presenting a paper addressing the Black Sea dynamics at the Joint Assembly of the International Association for the Physical Sciences of the Oceans (IAPSO[8]) and the International Association for Biological Oceanography (IABO) in Mar del Plata, Argentina, from October 21–28, 2001. Walter Munk was then over 80 years old, but with an extremely sharp mind. During the Assembly, IAPSO awarded him the Prince Albert I medal.[9] Walter Munk was one of the "fathers of oceanography", and no one was more deserving than him to receive this award for the first time. What I presented included the theory of wind-induced circulation and its validation in the Black Sea using altimeter data.[10] After my report, he asked me a question, and in the break between two sessions unexpectedly approached me. He was very excited to learn that the theory of general ocean circulation, to which he largely contributed, is applicable to the small Black Sea as well, and that the methods of satellite altimetry were bearing fruit

[8] At the Constituent Assembly of the newly established International Research Council (IRC), in July 1919, it was decided to establish the International Union of Geodesy and Geophysics (IUGG) as a union of six sections. One of the sections was that of Physical Oceanography (IAPSO). In later years, the IRC was known as the International Council of Scientific Unions (ICSU), an international non-governmental organization dedicated to international cooperation in the advancement of science. Its members are national scientific organisations and international scientific unions. In July 2018, ICSU merged with the International Social Science Council (ISSC) to form the International Science Council (ISC).

[9] This medal is named in honor of the late Prince Albert I of Monaco who, in 1919, organized the Oceanography Section of the IUGG. He also served as first president of that section. The medal is awarded biannually by IAPSO at its Assemblies.

[10] Satellite altimetry measures the topography of the sea surface, with high-resolution, global coverage and short revisit time (International Altimetry Team 2021).

Fig. 5 Walter Munk (1917–2019). *Credit* Photo by Jeff Cordia, courtesy of AIP Emilio Segrè Visual Archives, Gift of Dr. Munk

not only in the global ocean. It is a pity that research on regional seas sometimes goes unnoticed by much of the oceanographic community. This was perhaps the case of my work with my first Ph.D. student Nikolai Rachev, where we were investigating the propagation of Rossby waves in an enclosed basin, in our case the Black Sea (Rachev and Stanev 1997). My discussion with W. Munk on the basin oscillations was also very stimulating.

Wind torque not only determines the nature of the circulation, but it also shapes the horizontal distribution of temperature and salinity. Drift currents in the surface layer displace water masses in the Black Sea from the basin interior into the coastal zone. Hence, deep and denser water upwells in the basin interior and less dense water accumulates in the coastal zone. This specificity of the horizontal and vertical circulation explains the fact that the isopycnic surfaces rise in the basin interior and sink in the coastal zone; they mirror the shape of the sea level.

The dynamics of the Black Sea, like in many ocean systems, is strongly dependent on the bottom relief. The slope of sea level is greatest along the continental slope (Fig. 1), so the strongest currents (the main Black Sea current, also known as the Rim Current) occur there (Fig. 6). Surface velocities of this current are typically 0.3–0.5 m/s, reaching sometimes 1 m/s; currents in the deeper layers reach several cm/s. A complex competition and complementarity occurs between effects related to the Earth's rotation, topography, friction with the coast and bottom, and vertical stratification.

While horizontal circulation in ocean basins can be determined from observational data, the equally important vertical circulation remained unknown until the advent of numerical models. This is because vertical motions are so weak that they cannot be directly measured. Therefore, it is not possible to determine the fundamental hydrodynamic balances accurately and completely unless numerical models are used.

The numerical simulations of Stanev (1990) showed that the amount of vertically transported water (upwelling) in the Black Sea is ~ 10^5 m^3/s, and this flow is mainly concentrated in the upper 200–300 m layer. In the horizontal direction, the currents transport ~ 5×10^6 m^3/s, i.e., about fifty times the vertical transport.

The vertical circulation includes downward movements in the coastal zones and compensatory upward transport in the interior of the basin. The downward movements are structured in separate anticyclones between the main Black Sea current and the coast (Staneva et al. 2001, Fig. 6). Vertical circulation is further amplified as a result of entrainment of Black Sea waters by the sinking denser Marmara Sea waters. Interestingly, the process of sinking is blocked by strong stratification; hence, there are multiple feedbacks between the sinking rate and the sharpness of vertical stratification. Thus, by combining downward movements in the coastal zone (caused by descending Marmara Sea waters and coastal anticyclones) and upward movements in the interior of the basin (the latter mainly related to wind forcing and the Ekman drift), vertical circulation regulates the impact of different dynamic factors on stratification, which is a unique element of the Black Sea system.

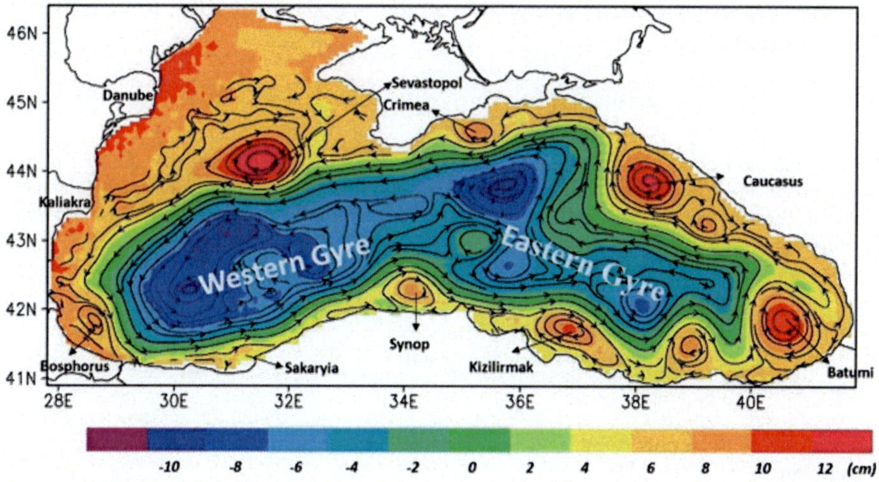

Fig. 6 A snapshot of the sea level in the Black Sea simulated by a numerical model. Arrows and lines show surface currents. The main circulation elements are the eastern and western circulation gyres. Eddies in front of Batumi and Sevastopol are almost always observed. Small anticyclonic eddies are located between the main current and the coast, usually bearing the names of nearby geographical features. The figure is from Stanev (2005) and is made available under the terms of the Creative Commons Attribution 4.0 International License (https://creativecommons.org/licenses/by/4.0/)

1.3 Thermohaline and Biogeochemical Properties

One of the most important thermohaline features in the Black Sea is the cold intermediate layer (CIL), a water mass of minimum temperature sandwiched between the seasonal thermocline[11] and the permanent halocline[12] (Fig. 7). The CIL is formed in winter as a result of strong surface cooling. The extremely stable halocline limits the depth of convective[13] mixing and sets the depth of penetration of the seasonal signal. The thickness of the CIL is ~ 20–150 m; the depth of its core is ~ 40–80 m. Following convection events, cold waters spread along isopycnal surfaces[14] over the entire basin. The strong stratification in the seasonal thermocline tends to reduce vertical mixing, shielding the CIL from the warm surface water in summer.

The variance in the density profiles depicts the small range of variability in potential density[15] (Fig. 7c) at the base of the upper mixed layer[16] (~ 40 m). Above this depth, the relatively large range of density variations is explained by the large range of potential temperature (Fig. 7a) due to the seasonal signal in the upper mixed layer. Below this depth, there is a large range of salinity in the halocline (Fig. 7b), which explains the range of density variations in the pycnocline.[17] This large salinity range is due to the strong horizontal gradients in the major frontal zone, to be presented later in this section.

---

[11] Thermocline ("cline", from ancient Greek κλίνειν klinein "incline") refers to the transition of water layers of different temperatures (sharp temperature jumps of relatively small thickness). Seasonal thermocline develops in the oceans in summer at relatively shallow depths due to surface heating.

[12] Halocline is the layer where the vertical salinity gradient is greatest.

[13] In fluid dynamics convection is often understood as process in which fluid parcels (fluid elements or volumes) move in a vertical direction, whereas by advection it is understood that they move in the horizontal direction. What is meant in the specific case here by convection is mostly caused by vertical mixing.

[14] An isopycnal surface is a surface with constant density inside a fluid. Other known isosurfaces are isobaric or isothermal surfaces, which describe surfaces with constant pressure or constant temperature.

[15] In situ temperature T is the temperature that the water parcel has at the measurement depth. Potential temperature θ is the temperature that the same water parcel would have if it were raised adiabatically (no heat exchange with the surrounding water) to the surface. The potential density σ_θ of a fluid parcel is the density the parcel would acquire if adiabatically brought to the surface.

[16] The upper mixed layer is a layer where turbulence generated by winds or surface heat fluxes tends to homogenize some range of depths.

[17] A pycnocline is the layer where the vertical density gradient is greatest. In the deeper parts of the Black Sea, pycnocline and halocline almost coincide because the role of temperature is small.

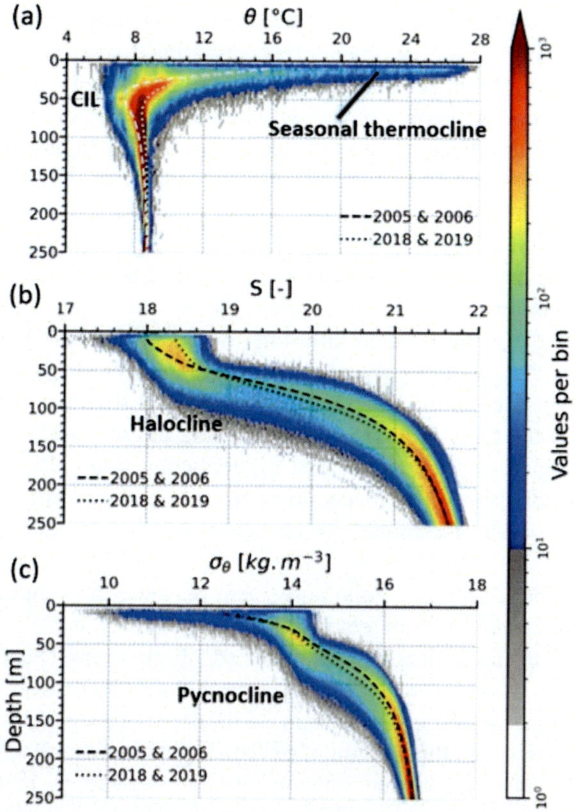

Fig. 7 Vertical distribution of **a** potential temperature, **b** salinity, and **c** potential density in the upper 250 m presented as a function of the number of samples per individual bin. The probability distribution would look the same; however, we prefer to present the number of profiles per bin to demonstrate the size of the observations. The samples, more than 6000, which have been taken by automatic ocean sensing robots (see Sect. 1 in chapter "New Observing and Modeling Systems and New Scientific Applications"), are described by Stanev and Chtirkova (2021). A gray color is used where there are less than 10 observations per bin. Bins, in which there are fewer than two observations, are ignored (white in the color bar). Dashed and dotted lines are the means for 2005–2006 and 2018–2019. From Stanev and Chtirkova (2021)

Unlike the neighboring Sea of Marmara, which has relatively "young water",[18] the deep water of the Black Sea is characterized by a slow renewal (perhaps thousands of years), therefore there are speculations that this water mass has been formed under climatic conditions different to today's. As has been shown in the numerical studies of Stanev et al. (2003), in modern conditions, ventilation of water masses from the surface occurs only during certain parts of winter. This is in contrast to the situation in the ocean, where ventilation continues throughout the year, with only the areas of ventilation changing their location. In other words, the ventilation of the intermediate

[18] Water whose properties are renewed relatively quickly, for example, within one or several years.

waters in the Black Sea is maintained by seasonal changes in its surface temperature, rather than by the presence of permanently outcropping thermal structures,[19] as is the case in the ocean. In brief, seasonal changes in the Black Sea make outcropping of isopycnic surfaces possible only in winter, whereas in the ocean there is continuous outcropping, only the places where this occurs move seasonally south and north. What was not understood prior to the studies of Stanev et al. (2003) was that water mass formation is localized predominantly over a very narrow band along the shelf edge of the Black Seas' West Coast.

The numerical experiments of Stanev et al. (2003) helped estimate the rates of water mass formation by convective cooling. The results in Fig. 8a are presented as the thickness of the water column intruding the pycnocline every year. Most cold water penetrates the CIL along the northwestern slope area (Fig. 8a). The advective cooling of the water column in these areas preconditions the penetration of surface cooling down to the CIL. Furthermore, the mixing of cold shelf water with saltier open sea water increases instability and triggers convection along the continental slope. The area of most efficient cooling acts as a small (compared to the basin surface) "throat" where the model predicts maximum penetration of cooled water into the CIL.

Figure 8b demonstrates the good agreement between the simulated interannual variability of the cold-water content between isopycnal levels σ_t[20] = 14.4 and 15.6 estimated from the numerical simulations and observations (Stanev 2005). This density interval is where most of the cold intermediate water is found.

Vertical movements of Bosporus origin are the cause of deep intrusions of oxygen-enriched water. However, the amount of oxygen input is too small to fully oxygenate the deep layers. The Black Sea is therefore characterized by sulfidic (hydrogen sulfide, highly toxic) deep waters (Fig. 9). The data in this figure are presented in density coordinates to reduce the scatter of observations caused by the strong spatial and temporal variability of the Black Sea state. Therefore, the analyses in density coordinates allow us to draw conclusions about the basin-wide changes in the physical and biogeochemical (BGC) state of the Black Sea. As seen in this graph, oxygen and hydrogen sulfide show almost opposite profiles. These two BGC properties are separated by the suboxic layer, where their respective concentrations are below the measurable values. At the upper boundary of the suboxic zone is the maximum of nitrates (NO_3), which provides nutrients for the BGC processes in the upper layers.

19 In the World Ocean, surfaces of constant density (isopycnal surfaces) have different depths. In the polar regions, some isopycnal surfaces reach sea surface (outcrop). This is due to the strong cooling there, which tends to increase water density, water stratification becomes unstable, and convection homogenizes vertically water masses. Outcropping positions change during the year (move to the south in winter and to the north in summer).

20 σ_t is a quantity used in oceanography to specify the density ρ of seawater at a given temperature T and salinity S. It is defined as $\rho(S,T)$-1000 kg m^{-3}, where $\rho(S,T)$ is in situ density. For example, a water sample with a density of 1026 kg m^{-3} has a σ_t value of 26.

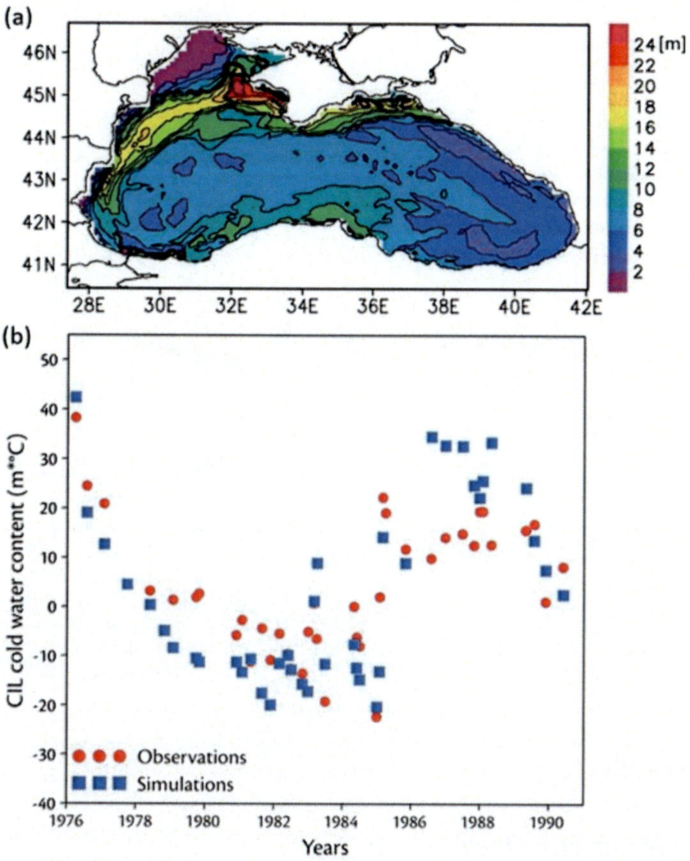

Fig. 8 Cold-water mass formation in the Black Sea by convective cooling estimated using an ocean circulation numerical model with a horizontal resolution of 5′. **a** Shows the geographical distribution of the intensity of water mass formation (see text), **b** shows temporal variability of the cold-water content between isopycnal levels $\sigma_t = 14.4$ and 15.6 estimated from observations and numerical simulations. The figure is adapted from Stanev (2005), which is made available under the terms of the Creative Commons Attribution 4.0 International License (https://creativecommons.org/licenses/by/4.0/)

BGC processes in the deep Black Sea layers are sustained by chemosynthesis. This is a process in which organic matter is not produced by sunlight, as in photosynthesis,[21] but the oxidation of inorganic molecules, such as hydrogen sulfide, serves as the energy source. Here I recall that the Black Sea is the largest enclosed anoxic (oxygen-free) basin in the World Ocean and therefore the interest in its biogeochemical processes is enormous.

[21] Photosynthesis is a process for producing energy-rich biomolecules from lower-energy substances using light energy. It is carried out by plants, algae, and some bacteria, which convert light energy into chemical energy with the help of light-absorbing pigments such as chlorophyll.

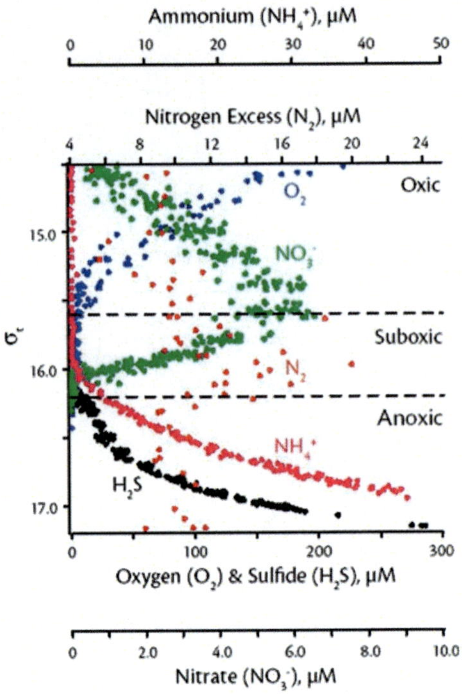

Fig. 9 Vertical biogeochemical structure of the Black Sea water presented in density coordinates. The figure is from Konovalov et al. (2005) and is made available under the terms of the Creative Commons Attribution 4.0 International License (https://creativecommons.org/licenses/by/4.0/)

2 Polymode

2.1 Eddies in the Ocean

2.1.1 What Are Eddies?

The years 1977–1978 will go down in the history of oceanography as the years of Soviet-American cooperation in the framework of the POLYMODE oceanographic experiment. In 1977, I was invited by the Shirshov Institute of Oceanology of the USSR Academy of Sciences to participate in the 27th cruise of the research vessel "Academician Kurchatov" under the POLYMODE program. After the first crewed international space mission Soyuz-Apollo, this was the second major scientific cooperation of the USSR and the USA. The research program was dealing with the rotational motions of ocean water (see the text in the following box).

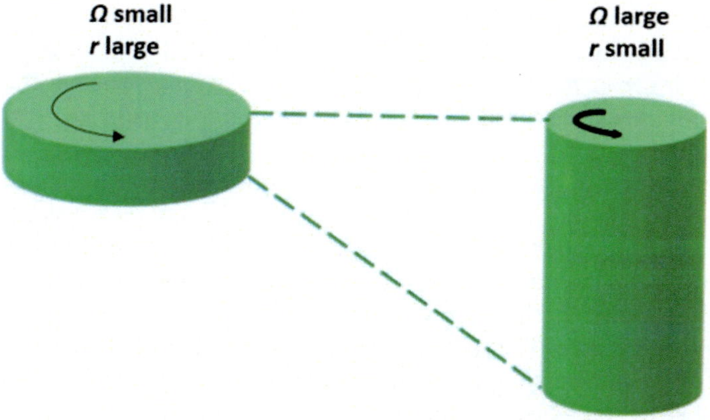

Fig. 10 To the explanation of the law of conservation of angular momentum. Courtesy of S. Hartmann

Conservation of Vorticity

Rotating bodies have the property of conserving their angular momentum K. The latter is proportional to the angular velocity of rotation ω and the square of the radius r of the rotating body, $K \sim \omega r^2$. If the rotating body is a cylinder of height H, its volume is $V = \pi r^2 H$, and $K \sim \omega V/H$. If $V = \text{const}$ (volume is conserved), it follows that $K \sim \omega/H$. Figure 10 illustrates the fact that if the height of the cylinder decreases, this will result in a reduction in its angular velocity, which we know from spinning ice dancers. This simplistic reasoning can be carried over to fluids. The difference is that on a rotating Earth, the horizontal component of the Earth's local angular velocity is added to the proper angular velocity of rotation (in fluids, we call it vorticity). The local angular velocity at latitude φ is $\Omega\sin\varphi$, where Ω is the Earth's angular rate of rotation. The local angular vorticity is $f = 2\Omega\sin\varphi$, known as the Coriolis parameter or planetary vorticity. If we formally use the angular momentum conservation law $K \sim \omega/H = \text{const}$ for fluids on the rotating Earth, we can add f to ω and write $(\omega + f)/H = \text{const}$. This equation, which is called the equation of conservation of potential vorticity, states that the two parts of vorticity, the one due to fluid rotation (ω) and the planetary vorticity (f) compensate each other. This mutual balance has very broad applications in ocean dynamics. For example, if for some reason a parcel of water changes latitude, its own vorticity ω has to change in order to compensate for the change in f (f is a function of latitude).

The ocean is full of mesoscale[22] eddies similar to cyclones and anticyclones in the atmosphere. Hence, one can say that eddies set the "weather" of the ocean.

[22] Mesoscale, of medium size or extent, such as ocean cyclones and anticyclones.

Temperature anomalies make the eddies visible on satellite images because eddies are three-dimensional structures, in which waters rise up to the surface, or descend deep into the ocean, depending on the sense of rotation. In case of upwelling, surface temperature usually decreases because deep water is cold, and vice versa in case of downwelling.

Eddies can be thought of as huge, with sizes from kilometers to tens of kilometers, "molecules" or volumes of water that pass from one area of the ocean characterized by certain properties (temperature, salinity, or biological properties) to another. They thus contribute to the exchange of water between polar and tropical latitudes and to the supply of nutrients to the surface layer of the ocean, causing plankton blooms. This exchange process due to eddies, unlike viscosity at the molecular level, is a function of the flow regime. The peculiarity of these "macromolecules" is that they are in a state of continuous rotational motion.

Eddies

The ratio between vorticity ω and Coriolis frequency f, known as the Rossby number (Ro), measures the relative importance of rotation of water parcel against the planetary rotation. At small Rossby numbers, vorticity is negligible and the motion is almost straight (geostrophic currents), while at large Rossby numbers (cyclones and anticyclones), the rotation of water parcels is comparable, or larger than the planetary rotation.

In geophysical hydrodynamics, the science of fluid dynamics of the Earth and other rotating planets, there is a fundamental scale called the Rossby radius of deformation (L_R), which gives a measure of the size of eddies. L_R increases with the increase of stratification and decreases with the increase of the Coriolis parameter. While f is only a function of latitude, the vertical stratification of water varies from one geographical region to another. In this way, conditions are created for the generation of eddies of different sizes depending on the local stratification. For example, eddies in the Arctic, where the stratification is weaker and the Coriolis parameter big, are many times smaller in size than eddies in tropical latitudes (in the tropics the stratification is stronger and the Coriolis parameter smaller). However, small ocean eddies can also be kilometer-sized. The scale dependence of dynamics of the atmosphere and oceans (in our simple example the size-dependence of Ro) results from the fact that the balance between the acting forces changes depending on the spatial size of motion.

Eddies (see the example in Fig. 11) are created as a result of the instability of ocean currents, such as the Gulf Stream. According to the Soviet scientist Andrei Kolmogorov, turbulence in fluids develops as follows. Once a jet or current loses stability, a separation of large turbulent eddies begins. These can also be unstable and break up into smaller ones until the energy dissipates (i.e., until it turns into heat). This cascading process is the basis for the conversion of jet energy to vortex

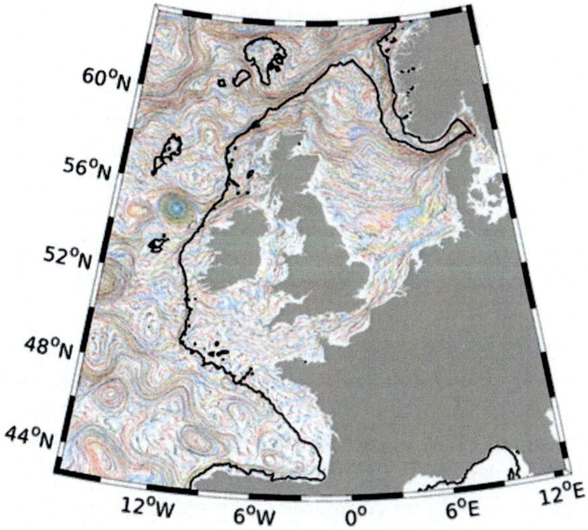

Fig. 11 Eddies in the Eastern Atlantic. The numerical model that reproduces the currents is described in Stanev and Ricker (2020). In the figure, ocean eddies are visualized as particle trajectories.[23] Particles are uniformly "seeded" at computational grid nodes and tracked over time. The eddies are well manifested in the deep ocean (the solid line is the 200 m isobath), whereas the shallow North Sea does not support their generation. Courtesy of M. Riker

energy and then to heat. In 1970s, it was not pretty clear how this theory worked as applied to large-scale motions in the ocean. What was also not very clear was how the atmosphere transmits mechanical energy to the ocean and how it is further distributed in space and time. Atmospheric processes are known to be highly variable; the atmosphere is continuously exciting the otherwise "calmer ocean". This could present a mechanism contributing to the generation of ocean eddies.

Vertical stratification of water masses in the ocean is much weaker than in the atmosphere. Therefore, eddies in the ocean are much smaller (typical horizontal scales are less than 100 km), and their timescales (about a month) are much longer than those of cyclones and anticyclones in the atmosphere (about a week). Here is an example: a horizontal scale in the ocean of 100 km would correspond to an atmospheric scale of 1600 km, and an oceanic period of thirty days would correspond to two atmospheric days. Because of the small scales and long characteristic times, complex and long-term measurements are needed to establish exactly how "ocean weather" works.

[23] In fluid dynamics, there are two approaches to study the motion. In the Lagrangian approach, the observer follows an individual fluid parcel as it moves through space and time. Plotting the position of an individual parcel provides its trajectory. Lagrangian particles are characterized only by their positions. They do not have mass and do not interact with the fluid (virtual particles). Following their propagation is referred to as particle tracking. In the Eulerian approach the observer looks at fluid motion in specific locations over time (e.g., a moored instrument).

2.1.2 The Scientific Background of POLYMODE

The small size of ocean eddies is one of the reasons they were discovered much later than their atmospheric homologues. Their existence was first demonstrated experimentally in 1970 during the Soviet experiment "Polygon 70" [24] conducted in the Atlantic Ocean north of the Cape Verde Islands. The observing system included 17 moored buoys placed on a cross measuring 240 × 240 km in zonal and meridional direction centered at 16.5° north latitude and 33.5° west longitude. This experiment lasted more than seven months and revealed an anticyclonic vortex with a diameter of approximately 200 km. This vortex was large because of the small value of Coriolis force between the tropics and the equator.

From March to mid-July 1973, the Mid-Ocean Dynamic Experiment (MODE-1, mainly USA, UK) was conducted to study eddies in a 600 km diameter region southwest of Bermuda. How representative this sampling was in order to provide a credible record for eddies can be understood from the following comparison with the atmosphere. MODE-1 could be compared to an experiment aiming to study atmospheric weather, which continued for about a week over an area of 1600 km. Recalling that cyclones and anticyclones have characteristic spatial scales of ~ 1000 km and temporal scales of about one week, we can then ask what the chance of a well-defined cyclonic or anticyclonic event being recorded in the atmosphere during this short period is, and in such a small survey area. It became clear in 1970s that multi-year measurements were to be made in different regions of the World Ocean. For that, the efforts of one country alone were insufficient. In the 1970s, the USA and the Soviet Union joined forces in an experiment called POLYMODE (the name was a combination of "Polygon" and "MODE") to advance the science of eddy motion in the ocean.

About POLYMODE

Issue 39 of POLYMODE news of 25 November 1977 formulated the main experimental and theoretical problems of POLYMODE:

1. To carry out field observations and experiments, primarily in open ocean regions of the western North Atlantic, designed to advance our knowledge of the kinematics and dynamics of the variability in that region and to determine their role in the circulation of the North Atlantic subtropical gyre,
2. To pursue theoretical/numerical modeling of the phenomenon and to apply theory to the design and rationalization of the POLYMODE field data via both local forecast-process numerical models and high-resolution numerical models of the North Atlantic gyre circulation.

Of major interest was the elucidation of the three-dimensional spatial and temporal structure of the eddy field, and the determination of the prevailing dynamical balances that govern the motion and interaction of eddies with

[24] In geometry, a polygon is a plane figure, which is enclosed by a finite number of connected line segments. What is rather understood here under "polygon" is a survey area.

motions of larger or smaller scales. An important milestone in the joint collaboration was the agreement to mutually make the results and data available. POLYMODE has shown the way forward for the rapid development of global science.

At that time I was taking my first steps in the field of numerical ocean modeling. Limited computational resources and insufficient global experience in this field of oceanography hampered the simulation of global eddy dynamics. The main difficulty stemmed from the fact that models resolving eddies are computationally heavy, as they need to operate with a fine spatial grid. This is imperative in order for them to capture small eddies (e.g., scales of ~ 10 km that are much smaller compared to the radius of the globe). This requires huge volumes of information to be processed, and this is where the conflict between available computing resources and the small size of ocean eddies requiring huge computing resources comes into play. Figure 4 in chapter "In Search of a Destination" gives an idea of the increase of computational performance in the last thirty years.

2.2 *On the Kurchatov Vessel*

The cruise in which I took part, from July 27 to October 15, 1978, was the last of the Soviet POLYMODE cruises. Along with the scientific objectives, we had the task of completing the last stage of the observational work and retrieve the equipment operating autonomously in the sea. The research area was not far from Bermuda.

I worked in the group "Processing the measurement data" with the mathematicians from Novosibirsk Vladimir Zalesny and Viktor Kuzin, former graduate students and collaborators of the President of the Siberian Branch of the USSR Academy of Sciences Gury Marchuk, future President (1986–1991) of the USSR Academy of Sciences. The head of the cruise was Konstantin Fedorov of the Shirshov Institute of Oceanology of the Academy of science of USSR (Fig. 12). His main scientific activities were in the field of observations of ocean fronts.[25] From 1963 to 1969, he was Secretary of the Intergovernmental Oceanographic Commission (IOC) of UNESCO.

I will characterize K. Fedorov by recounting the following case, which is characteristic for curiosity as an important scientific driver. At some point during the expedition, he began to fear that hot weather might bring us a tropical cyclone, or as they call it in the Atlantic, a hurricane. He guessed correctly; the US Hurricane Watch reported that a hurricane was forming in the immediate vicinity of our survey

[25] A front is a boundary between two different water masses. Fronts can be as narrow as a few hundred meters and as wide as tens of kilometers. Some fronts can persist for long periods of time.

Fig. 12 Portrait of Konstantin Fedorov outdoors with a camera. *Credit* Photo by Warren Wooster, School of Marine Affairs, University of Washington, courtesy of AIP Emilio Segrè Visual Archives

Fig. 13 A huge cylindrical body of light material is anchored at a depth of several thousand meters. Between it and the anchor automatic devices measuring the speed of ocean currents are fixed. A metal reflector is located at the top, facilitating its locating

area on 30 August. Initial information was controversial, and the direction of its movements not entirely clear. The situation was not safe, as the hurricane was raging at a distance of several hundred kilometers from us. The next day, 31 August, the picture became clearer. The diameter of the hurricane was 80–100 km, the wind speed, 140–160 km/h. It moved across the observational array with a speed of 13–15 knots.[26] The maximum speed of the ship was 18 knots. We were no more than 200 km from the hurricane and this was felt. The sea and sky had the same leaden gray color, it was pouring rain in the evening, and the waves were tossing the 7000 tons heavy ship.

We were made to close the iron bumpers on the portholes of the cabins, and the crew was fastening everything that could move. According to reports from Miami, the hurricane was heading for an area where three autonomous stations were operating. Konstantin Fedorov and the captain made the decision to move behind the hurricane in the direction of those stations. The scientific curiosity was stronger than fear of the elements. Although there was a danger that the hurricane would change its course, which by the way often happens, the "golden chance" to make unique measurements was not to be missed. Some organize special expeditions to study hurricanes, and we received such a gift from nature. The data we obtained allowed us to study the processes of interaction between the atmosphere and ocean in extreme situations, which is very important for science. This is because a significant part of the energy they exchange occurs during the short lifetime of hurricanes. It turned out that the powerful winds swept huge masses of water from the center of the hurricane. This, together with the strong cooling due to evaporation, led to a sharp drop in water temperature, which was amplified by the rise of colder water masses from the deep layers. Measured data showed that the hurricane left behind a trail of cold water, Fedorov used to say "a street of cold water". He was very enthusiastic and discussed the new results with youthful zest. Calculations showed that for the time the hurricane beat over the survey area, the heat input to the atmosphere from each square centimeter amounted to 5500 cal. For the whole area of survey, this represents a colossal number of 1.32×10^{15} kcal, in other words, equivalent to burning 100 million tons of oil, which is comparable to the world's oil consumption over a two-week period!

The work done during the voyage finalized the work done by dozens of scientists during fourteen months of continuous activities. Toward the end of the cruise, the intensity of the work of the group to which I belonged increased because all the raw data from the autonomous stations were available and we had to carry out their preliminary processing and prepare them for storage and further use. Some initial interpretation of the results was also necessary, i.e., a preliminary analysis of the ocean weather in the study area over the last three-month period. The maps of ocean eddies that we produced revealed complex temporal and spatial variability in temperature, salinity, and currents. These went into the cruise's final scientific report, which was more than a thousand pages long. The oceanographic observations and new knowledge we obtained can only be seen as one brick in the edifice of eddy

[26] A knot (kn or kt) is a unit of speed equal to one nautical mile (1.852 km) per hour.

science. In less than five years, the role of eddies in ocean dynamics has become much clearer (Robinson 1983).

3 In Search of the Best

Marine science, in the former socialist countries, was organized along the lines of the COMECON under the theme "World Ocean". My laboratory was working on the task "Numerical modeling and prediction of hydrological fields and processes in the most important regions of the World Ocean". Artem Sarkisyan, who became in 1992 an academician of the Russian Academy of Sciences, was the first who "took control" of the emerging Bulgarian theoretical oceanography. My work with the numerical model that A. Sarkisyan made available showed me that this model was a relatively simple system for diagnosis of ocean currents. Given a set of temperature and salinity measurements, the model could recover the currents corresponding to these fields. However, it had problems to solve the more general problem, namely to compute from given atmospheric data the temperature and salinity, as well as the corresponding currents and the evolution of the ocean system over time and in space as a result of changes in the state of the atmosphere and the internal dynamics of the ocean.

One day, in the early 1980s, S. Panchev entered my office saying: "Look at this report I got today". He was well known among foreign colleagues and many scientists used to send him their latest works. The institutes that were disseminating the results of their science had included him in their correspondence lists, and along with him I was also informed about new developments. He left me the report that turned out to be one of the most helpful documents I have ever received.

Back in the late 1960s, K. Bryan, from the Geophysical Fluid Dynamics Laboratory (GFDL) at Princeton University, had developed a model of ocean circulation (Bryan 1969) that he later refined with his collaborators Michael Cox and Bert Semtner.[27] What they were doing was the best of its time in the field of numerical ocean modeling, but my lack of sufficient self-confidence prevented me from asking K. Bryan to provide me with his model. The report that S. Panchev left me, (Semtner 1974), was the complete description of the model. The theory was known to me from the papers I had read, but as I looked through it, I found out that the second part of the report was the computational program of the model written in FORTRAN, and at the end there were the results of a test experiment. The first thing I did was to transfer the program to punch cards, which at that time we used as a communication medium with the computer. Next, I cleaned up the syntax and other errors until I reproduced the results of the published test experiment. Once you get an identical result, you are sure that you have a reliable tool for the job. And your work only now

[27] For brevity we will refer to this model as the GFDL ocean model, knowing that there are other ocean models developed in this laboratory, one of them being the Princeton Ocean Model (POM).

begins, because the test was for an idealized rectangular ocean.[28] I choose the area I am interested in, for example the Black Sea, get data on its bathymetry, data on the initial distribution of temperature and salinity in the sea, atmospheric data to force the model, for example, wind, air temperature, precipitation, etc. The overall result was that I was able to use the best for that time numerical modeling framework.

Years later, I developed personal relationships with colleagues from the GFDL. They were all amazed when I told them about how I started to use the GFDL ocean model, instead to directly asking its authors to send me the code on a magnetic tape. Perhaps there were not many in the world who had done the same… and that because of an iron curtain that was not completely impenetrable after all.

From October 2–10, 1983, the First International Symposium on "Integrated Global Monitoring of the World Ocean" took place in Tallinn, Estonia. The short name was MONOC, i.e., "monitoring of oceans". I was already advanced with the GFDL ocean model and presented a paper on "Numerical studies of seasonal variability of ocean currents" (Stanev and Roussenov 1985). This study evaluated the overall energy balance of the Black Sea using a model that was consistent in terms of energy conservation. It gave an assessment of the exchange of energy between the atmosphere and the sea and of its conversion between different types of motion. The vertical stratification of the sea was successfully reproduced by the numerical model, because we paid particular attention to the role of seasonally varying heat fluxes at the Black Sea surface in shaping the vertical and horizontal structure of seawater characteristics. The qualitative agreement between the simulated circulation, illustrated in Fig. 14 by sea surface height, and the circulation produced by a more refined model (Fig. 6), shows the reliability of these early modeling results. The work was very well received. [29]

In Tallinn I made two connections that have accompanied me throughout my scientific career. The first was with Jürgen Sündermann (Fig. 15) from the Institute of Marine research, Hamburg. It is important to note that he was the driving force behind the collaboration of German numerical modelers with the Siberian Division of the USSR Academy of Sciences. His joint works with Polish colleagues contributed to the rapprochement of oceanographers from the two neighboring countries. Perhaps the fact that his birthplace was in present-day Poland played a role. But he also went much further east, deepening contacts with Chinese oceanographers.

Another new colleague I met in Tallinn was Gennady Korotaev from the Marine hydrophysical institute in Sevastopol. After graduating from the Faculty of Mechanics and Mathematics of Moscow State University, he worked in Sevastopol, mainly in the field of Black Sea research. Since the time of political changes in the 1990s, Gennady and I have been working on similar problems, and I dare say that we enrich each other in a friendly competition.

[28] In many theoretical studies, the ocean is presented as a simple rectangular basin. Experiments with such idealized oceans are known as academic experiments.

[29] Sea surface height is the local elevation of the ocean relative to its resting level (called the geoid). Variations in sea surface height reflect dynamic phenomena and structures in the ocean, such as eddies or large-scale currents.

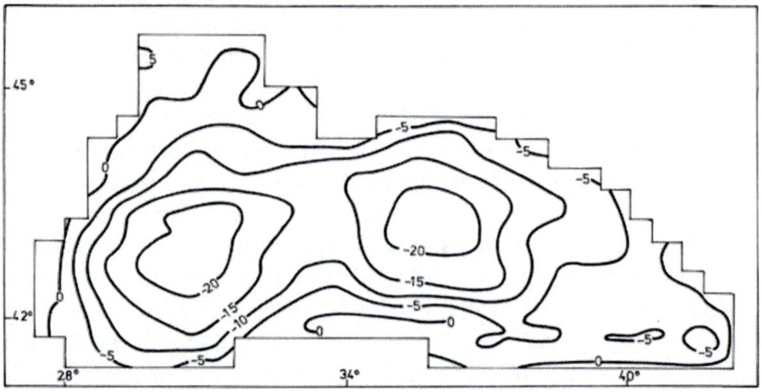

Fig. 14 Sea surface height in cm in winter. The numerical simulations are carried out with the GFDL ocean model with horizontal resolution 1° × 0.5°. From Stanev (1989a), with permission from Elsevier

Fig. 15 Jürgen Sündermann. From Frins (2019), photo: J.-R. Lippels, credit to J. Sündermann

4 The Chernobyl Disaster

For me, part of the story of the Chernobyl disaster is related to my work with the Moscow State University on a topic aimed at "Using numerical models to estimate currents in complex regions of the World Ocean and to produce maps of the climatic state of the hydrophysical fields in the Black Sea". On Monday, April 28, 1986, I arrived in Moscow to visit the Department of Oceanology of the Moscow State University and the Shirshov Institute of Oceanology of the USSR Academy of Sciences. Two days before, an accident had happened at the Soviet Chernobyl nuclear power plant. In the evening, after arriving at the hotel, I turned on the TV and saw the first announcement about the accident. On the background of a very unclear picture, the speaker reported that something had happened, but the tone was as if it was not dangerous. The announcement came after Swedish stations detected elevated radiation. The accident caused a cloud of radioactive waste to pass over parts of the USSR, Eastern Europe, and Scandinavia. Nearly 60% of it fell on the territory of Belarus. There were no official reports in Bulgaria until the third day, and later it was claimed that there is absolutely no danger to the country.

Because of the poor information, no one was aware of the danger. My kind hosts Alexey Kosarev, Alexey Blatov, and Valentin Tuzhilkin organized a picnic at a dacha near Moscow soon after my arrival. It was a rainy, wet and cool April day, and we were having a barbecue in the open air. Big drops were falling from the wet pines and larches. I later found out that at some time the radioactive cloud had turned north toward Scandinavia. What kind of radiation had we absorbed…

When I returned back to Bulgaria on 11 May, I learned that the situation was serious. This time, I was well informed by the colleagues from the Faculty of Physics. The radioactive cloud had reached Bulgaria on 1 May. The head of the Department of Nuclear Physics, Prof. Tsvetan Bonchev, a man of good intuition and flair, had mobilized his coworkers. Anyone who could measure anything to do with Chernobyl radiation was to do so. My colleagues from this Department discovered that Chernobyl contamination contained so-called hot particles, i.e., micron-sized particles that have high activity. I remember on the tiles and asphalt around the faculty building, colleagues had marked with chalk the places where they found "hot particles". A new problem was also discovered: the concentration of radioactive elements in the body of Bulgarian citizens was constantly increasing.

After the political changes in the Eastern Europe that came soon after the accident, I developed professional links with Hugh Livingston, since 1998 the Director of the Marine Environment Laboratory in Monaco at the International Atomic Energy Agency, and with Ken Buesseler, since 2013 the Director of the Center for Marine and Environmental Radioactivity at the Woods Hole Oceanographic Institution (WHOI) in Massachusetts. These world-leading radiochemists had realized that the experimental study of the intrusion of contaminated seawater together with numerical modeling of this process represented a powerful means of improving the quality of numerical models, making it possible to recover the spatial and temporal variation of concentrations. The work on caesium and strontium penetration into the deep layers of the Black Sea (Stanev et al. 1999c; Staneva et al. 1999) became later the basis for further research, supported by the Atomic Energy Agency, on "The application of tracer techniques in the study of processes and pollution in the Black Sea". It so happened that something positive in scientific terms came out of the huge accident and the disaster that followed, because the measured data contributed to the testing and improvement of our models. I recall that the founding (1993) coordinator of the Global Environment Facility Black Sea Environment Program in Istanbul, Laurence Mee, expressed a mistrust of models at the beginning of our communication, which was gradually replaced by admiration for what they were able to contribute to science.

5 At the Dawn of Marine Services

There is today one research and development (R&D) field that is particularly revered within the European research programs, namely "Applied Oceanography and User Services". I will discuss it in more detail in the next sections. However, when I try to reconstruct what was going on 40 years ago, I recall that, in Bulgaria, there was a distant precursor of what is being done in Europe today. The Ministry of Education, and in particular the Council for Higher Education, looked very favorably on our developments. Some of these developments were in the field of applied oceanography and used numerical predictive models, which allowed us to quantify the three-dimensional structure of temperature, salinity, currents, or the concentration of various pollutants in the marine environment. Although these reconstructions were model-based, i.e., they did not necessarily represent the situation in the sea completely and accurately, they were very useful, at least for providing preliminary estimates. The most valuable result was that the models gave us an enormous amount of information, incomparably more than what the most modern measurements of the time could provide. Our other advantage was that we could play out scenarios, something we cannot afford to do with nature.

We were sought after for consultations by the National Committee for Environmental Protection and the Hygienic and Epidemiological Inspectorate in Varna, which were interested in the problems of discharge of wastewater. One issue was that during the operation of the refinery in Burgas, the level of pollution of Lake Vaya had increased to an unacceptable level and the bottom was covered with substances poisonous to the ecosystem. The question was put like this. Is there an alternative for what to do with the wastewater? If so, can we discharge it into the sea? If we do so, where should we discharge it and in what quantities and concentrations?

The scientific problem is to prove with sufficient precision what will happen to the waste water that ends up in the sea. Will it ascend to the surface? This is possible since, although contaminated, this water has low salinity and its density is expected to be less than that of the surrounding fluid. However, if for some reason, it remains at the bottom, how thick will the contaminated layer be? Over how big an area will the plume spread? So far everything can be solved with the tools of numerical ocean modeling. The more difficult question was how to mathematically describe the BGC processes of interaction of the contaminated water with the surrounding fluid. Will these processes clean up the polluted water, or will the polluted water destroy the ecological system? If yes, will that be an irreversible process?

In the early 1980s, our research interests shifted to more coastal work in the frame of a project on numerical modeling of the dynamics of the Bay of Burgas and the distribution of pollutants in it. We used a model with a resolution of 1 km, which was very fine for this time, and developed it further in order to use it to analyze the sewage discharge and its propagation in this area. Here, it was essential to address the issue of defining the diffusion regime of pollutants in the models.

In 1981, starting the work on the topic "Numerical modeling of physical and chemical-biological processes in the Black Sea", I had to become literate in the field of Black Sea biogeochemistry. Not only were the new scientific aspects a high hurdle to overcome, but such seemingly simple things as units of different properties also posed problems. There was a great absence of coherence, some scientists were presenting their measurements in grams per liter, others in molar units, however, in a model all quantities have to be represented in the same units. I remember the Harvard University professor Allan Robinson saying, "The hardest thing for physicists and biologists to work out is a common language".

6 Bulgarian FerryBox

The Bulgarian Hydrographic Service (HS) of the Navy provides the hydrographic and hydro meteorological support of the military fleet and civil navigation in the territorial and internal waters of the country. It has the exclusive right to issue and distribute nautical charts and aids to navigation. It is also responsible for lighthouses and buoys in Bulgarian waters. Because both hydrographers and oceanographers, explored the sea, we had common interests that led us to some institutionalization of our works. Our collaboration involved coordinating work plans, hosting graduate students and giving advice, helping and assisting with data processing, giving lectures, and developing instruments. There was a commitment on the part of HS to include us in their expeditions, to provide us with data, and to make their facilities, including a hydrographic vessel, available within their means for the training of our students.

Hydrographers complained that there were some inconsistencies in their chemical observations and feared their poor quality. In the beginning of the 1980s, they were still using the classical measurements of seawater salinity based on a chemistry method called titration. This method estimates how much of a reactant is required for a particular chemical reaction to take place. The work is done manually; water

Fig. 16 Author in discussion with the officers of the Bulgarian hydrographic ship

samples are brought to the lab where a marine chemist determines the salinity. One drawback is that each measurement needs a long time to be completed.

In the beginning of 1980s, fellow geophysicists in the Department of Meteorology and Geophysics of the University of Sofia had developed a sensor for geological surveys that measured the salinity of drilling solutions. In practice, it measured the electric current flowing through an electrolyte, which depends on the concentration of the solution. This principle ensures that measurements are completed within a short time period. Similar sensors were already being used in western countries for ocean measurements, but in Bulgaria, it seems, the money to buy such equipment was not enough, and perhaps the information about the sensors was insufficient. It did not take long and we convinced the hydrographers of the need to conduct parallel measurements to test the quality of the data with our salinity sensor.

My colleague Atanas Geshev, who was one of the main designers of the salinity sensor, joined one of the cruises of the hydrographic vessel. We said that we wanted to make parallel measurements, but I personally wanted to check the precision of the lab technician's work. For the lab technician and Atanas, the samples they were measuring were just numbers. With Atanas, the measurement was completed in seconds, whereas with the lab technician it took many minutes. I repeated some samples without both of them knowing, whereby the measurement of Atanas gave the same results on repetition, while the ones of the lab technician often differed a lot for the same water.

This research cruise gave birth to an idea of which I am very proud. The ship's engineer (the one who looked after the engine work), Capt. III rank, Trifon Trifonov, showed me around the engine room. He was a very sociable and nice person. Trying to overpower the rumble of the machinery, he explained to me the details of the engine. I learned that it is not a problem to pump seawater continuously, so that at any moment or at any point of the ship's location it is possible to measure temperature and salinity. This is the working principle of the so-called FerryBox system.[30] Such a system started to be actively used in oceanographic practice (Petersen et al. 2006), but a long time after we put it in place on the HS vessel. I have to admit that, apart from temperature and salinity sensors, modern FerryBox systems use many other sensors, such as for oxygen, suspended sediment, etc. Before the next expedition on the hydrographic vessel, V. Roussenov, who was very keen on programming, developed a software to facilitate the continuous measurements and to analyze and visualize the data in real time.

[30] FerryBox is a flow-through system installed on board a vessel (ferry, cargo ship, or research vessel) for automatic and continuous measurement of marine parameters.

7 Mediterranean Sea and Baltic Sea

7.1 *First Steps*

My confidence in working with the GFDL ocean model arouse my curiosity to move beyond the Black Sea. My graduate student Stoyan Botev and I configured the model to work in the conditions of the Mediterranean (Stanev et al. 1989). This model turned out to be the first model for the Mediterranean Sea that accounted for the vertical distribution of temperature and salinity by discretizing the sea volume, introducing many layers in the vertical direction (Pinardi and Navarra 1993).

Mediterranean Sea Modeling, 30 Years of Continuous Progress

Stanev et al. (1989) used the GFDL ocean model to demonstrate the significance of seasonal forcing for the maintenance of the large-scale circulation in the Mediterranean Sea The model had a very coarse resolution compared to the present-day models, 1° in zonal and 0.5° in meridional directions, and only seven layers in the vertical (Fig. 17). At the sea surface, climatological data was used as atmospheric forcing. Although very coarse, the model was able to reproduce some fundamental features of water masses typical for the Mediterranean Sea. Among these are the formation of Levantine water and its westward spreading, as well as the deep penetration of surface water due to winter cooling in the northern part of the Liguro-Provençal basin.

The inset of Fig. 17 is an image from the simulations using the operational model for the Mediterranean Sea of Clementi et al. (2021). This model has horizontal grid resolution of 1/24° (ca. 4 km) and 141 unevenly spaced vertical levels. It has approximately five thousand times more grid boxes than the first model of Stanev et al. (1989). Horizontal resolution, which is ~ 18 times finer, necessitates using an 18 times shorter time step. This means that for the same period of model integration, about 100,000 more computational resources are needed to run the model of Clementi et al. (2021). This reflects approximately the power of present-day computers compared to the ones we used in the 1980s (see Fig. 4 in chapter "In Search of a Destination").

My interest to widen my expertise beyond the Black Sea led to my next steps toward the Baltic Sea (Fig. 18). What the Black Sea and the Baltic Sea have in common is that both have low salinity due to the large inflow from the rivers and the relatively small exchange with the Mediterranean and the North Sea, respectively. In both, a strong stratification of the water masses is build, making vertical exchange difficult. The amount of oxygen introduced with the sinking denser waters from the straits is insufficient to oxidize the organic matter in the deeper layers of two seas and the water there is anoxic. This explains the interest in studying the Black Sea and the Baltic Sea in parallel. In Sweden, our work on modeling processes in the

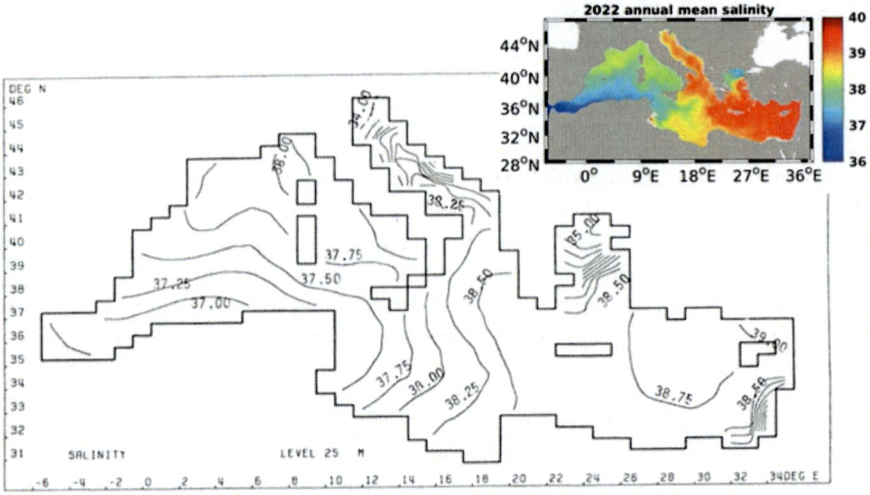

Fig. 17 Salinity distribution at 25 m depth in the model of Stanev et al. (1989), with permission from Elsevier Masson. The inset shows annual mean salinity in 2022 from the operational model of Clementi et al. (2021). Courtesy of K. Wahle for providing the plot in the inset

Black Sea was known, and V. Roussenov and I were invited to attend the conference on the dynamics of anoxic basins organized by the University of Gothenburg. My paper "Numerical modeling of the dynamics and distribution of oxygen and hydrogen sulphide in the Black Sea" (Stanev 1986), published in the conference proceedings, turned out to be the first attempt to model interactions between oxygen and hydrogen sulfide in three-dimensional space. The results of the numerical experiments gave clear indications of the strong correlation between physical and chemical fields, in particular on the role of vertical motions and horizontal circulation in shaping the temporal and spatial variability at the interface between oxic and anoxic waters.

7.2 *Water Masses and Topography Routing*

In what follows, I will briefly present some basic ideas about the thermohaline state of the Mediterranean and Baltic Seas. As an illustration of the similarities and differences between the individual semi-enclosed European seas, I have chosen the question of the long-term variability of the thermohaline fields (Stanev and Xi 2013 present an extensive review on diverse similarities and differences between the European semi-enclosed seas).

Differential heating through the sea surface causes density differences, which in the presence of the gravitational field drive the thermohaline circulation in the ocean and tends to create specific structure of water masses. In regional ocean basins, thermohaline patterns are also strongly affected by topography routing. The situation in the Black Sea, which was briefly presented in Sect. 1.3, is slightly simple, because the convective signal from the sea surface rapidly fades and no substantial intrusions of surface waters are observed in more than two-thirds of the water column. The situation is different in the relatively shallow Baltic Sea and in the weakly stratified Mediterranean Sea.

The complex bathymetry and geometry of the Baltic Sea and the Mediterranean Sea largely shape the geophysical flows, making them strongly dependent on the regional geometries. The principal difference between these two basins is that the Mediterranean Sea is a deep basin (mean and maximum depth of 1500 m and 5267 m, correspondingly), while the Baltic Sea is very shallow (mean and maximum depth of 55 m and 459 m, correspondingly. However, there are some similarities. The connection of the sub-basin known as the Gulf of Bothnia (Fig. 18) with the central Baltic Sea is restricted by the Archipelago Sea, which leads to the formation of contrasting water masses. This could be considered as a similarity with the Adriatic Sea, which is part of the Mediterranean Sea, in the sense that both regional basins are separated from the deep part of the sea by sills (with different depths) and that they have the potential to provide an important source for cold-water masses penetrating into the deeper layers of the main body of the basin.

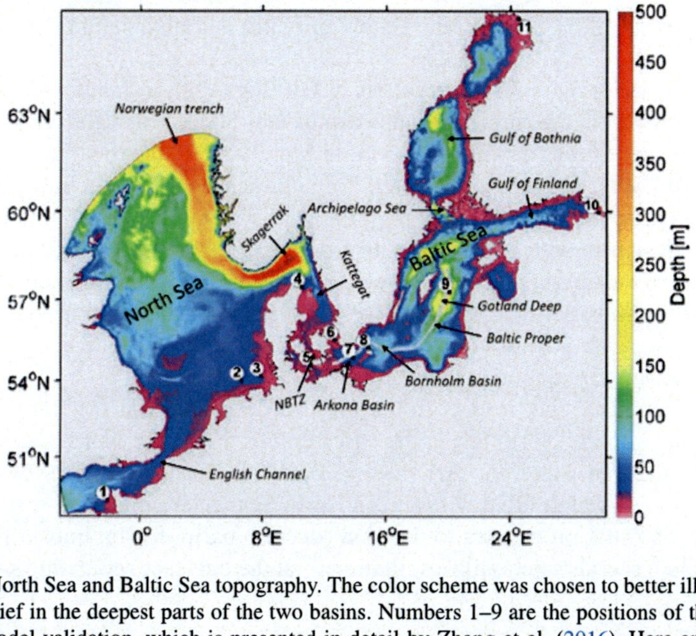

Fig. 18 North Sea and Baltic Sea topography. The color scheme was chosen to better illustrate the bottom relief in the deepest parts of the two basins. Numbers 1–9 are the positions of the stations used in model validation, which is presented in detail by Zhang et al. (2016). Here analyses are presented at location (9) and along the transect line starting in the western Arkona Basin and ending in the northern Gotland Deep (gray line). From Zhang et al. (2016), with permission from Elsevier

7.2.1 Baltic Sea

In the following, some results from the numerical simulations by Zhang et al. (2016), who validated their results against data from tide gauges[31] and observed profiles of temperature and salinity in the locations shown in Fig. 18, will be presented. The good agreement between numerical simulations and observations demonstrated the credibility of model estimates and gives a strong justification to use numerical data for understanding thermohaline dynamics, something that cannot be easily measured basin-wide.

The vertical stratification of waters in the Baltic Sea, like in the Black Sea, is dominated by vertical differences in salinity (Fig. 19b). The water in the central Baltic Sea is strongly stratified with upper layer salinity of about 6–8 and a deeper layer salinity of about 10–14. These large differences provide enormous stability, decrease vertical diffusion (Stigebrandt 1987; Meier 2001; Stanev et al. 1997), and therefore prevent the permanent halocline from being destructed by the winter cooling. The vertical salinity gradient below 100–150 m changes slightly, mostly in winter, as a result of the convective cooling. The small-scale oscillations at the interface, separating surface and deep waters (Fig. 19b), reveal the model representation of the oscillations of pycnocline. After day 435 from the start of the numerical simulation, this layer collapsed, as seen in the observations and numerical simulations presented by Zhang et al. (2016), due to the inflow of warm and high-salinity surface water from the straits.

The cold-water mass formation starts in fall (Fig. 19a); in January–February the cooling signal from the surface reaches depths of ~ 50 m. This value varies in space reaching in some areas depths of ~ 65 m. In March–April, the re-stratification starts and the cold water is overlaid by waters from the seasonal thermocline. As the time progresses after the maximum cooling, the cold-water content of the CIL starts to decrease, which is seen in Fig. 19a as a decrease of its thickness. The refill and thickening of the CIL in winter and the decrease of its cold-water content in summer repeats in an almost periodic way every year. These results demonstrate that the model replicates adequately features and processes associated with the temporal evolution of thermohaline circulation in the Baltic Sea.

Very interesting with respect to the refill of deep layers of the Baltic Sea with salty water from the North Sea is the cascading of the mixed North Sea water into the Gotland Basin. Here, the Arkona and Bornholm basins play the role of reservoirs for denser water, filled with mixed North Sea water up to their respective sill depths. This water propagates to the next (deeper) basin. During most of the time, the halocline reveals seiche-like oscillations[32] at the interface separating surface and

[31] Tide gauges are instruments, usually along the coastline, measuring water level.

[32] A seiche is a standing wave in an enclosed basin that occurs when the water level rises at one end of the basin and drops at the other.

bottom waters. Under strong wind conditions, these denser waters overshoot the sill and cascade into the next basin on its way to the Gotland Deep (Fig. 20).

The deepest part of the central (western) Baltic Sea is ventilated occasionally during strong inflow events. These events are known as the Major Baltic Inflows (MBI) and can be identified by the sporadically increased salinity in the subsurface layers. Some of the MBIs, e.g., the one in 1993, which have been well documented in the Baltic Sea literature (Lass and Mohrholz 2003; Mohrholz et al. 2006), were accompanied by an extremely strong inflow of Atlantic water resulting from very specific atmospheric conditions. Such events occur usually in winter and do not show clear periodicity. The simulated characteristics of the propagation of North Sea mixed water have much in common with the MBI.[33] The situation shown in Fig. 20a happened in February 2012 and is consistent with the overall occurrence of the inflow events. The analysis of the 500-day simulations (Fig. 20b) indicates that the inflow events repeat in a non-regular way many times during the analyzed period but usually with a lesser intensity than 1993-MBI.

With the increasing distance from the straits, the inflow increasingly resembles the gravity flows. Unlike the gravity current in the deep Black Sea, which usually does not reach depths larger than 700 m, that is it detaches from the bottom, the mixed North Sea water penetrates in most cases down to the deepest parts of the basin, and it provides a mechanism for ventilation down to the bottom.

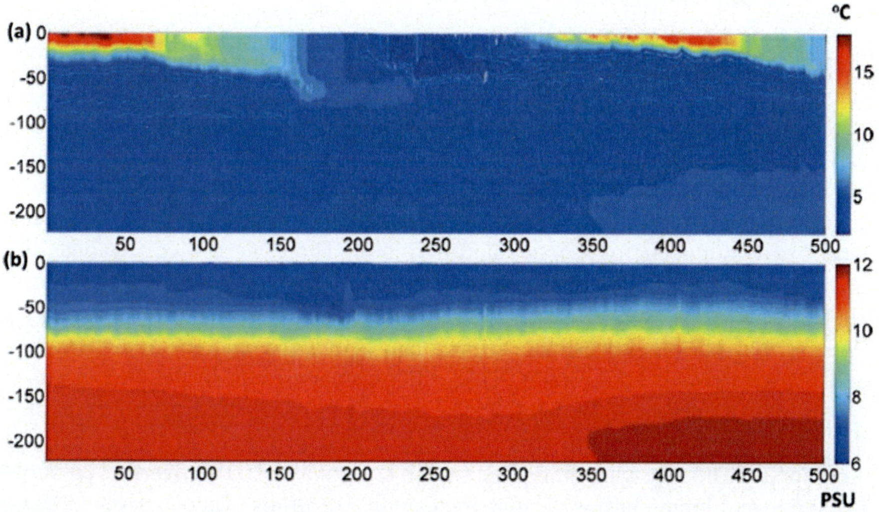

Fig. 19 Time versus depth diagrams of temperature (**a**) and salinity (**b**) at the Gotland Basin (location 9 in Fig. 18). The isolines in **a** represent 3, 3.5, 4, 4.5 °C. From Zhang et al. (2016), with permission from Elsevier

33 Major Baltic inflow events transport large amounts of saline water into the Baltic Sea, being the sole source for deep water. Observations indicate that they occur rarely, at about decadal time scales.

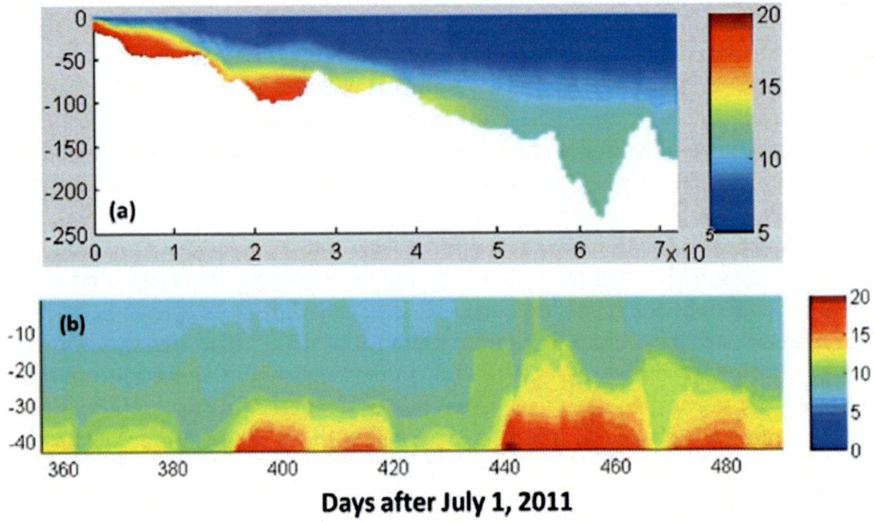

Fig. 20 **a** Salinity in the Baltic Sea along the section shown in Fig. 18 on February 29, 2012; **b** time versus depth diagrams of salinity in the Arkona Basin. The individual maxima at the bottom indicate inflow events. From Zhang et al. (2016), with permission from Elsevier

7.2.2 Mediterranean Sea

There are numerous specificities of the water masses of the Mediterranean Sea, among them the Eastern Mediterranean Transient (EMT), which is characterized by very slow changes, will be addressed in the following. This feature of water masses is quite interesting because it illustrates well the long-term dynamics of the thermohaline state of this sea, as well as the fact that the deep waters in the Mediterranean Sea are also strongly constrained by straits and sills (Cretan Arc straits, Cretan Passage, Straits of Otranto, Sicily, Corsica and Sardinia). The eastern Mediterranean deep layers are occupied by the Eastern Mediterranean Deep Water (EMDW). Before the 1990s, the EMDW was of Adriatic origin; however, the discovery of the EMT by Roether et al. (1996) suggested possible changes of thermohaline circulation (Klein et al. 1999; Roether et al. 2007). This major climatic event in the circulation and water mass properties of the Mediterranean Sea was recognized at the beginning of the 1990s, when the EMDW was not formed anymore in the Adriatic Sea but in the Aegean Sea. The new water mass (the Cretan Deep Water, CDW) overflowed into the Levantine and Ionian basins through the Cretan Arc straits (Theocharis et al. 2002). It then continued to flow along the continental slope in the Ionian basin (Roether et al. 2007).

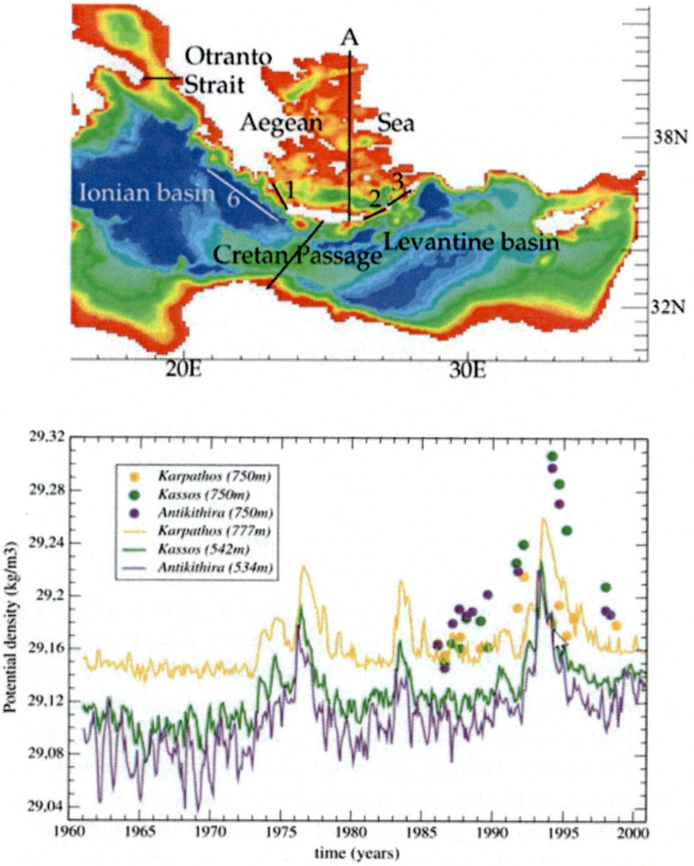

Fig. 21 Top: Bathymetry of the Eastern Mediterranean and some straits and analyzed locations. 1: Antikithira Strait. 2: Kassos Strait. 3: Karpathos Strait. 6: Hellenic Trench. Bottom: Monthly potential density (in kg/m^3) simulated on the bottom of the Cretan Arc straits (lines) and observed just south of the sill channels (circles): Karpathos (yellow), Kassos (green), and Antikithira (purple). The depth of each sill in the model and of each measurement is indicated in parentheses. From Beuvier et al. (2010)

What is the Cause for the EMT?

Existing hypotheses concerning the preconditioning of the EMT were recently summarized by Beuvier et al. (2010) as: (1) a change in the surface circulation, which could have prevented the Atlantic water flow toward the far eastern Mediterranean Sea, (2) the modification of the Levantine Intermediate Water path toward the Aegean Sea by the presence of three anticyclonic eddies south of Crete, (3) changes in the net surface evaporation and internal salinity redistribution, (4) a decrease in the Black Sea fresh water input due to a reduction of the river runoff, (5) intense winter convection in 1987 in the northern Aegean Sea, and (6) the occurrence of two successive winters (1991–1992 and 1992–1993) with strong and deep convection in all the Aegean Sea.

Although observations are crucial to understanding the thermohaline circulation, they cannot give a comprehensive four-dimensional description of the ocean. Here the development of numerical models helped greatly. In the long-term simulations of the circulation of the Mediterranean Sea, Beuvier et al. (2010) examined the contributions of the atmospheric and oceanic circulations during the late 1980s and early 1990s, namely the atmospheric buoyancy[34] loss, the change of the Atlantic water circulation in the Levantine basin, the occurrence of convection events in the Aegean Sea before 1992 and the Black Sea low salinity water discharge. Their results suggest that the key triggering factors of the EMT are the surface heat and water losses; hence, the buoyancy loss that occurred during the severe winters of 1991/92 and 1992/93, as already suggested by Josey (2003). The analysis showed that the heat loss was the major component of the buoyancy loss both in terms of mean value as well as in terms of time variations. The evolution of the volume of dense waters in the Aegean Sea (Beuvier et al. 2010) demonstrated that the newly formed waters were very dense ($\geq$ 29.2 kg/m^3), in agreement with observations of Theocharis et al. (1999). The numerical simulations reproduced the sequence of the EMT events also in agreement with the observations. The heat and water losses at the surface of the Aegean Sea during the winters 1991/92 and 1992/93 triggered strong winter convection, which was followed by a huge outflow of dense waters from the Aegean Sea to the Ionian and Levantine basins during the following two years (Fig. 21). With respect to the preconditioning, it was found in the simulations that the changing of the Atlantic water path in the Levantine basin together with a period of net reduced fresh water inputs over the far eastern Mediterranean Sea led to a salt increase of the Aegean Sea. The key triggering elements of the EMT are thus the surface heat and water losses which occurred during the severe winters of 1991/92 and 1992/93. The interannual variations of the Black Sea discharge in the surface layer of the northern Aegean Sea did not influence the trigger of the EMT, but rather impacted the chronology of the convection event during the previous decades, with intensification in the 1970s and weakening in the 1980s.

[34] Buoyancy is a vertical force exerted by a fluid that opposes the weight of an immersed object or fluid particle (Archimedes principle). Its magnitude is equal to the weight of the fluid that would otherwise occupy the submerged volume of the object, i.e., the displaced fluid.

8 A Step into the Atlantic and Arctic Numerical Modeling

In 1986, I received a scholarship from the A. von Humboldt Foundation. My acquaintance with J. Sündermann led me to the Institute of Marine Sciences in Hamburg. Although he was my official host, I worked with Dr. Hans Friedrich (Fig. 22), another user of the GFDL numerical ocean model at that time. This is explained by the fact that J. Sündermann was mainly concerned with the dynamics of shallow and shelf seas, while Hans and I were dealing with deep ocean basins. We both agreed that I would involve Hans in the Mediterranean Sea works, and that I will brighten my research focus adding the Atlantic and Arctic Ocean dynamics, something he was working on.

Waters in the Atlantic are mixtures of waters (in different percentages) that are formed in certain ocean areas (water mass formation areas). For example, in the region of the Labrador Sea, some of the coldest water masses in the World Ocean are created. They are denser due to the low temperature, sink to a certain depth, and then spread very slowly in the direction of the equator. Around Antarctica, a similar process occurs, but there, the sinking penetrates to a different depth. These waters, with temperature and salinity, which are different from those of the Labrador Sea water, spread northward. A third type of water mass forms at the outlet of the Mediterranean Sea (Fig. 23). These are saltier waters because the salinity in the Mediterranean is higher than in the ocean. Measurements have shown that these Mediterranean Sea waters spread westward as a salt tongue at a depth of about 800 m.

Until the mid-1980s, there were no numerical models of the Atlantic and Mediterranean, but only individual models of each basin. This did not allow the exchange of waters between the two basins to be properly included in the models. Hence, the salt

Fig. 22 Hans Friedrich. Courtesy of J. Sündermann

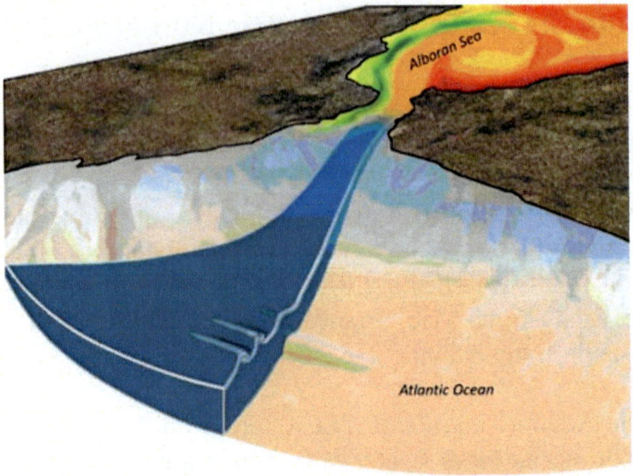

Fig. 23 Schematic representation of inflows and outflows. This figure gives an idea of only some of the features on both sides of the Gibraltar Strait, such as the clockwise rotation of surface current in the Alboran Sea, deflection to the right of the gravity current in the Atlantic Ocean and generation of instabilities on its interface in the regions of sills. The bathymetry and coastline are schematically shown. From Stanev and Xi (2013). Courtesy of Gerhard Gayer

tongue of the Mediterranean Sea water was missing in the numerical simulations. Rather than taking the step to link the two basins, I chose to solve a simpler problem by introducing the source of the salty Mediterranean waters using climate data. To the best of my knowledge, the new model of Atlantic water dynamics (Stanev 1992) was the first to reproduce realistically the distribution of Mediterranean water in the Atlantic (Fig. 24).

The other challenging topic that Hans and I started working on was coupling (joint modeling) of the Atlantic and Arctic Oceans. These two basins exchange waters in the relatively narrow Denmark Strait between Iceland and Greenland (see thick arrow in Fig. 25). This water exchange is very important for the dynamics of two oceans; hence, it is not appropriate to address them separately. This led us to developing a model, which included both of them. There were two problems to overcome. One was that in the coordinate systems we usually use in oceanography, the meridians converge at a single point at the pole that makes the resolution of the model there too fine. In order to provide a more uniform resolution in the ocean areas of interest, we rotated the coordinate system so that the pole of the new coordinate system was positioned in Canada. For this change, we had to only re-define in the new coordinate system the model representation of the Coriolis force, which depends, in the new coordinates, on latitude and longitude. Many use this original approach today. Unfortunately, we did not publish our study because of our dissatisfaction with some results and because Hans was excessively critical. The point was that due to insufficient computing power, our models worked with a coarse grid. We had two versions of the model, one with a horizontal resolution of half a geographic degree, which is approximately 50 km, and a coarser resolution version of one degree. With

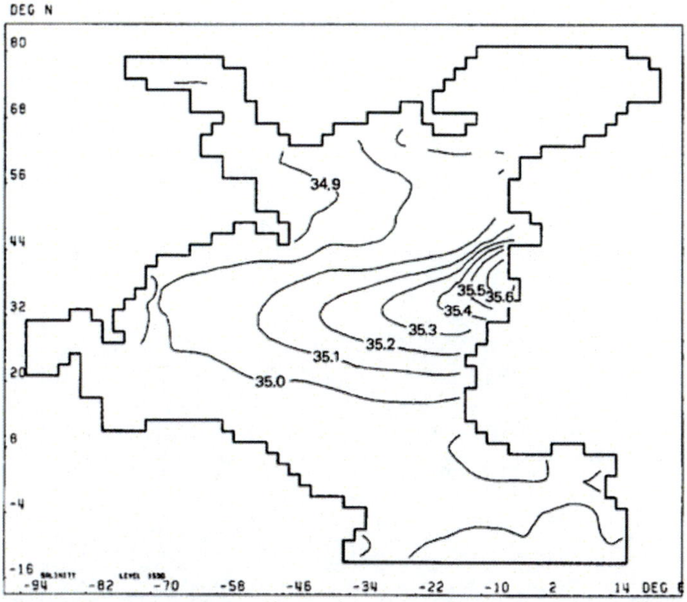

Fig. 24 Horizontal salinity patterns at 1500 m. Contour interval is 0.1. Horizontal resolution of the model is 2°, and the computations are carried out at 12 levels. From Stanev (1992), with permission from Elsevier

this resolution, it was not possible to adequately resolve the narrow deep part of the Denmark Strait and thus accurately calculate currents. If the area of the cross section of the strait in the model is to be kept as large as it really is, the model strait becomes too shallow, which precludes the possibility of having deep currents. If the maximum depth of the strait in the model equals the actual depth, the cross section of the strait becomes too large, making the exchange of water between connected basins unrealistic in numerical simulations. I made many attempts to overcome the problem by testing different parameterizations, but to no avail. For some problems, it is necessary to adequately represent key topographic features. In recent years, I have returned to the issue of interbasin exchange in the Black Sea and Baltic Sea Straits (see Sect. 7 in chapter "New Observing and Modeling Systems and New Scientific Applications"). Nowadays, in the area of these narrow straits, we use a raster below 100 m. In some global models, they use a resolution of 1/16 degree. This resolution is ~8(16) times finer than what we used at that time in our two models, which makes possible to sufficiently resolve processes in the Denmark Strait. Again, one can easily estimate how much computer power has increased in the last 35 years.

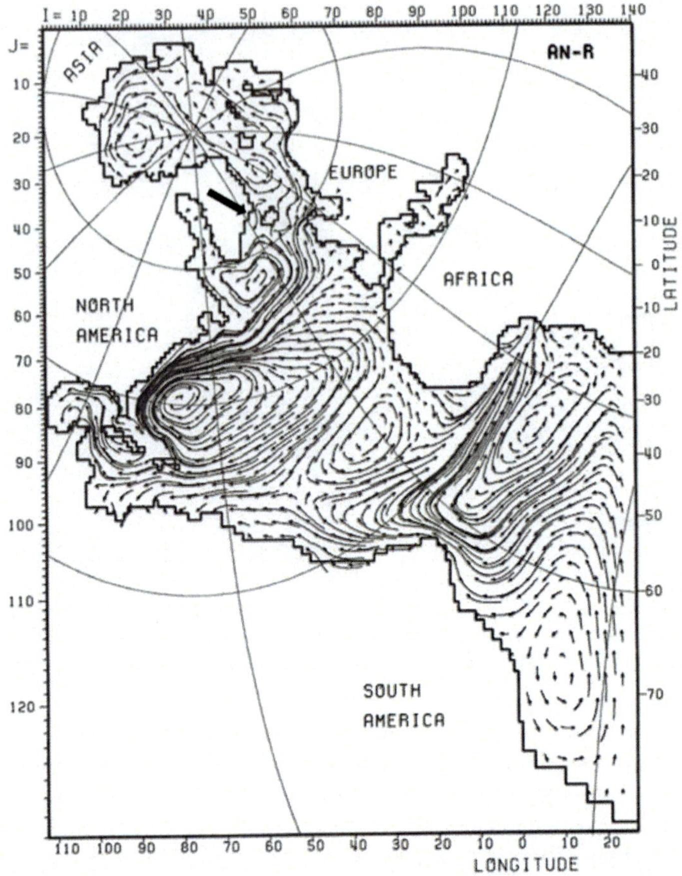

Fig. 25 Sea currents reproduced in a model with a spherical coordinate system but with offset poles. The parallels and meridians are also shown in order to illustrate the particular nature of the Arctic and Atlantic model in the new coordinate system (notice the numbers on the horizontal and vertical axes and the "distortion" of the shape of the continents, mostly in the south). The thick arrow shows the position of the Denmark Strait

9 Some Important International Conferences

From April 9–14, 1987, the Twelfth General Assembly of the European Geophysical Society (EGS)[35] was held in Strasbourg. I gave two presentations, the first one was on "Numerical modeling of the dynamics of the Black Sea and the distribution of

[35] EGS was founded in 1971; the first Assembly was in 1973. In 2002, the EGS merged with the European Union of Geosciences (EUG) under the name European Geophysical Union (EGU).

hydrogen sulphide and oxygen in it". This paper presented in a synthesized form results that I had already published in Soviet journals, as well what I had reported during the previous year in Gothenburg (see also Stanev 1989a). The second paper was on "Numerical experiment on the seasonal variability of the dynamics of the Mediterranean Sea". The session, at which I presented both papers, was attended by A. Robinson, Paola Malanotte-Rizzoli from the Massachusetts Institute of Technology (MIT), Nadia Pinardi, then from IMGA-CNR[36]-Modena, who was working in the same field as I and several other scientists involved in the international project Physical Oceanography of the Eastern Mediterranean (POEM). It became clear to A. Robinson and P. Rizzoli, who were cocoordinators of POEM, that I was a pioneer in the application of the GFDL ocean model to the Mediterranean, and they invited me to participate in the next year from 31 May to 4 June in the POEM workshop in Trieste.

The work in POEM was conducted under the supervision of the IOC, and in particular of Dr. Selim Morcos, a scientist who has long been interested in the Eastern Mediterranean and who was the architect and tailor of POEM. For this project, it was of paramount importance to (1) create an intercalibrated dataset; (2) make a synthetic scientific assessment of the current state of the Eastern Mediterranean; and (3) launch and plan future Eastern Mediterranean science.

According to the assessment of A. Robinson in the 53rd Report of the IOC, there were three major accomplishments during the first phase of POEM: data collection, model setup and calibration, and the establishment of an effective and robust collaborative science base. I highlight this information because POEM turned out to be a precursor to the Cooperative Marine Science Program for the Black Sea (CoMS-Black), a Black Sea project that started about five years later. In other words, POEM was developing patterns for international cooperation applicable to other areas of the World Ocean.

The EGS session in which I participated was also attended by Jacques Nihoul from Belgium, who was organizing the 19th International Liège Colloquium on Ocean Hydrodynamics from May 4–8, 1987, on the topic of "Small-scale turbulence and mixing in the ocean". This was one of the events of the SCOR[37] Working Group 69 "Small-Scale Oceanic Turbulence".[38] Jacques Nihoul was a visionary in the field of ocean hydrodynamics and a pioneer in interdisciplinary modeling of marine systems. He made a major contribution to the creation and cohesion of the

[36] Istituto per lo studio delle Metodologie Geofisiche Ambientali (IMGA), Consiglio Nazionale delle Ricerche (CNR), Italy.

[37] The Scientific Committee on Oceanic Research (SCOR), which is part of the ICSU, organizes working groups in various areas of oceanography. Working Groups consist of no more than ten permanent and no more than ten associate members and are tasked with discussing the problem they are investigating and preparing a review or other document outlining the directions in which the relevant scientific field should develop. The mandate of the group is limited to four years.

[38] SCOR Working Group 69 Small-Scale Oceanic Turbulence, Chair: K. N. Fedorov (USSR), Other full members: T. R. Osborn (USA), C. Cox (USA), C. H. Gibson (USA), J. Woods (FRG), M. Gregg (USA), C. Garrett (Canada), J. S. Turner (Australia), R. Ozmidov (USSR), and S. Panchev (Bulgaria).

European scientific oceanographic community. He had a wonderful sense of humor and was a true gentleman. "I hear Panchev will not be attending", he said. "He has other pressing commitments at the same time", I replied. "The funds for his business trip are available, why don't you attend instead of him?", continued J. Nihoul. "I cannot represent him, I am not a specialist in turbulence", I replied, "It doesn't matter, you're a specialist in something I'm much more interested in. We are also getting into numerical modeling; the Black Sea and the Mediterranean are interesting to us and you are the one I want to collaborate with".

In less than a month, I left Hamburg for Liège. There the leading scientists in the field of marine turbulence had gathered. For me, it was a great university. There I met Jean-Marie Beckers, a final year student of J. Nihoul, developing his own model of the currents in the Mediterranean. The other outcome of my visit was that I received an invitation to participate next year in the Liège colloquium, dedicated to "Mesoscale Coherent Structures in Geophysical Turbulence". There I was no longer just a substitute for S. Panchev, because I presented my own work on the study of eddies in the Black Sea triggered by seasonal atmospheric forcing (Stanev 1989b). The model I used in this study was much finer than the previous versions of my Black Sea models. It had a horizontal resolution of $1/3° \times 1/6°$ in the meridional and zonal directions, respectively, which was good for that time and permitted large eddies to appear in the numerical simulations. These simulations showed that the seasonality in the atmospheric conditions contributes to fundamental changes in eddy processes, i.e., a tendency for eddy energy patterns to self-organize.

Before the end of my stay in Germany, I attended from August 23–31, 1988, the Fourth General Oceanographic Assembly, which was held in Acapulco, Mexico. There I met K. Fedorov for the last time as he died the following month in Moscow. The Chinese scientist Jilan Su also attended the conference. During the official dinner, we sat next to each other and he told me a little about his interesting biography. He was born in mainland China, graduated from university in Taiwan in 1957, and was one of the first Chinese oceanographers to receive advanced training and specialization in the USA. He received his master's and doctoral degrees there and worked at several leading institutes until 1977. After his research career in the USA, he returned to China, holding various leadership positions. He was elected an academician of the Chinese Academy of Sciences in 1991. I met him again in 2017, about thirty years after our first meeting, at his Second Institute of Oceanography (SIO) of the State Oceanic Administration in Hangzhou. I led a German-Chinese scientific project on the German side. My research partner was Fei Chai, who was sharing the professorships positions at the University of Maine (USA) and at the SIO. On the occasion of frequent and sometimes unexpected meetings with scientists, I remember the words of my friend V. Kuzin: "Scientists meet often because they have one less dimension. They move in a space limited by their scientific interests, which increases the probability of meeting". His theory provides an "explanation" for repeated and sometimes unplanned meetings with colleagues.

10 In Belgium

On January 3, 1990, I left Sofia for Liège, where I was to spend a year as a visiting professor at the invitation of J. Nihoul. I shortened this time a little and divided it into two parts, asking V. Roussenov to cover the remaining four months in Belgium. The main purpose of the invitation was to help J.-M. Beckers in his work, which dealt with the problems of currents in the Mediterranean and the Black Sea. He had developed on his own a numerical model with a very specific, but also adequate representation of vertical coordinates as a function of depth. This model known as GHER[39] model eliminated some of the shortcomings of the then existing models, but it was yet to be exploited in real conditions in order to prove its quality. The Black Sea being relatively small, but with a wide variety of bathymetry and a specific horizontal and vertical distribution of seawater density, was an ideal testing area.

The novelty of our study was the investigation of barotropic and baroclinic oscillations[40] in the Black Sea using numerical modeling. Prior to our first experiments, numerical models used the so-called rigid lid boundary condition for the sea surface, which effectively excluded the propagation of surface waves. The GHER model, which has an active free surface, made it possible to simulate barotropic and baroclinic waves at different scales. Furthermore, a parameterization was developed that allowed to realistically simulate the transport in the Bosporus Strait and its dependence on sea level dynamics (Stanev and Beckers 1999a). It was found that sea surface oscillations interact with oscillations in the pycnocline, contributing to a change in vertical stratification. Vertical stratification, in turn, in conjunction with the bottom topography causes different manifestations of oscillations over the continental slope (coastal waves) and in the interior of the basin. It appears that changes in stratification caused by the seasonal cycle of temperature and salinity in the Black Sea are an important factor modifying wave processes and the resulting internal mixing.

Our joint work, as well as our complementarity, not to forget our close friendly relations, have borne fruits (Stanev and Beckers 1999a, b). Jean-Marie defended his thesis in 1992 and continued to work actively in the field of data analysis, numerical modeling and data assimilation in hydrodynamic models. Today he is a full professor at the University of Liège, where he is Chair of Physical Oceanography. He continued the work of J. Nihoul as organizer of the International Liège Colloquium on Ocean Hydrodynamics (Fig. 26).

While I was working more actively with Jean-Marie, V. Roussenov began an interesting project with Pierre Brasseur, another student of J. Nihoul. This research was more focused on observational data analysis and later it evolved toward data assimilation in the ocean, where today P. Brasseur is one of the leading scientists. I will

---

39 GHER (GeoHydrodynamics and Environment research) is a research group of the University of Liège, directed J. Nihoul and later by J.-M. Beckers.

40 We can think of barotropic oscillations roughly as waves in an ocean of constant density (a homogeneous ocean). Baroclinic oscillations are such that density is an essential part of the wave dynamics.

Fig. 26 J. Nihoul (standing on the left) and J.-M. Beckers (seated in front of the computer terminal) explain their results during a poster session at the International Liège Colloquium on Ocean Dynamics. Courtesy of J.-M. Beckers

talk more about him, as well as about our work in the field of scientific development of European operational oceanography, later on.

11 Data Assimilation

You may have noticed that when I first encounter something new, meet a new person who leaves a trace in my scientific development, or when I do something new that has not been done before, I report it. This leads me to the assimilation of observational data (Cummings et al. 2009) in connection with my participation in the International symposium on the assimilation of observations in meteorology and oceanography organized by the World Meteorological Organization (WMO) in Clermont-Ferrand, France, from July 9–13, 1990. The method of data assimilation was first developed in the field of meteorology and was later transferred and adapted to the marine environment.

Data assimilation combines observations and models to (1) improve our understanding of ocean circulation over a wide range of temporal and spatial scales of phenomena; (2) monitor; and (3) predict the state of the ocean. Essentially, data assimilation is a combination of observations and model-reproduced ocean states. Measurements are typically local in time and space, so it is not possible to produce a complete picture of the state of the ocean based on observations alone. When observational data are inserted into numerical models, it becomes possible to provide a four-dimensional (three spatial coordinates and time) reconstruction of ocean fields.

The data assimilation method goes beyond the classical methodology where observations and theory are developed independently. The challenge is to extract the most relevant information from relatively sparse and noisy observations, and to feed this information optimally to numerical predictive models. Observational errors are due to instrumental noise, environmental noise, sampling problems, and possible misinterpretation of measurements. Ocean numerical models are not without errors either. Their errors come from incomplete (imperfect) descriptions of physical processes, insufficient grid resolution, or problems with atmospheric or hydrological forcing of the model. Even if there are "perfect" ocean models, over time, the estimates the models provide will continuously deviate from reality. This is known as a loss of predictability, a typical characteristic of the complex (nonlinear) fluid dynamics. The time limit of predictability depends on the type of geophysical fluids and the dominant processes. As atmospheric and ocean eddies are the major representatives of nonlinear large-scale dynamics, their characteristic temporal scales (life times) limit predictability times. For synoptic processes in the ocean, predictability is lost over timescales on the order of weeks to months, for the coastal ocean, where temporal scales are shorter, this time is on the order of days. The inability to predict over long time periods (such as predicting the motion of the planets of the solar system) is a natural characteristic of geophysical systems. This hinders accurate and long-term predictions and requires periodic adjustment of models using observations with appropriate spatial coverage. It is also important that observations are taken at intervals smaller than the predictability limit. Knowledge of the multivariate aspects of the model data and the observations to be combined is needed for the optimal process of data assimilation. A major problem is to know the errors of the observations and those of the model. Roughly speaking, the larger the measurement errors, the less weight the corresponding data will have in the process of model correction. Conversely, the larger the model errors, the more weight the measurement data is given in the assimilation.

Fig. 27 Acad. A. S. Sarkissian (left) and Acad. G. I. Marchuk. Courtesy of Juergen Sündermann

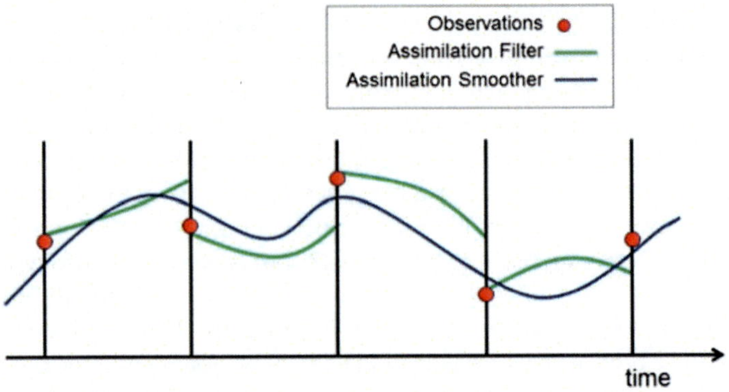

Fig. 28 Schematic representation of the trajectory of the model state during the data assimilation process in variational (blue curve) and sequential (green lines) approaches. The red dots symbolize observations. The idea behind data assimilation is to bring the model state close to the observed data. This can be done at certain times, e.g., daily, or "smoothly" as the model continuously follows the behavior of the observed data over a longer period. From Stanev and Schulz-Stellenfleth (2014)

The question of the optimal use of measurements had preoccupied me before the Clermont-Ferrand symposium, and I already obtained some results using relatively simple data assimilation methods that were worth showing. The works of Stanev and Friedrich (1991) and Stanev (1994) turned out to be the first on data assimilation in the Mediterranean and Black Sea using three-dimensional basin-wide models (one example of the simulated currents in the Mediterranean is given in Fig. 29). More important for me at the Clermont-Ferrand symposium was meeting Yoshikatsu Sasaki, Olivier Talagran, and several other leaders in the field, which gave me further motivation to continue to dive into the science of data assimilation.

Modern data assimilation in meteorology and oceanography uses optimal control theory (or variational approach), based on earlier works by the famous French mathematician Jacques-Louis Lyons, an active participant in the French Resistance, and even earlier works by the Soviet computational mathematician G. Marchuk. As I have already mentioned, G. Marchuk was involved in atmospheric physics and was the thesis advisor of my friends V. Zalesny and V. Kuzin when he was President of the Siberian Branch of the USSR Academy of Sciences. Gury Marchuk (Fig. 27) was the first to apply the theory of optimal control in environmental modeling. I remember, that in the early years of my scientific career, his coworkers and students used to visit the BAS and promote G. Marchuk's computational methods. The advantage of the variational approach (Fig. 28) is that it makes two things simultaneously possible that do not work with other approaches: (1) that the meteorological fields are consistent, i.e., they obey the equations describing the fluid dynamics, and (2) that the difference between the observations and what the model predicts remains minimal. Aware of G. Marchuk's work, in 1966 J.-L. Lyons invited him to visit France. The invitation was hand-delivered by the French President Charles de Gaulle during his visit to the academic city ("akademgorodok") of Novosibirsk.

At the Clermont-Ferrand symposium, the main papers were presented by J.-L. Lyons and Y. Sasaki, the latter also using a method similar to that developed by G. Marchuk.

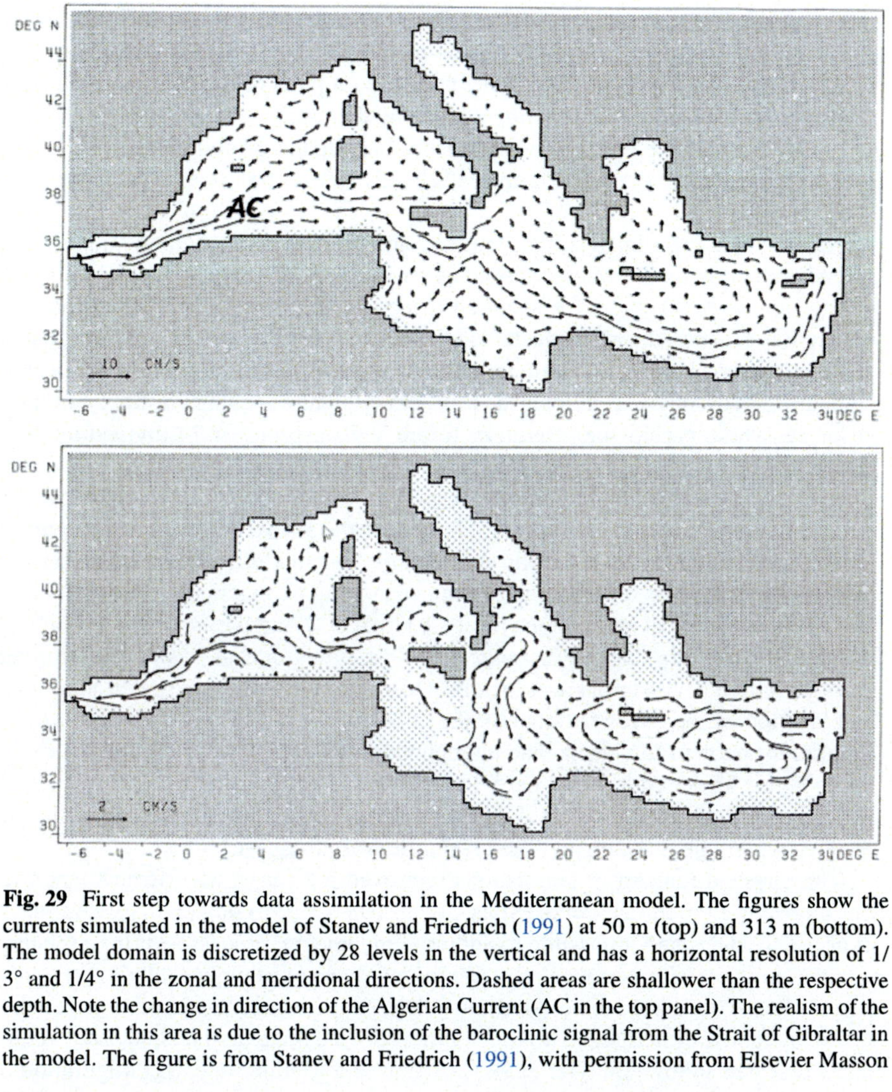

Fig. 29 First step towards data assimilation in the Mediterranean model. The figures show the currents simulated in the model of Stanev and Friedrich (1991) at 50 m (top) and 313 m (bottom). The model domain is discretized by 28 levels in the vertical and has a horizontal resolution of 1/3° and 1/4° in the zonal and meridional directions. Dashed areas are shallower than the respective depth. Note the change in direction of the Algerian Current (AC in the top panel). The realism of the simulation in this area is due to the inclusion of the baroclinic signal from the Strait of Gibraltar in the model. The figure is from Stanev and Friedrich (1991), with permission from Elsevier Masson

12 Further Developments of the Mediterranean Model

What was new to me in 1991 was the Unix operating system. The operating system is a fundamental piece of computer software that manages and coordinates the resources of hardware and software and serves the computer programs that are running. Another well-known operating system is Microsoft Windows. Unix was developed in 1969, and until the 1990s, it was mainly used in specialized applications such as workstations and servers, especially in universities and research institutions.

Computer codes were also evolving. Their versatility was extended by the use of more modern computer languages, and by the possibility of writing computer codes in such a way that "switches" could be used to turn parts of the code on and off, or to change its character in a direction desired by the user.

In 1991, I had the chance to expand my knowledge of computer systems and programming. Nadia Pinardi, whom I had already met several times at Mediterranean conferences, invited me to work with her for a year at her institute IMGA-CNR in Modena. We had similar scientific interests and used the same numerical model. In recent years, this model was published as a new version,[41] which we had to master during our work together. We also had to update the model block describing the interaction between the atmosphere and the ocean. The idea was to put in the hands of the Mediterranean computational community a new and modern instrument of computing hydrophysical fields. Due to teaching commitments in Sofia, I could not cover the whole period she offered therefore V. Roussenov and I divided the time equally. I visited Modena in the first and last quarter of the year, and Vassil worked there continuously for the remaining six months. The model we developed together with N. Pinardi (Roussenov et al. 1995; Pinardi et al. 1997) received wide recognition among the Mediterranean scientific community. This model was actively used for at least another 20 years for research and in oceanographic practice.

Our first numerical simulations presented in Sect. 7 and the ones which followed (Pinardi and Navarra 1993; Roussenov et al. 1995; Pinardi et al. 1997) demonstrated that the seasonal variability was manifested not only by the change in amplitude and location of the gyres but also by the appearance of seasonally recurrent gyres in different parts of the basin. The interannual fluctuations in the circulation of the upper layers appeared to be of particular relevance to the water mass formation. It was demonstrated by Pinardi et al. (1997) that the Eastern Mediterranean basin was the area where the interannual ocean response was strongest.

The physical realism of simulations improved a lot in about fifteen years after our pioneering modeling works (Stanev et al. 1989; Roussenov et al. 1995; Pinardi et al. 1997) and made it possible to address fundamental processes of thermohaline circulation of the Mediterranean Sea. The simulations of Beuvier et al. (2010) for the period 1961–2000 have been carried out with an eddy-permitting circulation model,[42] driven by realistic interannual high-resolution air–sea fluxes and realistic river runoff and Black Sea inflow. Beuvier et al. (2010) used a Mediterranean configuration of the NEMO (Nucleus for European Modeling of the Ocean) model (Madec 2022)[43] with a horizontal resolution of $1/8° \times 1/8°$, equivalently, 9–12 km. The exchange with the Atlantic Ocean was introduced by specifying a buffer zone in the Atlantic Ocean where the 3D temperature and salinity of the model were relaxed toward the

[41] GFDL scientists Ron Pakanowski, Keith Dixon, and Tony Rosati have rewritten the GFDL model into a modular form that allows an easier generation of different options and configurations and easier incorporation of new physical parameterizations. The model was published in 1990 and became known as the Modular Ocean Model v1.0 (MOM1).

[42] Eddy-permitting models are relatively coarse and can only resolve larger eddies. The horizontal and vertical resolution in eddy-resolving models is fine enough and allows to resolve the most important characteristics of eddy field.

[43] This was my first touch with this model, which in this study was operated by the French colleagues.

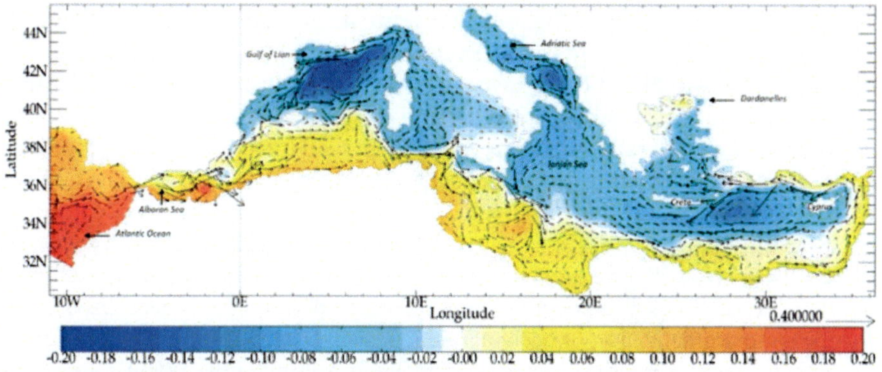

Fig. 30 Sea surface height (colored field, m) and currents at 35 m depth (vectors, m/s), averaged over 40 years (1961–2000). Each fourth velocity vector is plotted. From Beuvier et al. (2010)

climatological fields of temperature and salinity. The river inputs were explicitly added as a fresh water flux. The Black Sea discharge through the Dardanelles was modeled as a river runoff into the Aegean Sea. The dynamics of the Mediterranean Sea illustrated in Fig. 30 by the sea surface height and currents at a 35 m depth reveal the well-known general cyclonic path of the Atlantic water. The simulations replicated three intense cyclonic gyres (in the Gulf of Lions, in the southern Adriatic Sea, and the Rhodes gyre in the Levantine basin) in zones where winter convection generally occurs. The anticyclonic gyres were most pronounced in the Alborán Sea providing the origin of the meandering coastal current along the African coasts.

The typical Mediterranean thermohaline circulation (Fig. 30) determined by the simulations of Beuvier et al. (2010) is quite complicated. At the surface, the low density Atlantic water enters the western Mediterranean Sea at the Strait of Gibraltar and flows along the African coast toward the eastern Mediterranean Sea (Millot 1999). Winter convection forms dense waters in the following four main areas: (1) the Western Mediterranean Deep Water in the Gulf of Lions, (2) the Adriatic Deep Water (ADW) in the Adriatic Sea, (3) the Cretan Intermediate Water in the southern part of the Aegean Sea, and (4) the Levantine Intermediate Water in the Levantine basin (mainly in the Rhodes gyre). The Levantine Intermediate Water, which is the most important intermediate water mass, flows at a depth of about 300–500 m from the Levantine basin toward the western Mediterranean Sea.

13 CoMSBlack, a Collaborative Marine Science Program in the Black Sea

The Black Sea is the world's most stagnant marine system, where water layers are very difficult to mix. During the Cold War, this was also true for the system of scientific contacts, difficult exchanges, only formal collaborative or no scientific work between colleagues from the Black Sea countries. The political changes in Eastern Europe triggered an intense interaction between researchers from the neighboring countries.

American scientists oriented themselves toward the management of joint Black Sea research, actively assisted by Turkish colleagues with whom they had long-standing contacts. However, they did not have the necessary contacts with researchers from the other Black Sea countries, nor sufficient access to data other than Turkish data. The Russians and Ukrainians, on the other hand, held the bulk of the observational material and were all interested in making their data available for use. The idea was to harmonize oceanographic information, to build modern databases and to make the access to these databases easier. In terms of databases, the Americans had a good track record. They also had the experience gained during the POEM, which they sought to apply to future Black Sea cooperation. The situation was also characterized by the increasing marginalization of oceanographic research in the former Soviet Union.

In these times of political transition, Dr. David Aubrey of the WHOI came to the fore. Allan Robinson (Fig. 31) and P. Rizzoli supported him; the Turkish scientist Dr. Ümit Ünluata was actively involved in the management of the future work. Who were the Bulgarian actors? Through a good coincidence, the President of the BAS at that time was Acad. Blagovest Sendov (Fig. 6 in chapter "In Search of a Destination"). The vice-presidents were corresponding members of the BAS S. Panchev (since 1997 academician, Fig. 2 in chapter "In Search of a Destination") and Alexander Yankov (Fig. 32), both with very clear interests and expertise in marine research. I had the chance to be supported by this high level "troika" in the process on international restructuring of our Black Sea activities.

The International Workshop on the Black Sea was held in Varna from September 30 to October 4, 1991. A total of 125 scientists and managers discussed the health of the Black Sea, the possible causes of environmental degradation, and the state

Fig. 31 From left to right, Jean-Marie Bouquegneau, Anne Mouchet (from the back), A. Robinson and J. Nihoul during a cocktail reception at the International Liège Colloquium on Ocean Dynamics. Courtesy of J.-M. Beckers

Fig. 32 Alexander Yankov (1924–2019) was member of the UN International Law Commission (1977–1996) and its Chair in 1984. He participated in the United Nations Committee on the Peaceful Uses of the Sea Bed (1968–1973) as its Deputy Chairman. He chaired the Third Committee of the Third UN Conference on the Law of the Sea in all sessions from 1973 to 1982. The United Nations Convention on the Law of the Sea (UNCLOS) provides the legal framework within which the IOC acts. It influenced ocean research by establishing well-defined ocean regimes and rules for research. A. Yankov was also Judge of the International Tribunal for the Law of the Sea (1996–2011). Photo of Jordan Simeonov, from Yankov et al. (2006)

of ocean science in the Black Sea basin. These issues were at the heart of CoMS-Black, an interdisciplinary program that was driven by concern about the state of the Black Sea environment. CoMSBlack created the conditions for the highest quality oceanographic science and for effective and integrated management of the Black Sea resources. The information about the workshop, published in Issue 86 of the IOC Proceedings, argue for a deeper cooperation between countries in the study of the Black Sea.

Main Objectives of CoMSBlack

Ünlüata et al. (1993) specify the main objectives of CoMSBlack as:

- To provide an assessment of natural and anthropogenically induced environmental changes using historical data,
- To determine past and present fluxes of water, sediment, carbon, nutrients, heavy metals, hydrocarbons, and other selected materials from rivers, atmosphere, straits, and bottom sediments,
- To determine the fluxes of carbon, nutrients, organisms, and selected pollutants across the coastal and shelf seas, the shelf break, and within the basin,
- To understand the fundamental physical and BGC processes governing the transport, transformation, and fates of carbon, nutrients, suspended sediments, and selected substances,
- To provide a quantitative understanding of the physical and BGC processes and nutrient fluxes affecting primary productivity,[44]
- To assess man-made and natural influences on the ecosystem structure and function in the water column (phytoplankton and zooplankton[45]) and benthos (selected benthic organisms[46]),
- To develop multiple- and interactive-scale models for general and mesoscale circulation, ecosystem, and regional processes that will be applicable to the studies concerned with the ecology and biogeochemistry of the sea, and
- To assess space and time scales of general circulation and mesoscale features and their energetics and processes leading to the formation, spreading, and transformation of CIL.

Within CoMSBlack, several working groups were established. Together with the colleague from Turkey Temel Oguz, I was the rapporteur of the Working Group on Dynamics and Transport in Coastal and Offshore Waters, chaired by Ü. Ünlüata. The aim of the scientific program of this group was to arrive at a quantitative understanding of the circulation and dynamics of the Black Sea, including its coastal and shelf regions, as well as the interactions of these regions with the interior of the basin, fluxes across boundaries, transport and transformation of water masses and selected substances.

---

[44] Primary production is the synthesis of organic compounds from atmospheric or aqueous carbon dioxide. It occurs mainly through the process of photosynthesis.

[45] Microscopic plants (phytoplankton) and animals (zooplankton) in freshwater and seawater.

[46] The bottom layer of the ocean is known as the benthic zone. This is a term used in the ecological studies to identify the region at the bottom of the water column including sediment layers. The name comes from ancient Greek, βένθος (bénthos), meaning "the depths".

Purpose and Major Scientific Objectives of the Working Group on Open Sea Dynamics and Transport

As reported by T. Oguz and E. Stanev (see, Aubrey 1991), the purpose of the scientific program on Costal and Open Sea Dynamics and Transport[47] was to obtain a quantitative understanding of the circulation and the dynamics of the Black Sea. This included its coastal and shelf regions, and the interactions of these regions with the interior of the basin, fluxes through boundaries, and the transports, transformation and fates of water masses and selected substances. The major scientific objectives identified by the Working Group are:

- To establish a definitive phenomenology for the salinity, temperature, O_2 and H_2S structures, transport and dispersal characteristics, as well as suspended sediment distribution,
- To assess the relative importance of wind, buoyancy fluxes (from atmosphere, rivers, and straits) as well as the role of topography on the general circulation and dynamics,
- To assess space and time scales of general circulation and mesoscale features and their energetic,
- To investigate the processes of water mass formation, spreading, and transformation in space–time,
- To evaluate the existing database on climatic changes (including paleoclimatic data) and to elaborate the theoretical and experimental background for the analysis and forecasting of the possible climatic changes,
- To study the transport, sedimentation, and dispersal of terrigenous materials and some specific tracers,
- To investigate the characteristics, origin and driving mechanisms of the deep benthic boundary layer and its role in the overall thermohaline structure of the Black Sea,
- To identify the principal processes involved in the circulation and dynamics of the coastal and shelf seas and the shelf-deep sea interactions and fluxes,
- To develop multiple scale models with the components of general circulation and regional processes that will be applicable to the studies concerned with the ecology and biogeochemistry of the Black Sea,
- To establish capabilities for continuous monitoring systems, and
- To develop water quality control and management-oriented operational models for transport, spreading, and dispersal of contaminants.

The main scientific objectives we set were an attempt to make a complete definition of the essential tasks to be solved on the way to overcoming the critical problems in Black Sea science and marine environment management. A number of measures were envisaged in the field of education and training, procurement of modern equipment, development of data management systems, intercalibration and

[47] https://www.jodc.go.jp/jodcweb/info/ioc_doc/Workshop/w086.pdf, last accessed on 03.02.2023.

standardization, clarification of the need for compatible computer systems and platforms, as well as the need for international cooperation, including the ratification of the Convention on the Protection of the Black Sea against Pollution. Much of what we planned is still relevant today, more than thirty years later.

An ad hoc group was set up under the Steering Committee, including T. Oguz and Emin Özsoy (Turkey), Emil Stanev and Dimitar Yordanov (Bulgaria) and Leonid Ivanov and Sergey Demishev (USSR[48]). The international seminar was followed by the first observation campaign called HYDROBLACK '91. It took place from September 2–29, 1991, with the participation of five research vessels from three different Black Sea coastal countries. This was the first quasi-synoptic survey of the Black Sea (Oguz et al. 1994), in which measurements were made at nearly 300 hydrographic stations. Results revealed an eddy-dominated circulation with anticyclonic eddies around the periphery of the basin (see Fig. 6). The meandering Rim Current along the continental slope was separating this anticyclonic area from the predominantly cyclonic basin interior, exhibiting large filaments and dipole eddy structures. Observations also showed that quasi-persistent features of the general circulation were fairly coherent down to ~500 dbar.[49] HYDROBLACK '91 was followed in July 1992 by the CoMSBlack '92, a survey campaign, which involved five ships. This time, the measurements covered various aspects of biological oceanography and of the distribution of radioactivity associated with the Chernobyl disaster.

14 ONR and the Black Sea Research

I had received an invitation from the Office of Naval Research (ONR) to visit several scientific organizations in the USA in 1992: the ONR Headquarters at Arlington, the Naval Research Laboratory (NRL), which is located on the grounds of the National Aeronautics and Space Administration's (NASA) John Stennis Space Center (SSC) close to New Orleans, National Center for Atmospheric Research (NCAR) and the University of Colorado in Boulder, and finally the WHOI and Harvard University near Boston. The goal was to get to know each other and possibly to start collaborative work, therefore, I had to give seminars at each of the places I visited. At the ONR Headquarters, I met Manuel Fiadeiro in person, from whom I learned numerical schemes during my graduate student years. At NRL, I was surprised to see among the attendants of my seminar the great Soviet oceanographer and theoretician, Vladimir Kamenkovich, who had immigrated to the United States. He and I had met many years ago in Moscow. NRL scientists wanted to expand their numerical modeling

[48] The USSR dissolved on 26 December 1991.

[49] Bar is a unit of pressure, defined as equal to 10^5 Pa. The name is derived from the Greek $\beta\acute{\alpha}\rho o\varsigma$, meaning *weight*. One bar is approximately equal to the atmospheric pressure at sea level. The value of pressure in the ocean in terms of dbar is approximately equal to the value of depth in m. In the International System of Units (SI), the pascal (Pa) named after Blaise Pascal (1623–1662, a French mathematician, physicist, man of letters, and Christian philosopher) is used as a unit for pressure. It is defined as 1 Pa = 1 N m^{-2}. 1 atm = 101325 Pa = 1013,25 hPa = 1,01325 bar = 1013,25 mbar.

into the Black Sea, but it took us many years to start working together. That did not happen until 2004, the next time I visited NRL (see Sect. 6 in chapter "From Regional to Coastal Oceanography"). In Boulder, I visited NCAR, where my host was William (Bill) Holland, a scientist who had spent part of his career at GFDL. He was one of the main developers of ocean modeling at the time. My friend from Hamburg, H. Friedrich, highly recommended that I visit his good friend Bill.

From Boulder, I flew to Boston. This was my second visit there. The first time, in 1977 during the POLYMODE cruise, we stayed there for three days. Now, I was there at the invitation of A. Robinson, who is recognized as one of the pioneers and leading experts in modern ocean forecasting. Not far from Boston is the WHOI, where I visited some of the American leaders of CoMSBlack, H. Livingston and D. Aubrey. Hugh and I had befriended each other in Izmir, Turkey, where the CoMSBlack Steering Committee meeting was held on 21 and 22 April 1992. At that meeting, the decision was taken for the Sofia Workshop on Black Sea Ecosystem Modeling to be organized, which I will talk about in a moment. The direct result of this "marathon" was the expansion of my personal ties with the institutes I visited. The long-term result is the opening of Black Sea oceanography to more intensive international scientific collaboration.

15 NATO and the European Community Move Toward the Black Sea

NATO took an interest in the Black Sea and, using the Science for Stability (SfS) program, launched the project "Ecosystem Modeling as a Management Tool for the Black Sea: a Regional Multi-institutional Cooperation Program", known as TU-BLACK SEA.[50] As the only NATO country in the region at that time, Turkey, through the Institute of Marine Sciences, Middle East Technical University, Erdemli, stepped in as coordinator. This project was NATO's first major venture to establish scientific cooperation with Central and Eastern European countries. The aim of the project was to contribute to improving the health of the Black Sea using ecosystem models as a management tool, through capacity building and by fostering an interactive Black Sea science community.[51]

University of Sofia was asked, in my person, to organize a workshop on ecosystem modeling of the Black Sea in Sofia. The meeting took place from March 23–25, 1994. It was one of the main events of the NATO TU-BLACK SEA project. The meeting was at a high expert level, hence, only 31 selected scientists participated, among them the

[50] http://sfp1.ims.metu.edu.tr/TU-BlackSea/default.htm, last accessed on 03.02.2023.

[51] In the framework of the project, nearly 100 scientists from, a total of, 14 prominent oceanographic institutions (9 regional and 5 non-regional) collaborated on four main activities. They focused on the development of a database and data management system, the development of appropriate infrastructure, and the development and application of interdisciplinary ecosystem models. Last but not least, it had to be ensured that the products of the project TU-BLACK SEA could be used by managers, decision makers, and other regional programmes.

leading American researchers in the field Hugh Ducklow,[52] P. Rizzoli, James (Jim) Murray[53] and some others. The workshop clarified plans for joint activities as well as a number of details of the implementation of the tasks.

After five years of work in the Gulf of Lyon and the Western Mediterranean, where the European Community funded research on regional processes in river and marine systems, interests moved eastward. Scientists from Western and Eastern Europe decided to join forces to study the Danube Delta and the Northwest Black Sea in the framework of the pilot project "European River Ocean System: the interactions between the river Danube and the north western Black Sea—Pilot phase (1994–1995)", known as EROS 2000 (Lancelot et al. 2002a). This was the first Black Sea project funded by the European Community. The study area represented the best example of the degradation of coastal and marine aquatic environment. EROS 2000 can be seen as a pilot project for understanding the complex interactions between human activities and the marine environment. Interdisciplinary research on BGC processes associated with eutrophication,[54]pollution and erosion were therefore at the core of the project. The expectation was to identify measures to be taken to prevent a similar catastrophic situation in other semi-enclosed seas.

Specific Issues Dealt with in EROS 2000[55]

1. What would be the consequences of a reduction of the river input of nutrients for the eutrophication and the ecosystem structure of the Northwestern Black Sea?
2. What are the riverine and atmospheric sources, levels, and fates of key organic/inorganic pollutants in the Black Sea environment?
3. What are likely to be the short- and long-term consequences of the regulation of river discharge for sediment transport/erosion/deposition and water stratification?
4. What is the significance of the Black Sea source of important key greenhouse gases?

[52] H. Ducklow is one of the world leaders in ecosystem modeling. During the time I am referring to, he was a professor at the the Virginia Institute of Marine Sciences.

[53] James Murray is a professor from the University of Washington, Seattle with an important contribution to the Black Sea biogeochemistry. More details about his research are given further in the text.

[54] Eutrophication is the process by which water is enriched with minerals and nutrients, such as nitrogen and phosphorus.

[55] https://cordis.europa.eu/project/id/EV5V0501/fr, last accessed on 03.02.2023.

EROS 2000 was followed by EROS 21, within which a complex ecological model named BIOGEN was developed. This was a model with a high degree of representativeness of the ecological system (Lancelot et al. 2002b). My main work was on the direct (online) coupling of physical and biological processes (i.e., the coupling of the MOM1 and BIOGEN models). Experimentations with the coupled model were largely done by Joanna Staneva, a recent graduate student, who had joined my lab, and who coincidentally had the same family name as mine. The above publication analyzed the results of different scenarios of anthropogenic influence for the period since 1960 and the response of the Black Sea ecosystem to anthropogenic pressures. Quantitative estimates of the ability of highly polluted systems to recover, after a change in the agricultural regime of countries in the area, were very instructive. They demonstrated the ability of the Black Sea to recover, at least partially, the quality of its ecological system following a reduction in the inputs of fertilizer-enriched waters of the river Danube. This had indeed happened after the political changes when agriculture in some Eastern European countries was declining. My stay in Liège in 1995 gave me the opportunity to start working productively with Marilor Grégoire, another young member of the team of J. Nihoul, who was entering the field of ecological modeling.

The Bulgarian participation in EROS 2000 and EROS21 was encouraging. In the mid-1990s, the leadership of IO-BAS shifted toward ecology. The head of the Chemistry Section, Alexander Stoyanov, was respected by the project management. Snezhana Moncheva (since 2017 Director of the Institute), with whom in the past years we were the most active Bulgarian participants in the series of CoMSBlack workshops and subsequent NATO projects, was establishing herself as a valuable scientist in the EROS 2000 and EROS21 consortia (Fig. 33).

The scientific cooperation in Europe was constantly increasing bringing scientists from different regions together. From November 14–18, 1995, the NATO Advanced Research Workshop was held in Varna on the "Sensitivity of North Sea, Baltic Sea and Black Sea to Anthropogenic and Climatic Changes" (Özsoy and Mikaelyan

Fig. 33 Snezhana Moncheva (left) and Joanna Staneva during a Black Sea workshop

1997). The idea was to provide a forum for scientists from northern European and Black Sea countries to exchange information on the similarities and differences in the physical and biological systems of the North Sea, the Baltic Sea and the Black Sea.

The Workshop on the Study of the North, Baltic, and Black Seas

The exchange of water between the North, Baltic, and Black Seas and the ocean is weak, and therefore, their self-cleaning capacity is also weak. There has been a deterioration in their environmental quality due to increasing pollution in recent decades. The workshop dealt with the distinctive physical and ecological characteristics of the three seas in a comparative manner to identify the types of processes affecting biological productivity, nutrient cycling, physical transport, and mixing. A comparative study of these basins would allow us to better understand ecosystem sensitivity in these different environments. Presentations at the workshop highlighted the complexity of semi-enclosed seas related to the interactions between physical, chemical, and biological fields, as well as differences in temporal and spatial scales within each system. It was realized that there is a strong climate signal in these systems that manifests itself in interannual, interdecadal, and long-term variability. Part of the variability is likely related to background climate changes on the planet.

16 Science and Policy

The Inter-Parliamentary Conference on the Protection of the Black Sea Environment took place in Istanbul from July 9–12, 1996. It was jointly organized by the Parliamentary Assembly of the Black Sea Economic Cooperation (PABSEC) and the Parliamentary Assembly of the Council of Europe (PACE). PABSEC was established as a result of the major political changes, which occurred in the late 1980s. The countries of the region were trying to respond to the challenges of national development and European integration, enhancing the bilateral and multilateral relations. Parliamentarians from 24 countries, representatives of regional and international organizations, and distinguished international experts participated. Prof. Dimitar Syrakov from the National Institute of Meteorology and Hydrology (NIMH) of Bulgaria and I participated as Bulgarian experts.

The July 1996 conference, which was a step along the path that started with the adoption of the Convention for the Protection of the Black Sea against Pollution (Bucharest Convention, 1992[56]), was not without political background. The issue of pollution of the Black Sea was raised in the context of ever-increasing traffic through

56 http://www.blacksea-commission.org/Official%20Documents/The%20Convention/Full%20Text/, last accessed on 03.02.2023.

the Bosporus. There has been some friction between Russia and Turkey, the latter arguing that navigation should be restricted in order to reduce the potential danger of an environmental disaster. Many of the presenters talked about possible disasters in the Black Sea if a marine accident happened in the Bosporus. However, it is clear to anyone who is a bit familiar with the hydrological situation that the water flow in the Bosporus, at the surface where pollutants from tankers propagate, is from the Black Sea to the Sea of Marmara (Fig. 2). In case of such accident, the Sea of Marmara would be more endangered. Something like in Lafontaine's fable "The Wolf and the Lamb" was happening. During a break, shortly before my talk, my Turkish friend Ü. Ünlüata came to me and said: "With your report, you have to make a statement about how important it is that we restrict traffic across the Bosporus for safety". My report was on "The Vulnerable Ecological System of the Black Sea". I presented the physical preconditions for the formation of certain ecological states in the Black Sea, and the degradation of the Black Sea environment over the past decades because of eutrophication caused by the huge amounts of fertilizer-enriched waters entering the sea. I did not make the statement that Ümit wanted. This cooled our relations.

After a while, I realized that the 1998 recommendations of the Parliamentary Assembly of the Council of Europe on sustainable development of the Mediterranean and the Black Sea referred to the Istanbul Conference. In these recommendations, I find the statement that I was making with my report, namely that pollution coming from the rivers flowing into the Black Sea could affect even the Mediterranean. It has taken two years to realize that the current on the surface of the Bosporus flows in the opposite direction. I started to realize that the benefit of research results for politicians is limited by their capability to understand them and that the scientific community has a responsibility to communicate scientific data and results timely, perhaps also proactively, and in an understandable fashion. Obviously, increased understanding is needed on both sides, scientists and politicians, to optimally address issues of great socio-economic relevance.

17 1997, One Important Year

In 1997, I had a long journey ahead of me. The Joint Assembly of the International Association of Meteorology and Atmospheric Sciences (IAMAS) and IAPSO was held in Melbourne from 1 to 9 July. On this trip, I made two new friends, David Dietrich[57] and Malcolm Bowman[58] (Fig. 34). David passed away in 2018, and our mutual friend Malcolm invited those who had worked with David to write something of their experiences. Here is my text.

---

[57] At that time D. Dietrich was Senior Research Scientist at the Mississippi State University Center for Air Sea Technology, Stennis Space Center, MS 39529-6000.

[58] Professor of Physical Oceanography at the School of Marine and Atmospheric Sciences, State University of New York at Stony Brook.

"Dave was a great professional and an excellent scholar. I will never forget his enthusiasm and cheerfulness. There was something about him that reminded me of a young man in the shape of a grown person. I first met him in Melbourne, 1997. He was presenting his results on eddy formation and evolution in the East Australian Current region, very nice movies. I approached him and told him that there were many similar processes in the Black Sea that I wanted to simulate. A month later Dave sent me some of his results. He had done all numerical simulations himself; he had found data on the bottom topography and atmospheric forcing, and he had set the initial state of his model by taking hydrological data from the Sid Levitus Atlas (Levitus 1982). The salinity looked strange, but not because of problems in the model, but because at that time, S. Levitus had not paid much attention to the Black Sea.[59] The enthusiasm of Dave, my criticism and the curiosity of the two of us brought us together. This is how we started our joint work 20 years ago…"

D. Dietrich developed the model named DieCAST (Dietrich Center for Air Sea Technology) ocean model. This model uses higher-order treatment of the generally dominant terms in ocean dynamics, i.e., pressure gradient and Coriolis force. This leads to accurate and robust (stable with realistically small dissipation) models. It appears an excellent tool to carry out numerical studies of the Black Sea circulation (Staneva et al. 2001) because this model made it possible, for the first time, to resolve

Fig. 34 David Dietrich, (on the left) and Malcolm Bowman attending the International Liège Colloquium on Ocean Dynamics. Courtesy of J.-M. Beckers

[59] In this version of the Atlas, the Black Sea region was "contaminated" with salinity observations pertaining to the Mediterranean Sea.

all important mesoscale features of the circulation in the Black Sea. The simulations shown in Fig. 6 are produced using this model.

At the very beginning of the political transition in Bulgaria, my cousin Georgi Popov, a computer engineer, settled in Auckland, New Zealand. I decided to visit him. Coincidentally, Dave was also traveling to New Zealand. "Where are you going to settle down?" he asked, and when he found out my main destination was Auckland, he said, "You know, I have a friend there whom I'm going to visit. His name is Malcolm Bowman. He's a professor in the US, but he has a second job in Auckland. Why don't you visit him and give a talk there?" Between 1996 and 1999, M. Bowman, who is Scottish, New Zealander and American, took a sabbatical leave from Stony Brook University near New York and returned to his native New Zealand as head of the School of Environmental and Marine Sciences at the University of Auckland. It didn't take much for David to convince me, I contacted Malcolm and visited him at his institute. The friendships we made, which later developed into family contacts, continue to this day. Malcolm and I cross paths all the time, from Brazil to Canada and from Taiwan to Long Island. Together with him and Dave, and with my Ph.D. students from the 1990s, Joanna Staneva and Elisaveta Peneva, we published some results of our research in the Black Sea (Staneva et al. 2001; Stanev et al. 2003), which resonated well in the scientific community. In the same year, I married Joanna, with whom we continue to this day on our common scientific path.

18 A Revolution in Ocean Science in the Second Half of the Twentieth Century

18.1 Satellite Oceanography

Satellite oceanography is concerned with the application of aerospace electromagnetic remote sensing technology, which enables to study the oceans and provides a new tool in marine observational practice. You might ask why satellites? The oceans are vast and difficult to cover thoroughly with observational instruments. On the other hand, satellites provide a uniform sampling, and the data are taken quickly and over large areas. However, there are also limits to what can be observed from satellites as they see only the surface of the ocean or a very thin surface layer. Furthermore, the number of parameters of the marine environment that can be measured by satellites is limited.

There are two main measurement methods: passive and active. In the active method, the satellite sends a signal to the ocean, it is reflected and returned back to the satellite. In the passive method, the satellite detects a signal from another source, for example sunlight, that has been reflected from the surface of the ocean.

The properties of the signal received by the satellite are used to estimate the properties of the ocean in the area from which the signal was received.

In terms of orbits of satellites, there are two types: geostationary orbits and polar orbits. In the former, the satellite orbits synchronously with the Earth, staying or "hovering" in the same place. Satellites flying in such orbits enable a high temporal resolution (continuous observation of the same region of the ocean), but the coverage in space is limited by the area the satellite observes (Fig. 35). A polar orbit is one in which the satellite passes near both poles of the Earth (Fig. 36). On a rotating Earth, each orbit is offset from the previous one, and over a period of time the satellite can capture the entire Earth's surface. The problem with these orbits is the lower temporal resolution. Typically, geostationary satellites fly in high orbits, for example at a distance of about six Earth radii (about 36,000 km), while polar orbits are low—850–1000 km altitude.

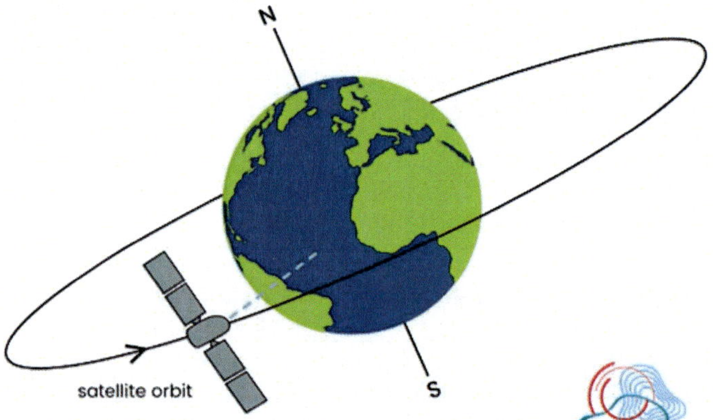

Fig. 35 Geostationary satellite. Courtesy of Beate Meiners

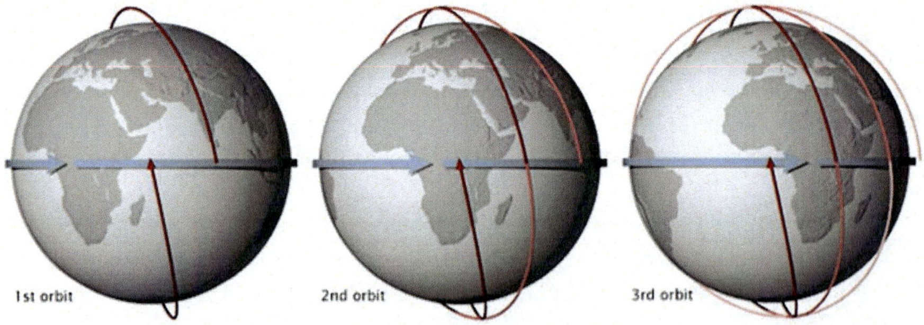

Fig. 36 These illustrations show three consecutive orbits of a sun-synchronous satellite. The satellite's most recent orbit is indicated by the dark red line, while older orbits are lighter red (NASA illustration by Robert Simmon)

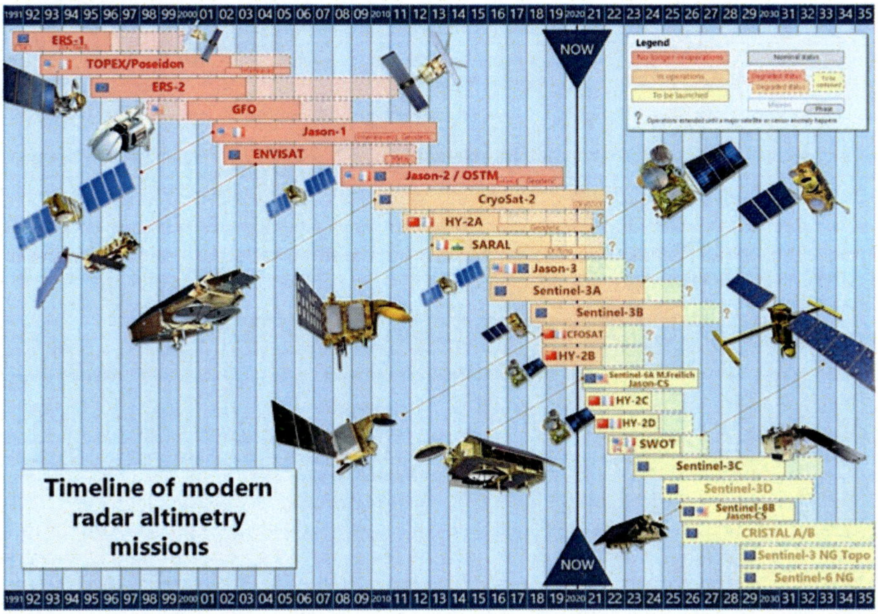

Fig. 37 Altimetry satellites timeline (International Altimetry Team 2021), with permission from Elsevier

The measurement of ocean color, by which the content of chlorophyll or other suspended matter is determined, is an example of a passive measurement. It measures the characteristics of light in the visible part of the spectrum scattered by the surface layer of the ocean and transmitted upward through the ocean surface. These measurements are particularly important for the study of the biological state of the ocean (Groom et al. 2019).

Another representative of a passive measurement is the measurement of surface temperature. Like any body, the Earth radiates infrared light, and the respective flux depends on the surface temperature. By measuring the intensity of the long-wave radiation with special infrared sensors called radiometers, the temperature can be determined (Donlon et al. 2009).

A specific example of an active observation is the measurement of ocean level (International Altimetry Team 2021, see also Fig. 37). This observation uses microwave pulses sent by radar mounted on satellites. The signal is reflected from the ocean surface and returned. The travel time of the signal determines the distance between the orbit and the ocean surface. The latter is not horizontal. Its slope can be used to estimate the magnitude of surface currents or get an indirect information about the density distribution of seawater. As a rule of thumb, the ocean level is higher in low density areas and vice versa. The accuracy of ocean level measurements being very high (errors of approximately 3 cm), the satellite can see even smaller irregularities or individual ocean eddies.

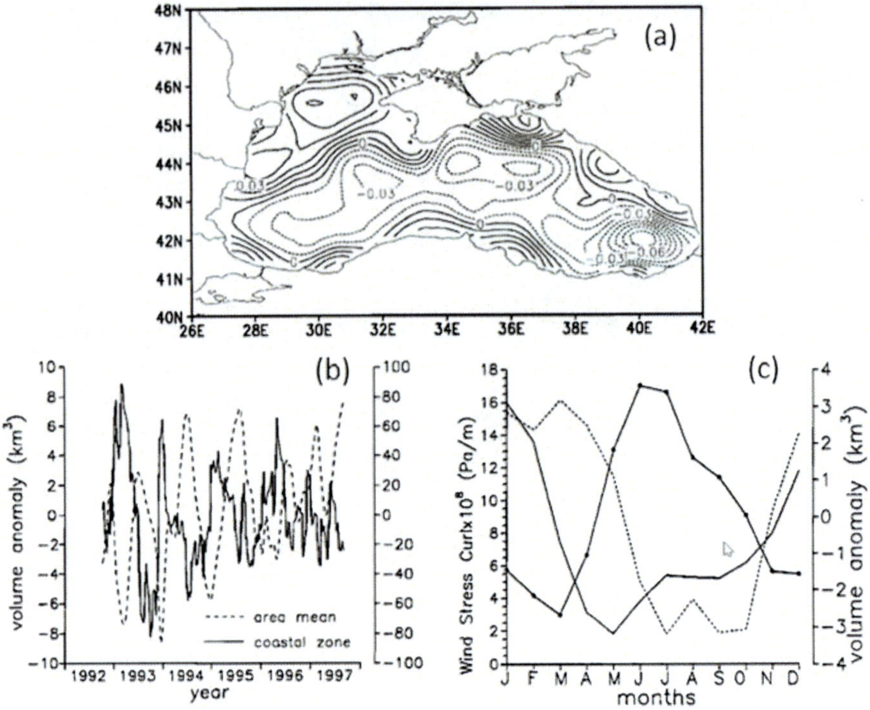

Fig. 38 Figure on the top **a** shows the first empirical orthogonal function (EOF) of grided altimeter data. From Peneva et al. (2001), with permission from Elsevier. **b** Solid line (left *y*-axis) shows the variability of the sea level anomaly in the coastal zone minus the basin mean sea level anomaly. The dashed line (right *y*-axis) shows the variability of the basin mean sea level anomaly. **c** Solid line shows the seasonal variability of the wind stress curl estimated from the climatic data of Staneva and Stanev (1998). The dashed line shows the sea level anomaly in the coastal zone minus the basin mean sea level times the area of the coastal zone. The curve with small circles shows the mean for the four-year period sea level anomaly corresponding to the solid line in **b** and scaled by 10^{-1}. From Stanev et al. (2000)

Another example of an active measurement method is the determination of wind at the sea surface (Ribal and Young 2019). The idea is that a microwave radar signal sent from a satellite is scattered differently by the sea surface, depending on different smoothness in different areas of the ocean or at different time intervals. Sea surface is rougher when the waves are higher. Since wave height is directly related to wind speed (see Sect. 5.2 in chapter "New Observing and Modeling Systems and New Scientific Applications), the latter can be determined with an accuracy to within $2\ \mathrm{m\ s^{-1}}$.

18.2 Winter Intensification of the Black Sea Circulation

By the end of the twentieth century, methods of ocean remote sensing had become an indispensable component of the ocean observing toolkit. In the 1990s, changes in sea level and ocean currents started to be measured in the frame of the French-American TOPEX/Poseidon program. The first satellite was launched on August 10, 1992, from the Kourou space center on an Ariane 42P launch vehicle.[60] On the French side, Pierre-Yves LeTraon[61] was responsible for much of the scientific program. The data were free to use, and he and I decided to start processing the first four years of measurements in the Black Sea area. E. Peneva joined the work (Stanev et al. 2000[62]).

In addition to presenting some general features of the Black Sea circulation in Sect. 1, I show below the results of a statistical decomposition of the altimeter dataset in terms of empirical orthogonal functions (EOF analysis[63]). The first EOF (Fig. 38a) accounts for the largest possible variance (here 24%). It shows the dominant underlying pattern of variability, in this case the opposing trends of sea level variability in coastal and open sea areas. Corresponding temporal changes reveal seasonal variability, suggesting intensification of surface currents. The analysis of atmospheric conditions in the Black Sea showed that the winter intensification of currents is related to the spatial variability of atmospheric winds, and in particular to the stronger wind vorticity in winter (Fig. 38b, c). The longer the measurements, the more reliable and comprehensive the results are. With the help of altimeter data, accumulated over about thirty years and results from numerical models, Peneva et al. (2021) proved recently the correctness of the findings of Stanev et al. (2000) about the predominating wind-induced circulation.

[60] Jason-1, the successor to the TOPEX/Poseidon satellite, was launched on December 7, 2001. Jason-2 is also known as OSTM (Ocean Surface Topography Mission). It was launched on June 20, 2008. The two satellites are deployed in circular orbits at 1340 km. Jason-1 and Jason-2 went out of service in July 2012 and October 2019, respectively. Their operations were continued by Jason-3, launched on January 17, 2016. Sentinel-6 Michael Freilich was launched in November 2020 and orbits 30 s behind Jason-3. Scientists have recently agreed that Sentinel-6 Michael Freilich should become the reference satellite for sea level measurements. See Fig. 37.

[61] In the 1990s P.-Y. LeTraon was vice director of the Space Oceanography Division (Collect Localisation Satellites, CLS).

[62] Almost at the same time, analysis of altimetry observations of the Black Sea level was also performed by Korotaev et al. (2001).

[63] EOF analysis is a decomposition of a signal or dataset in terms of orthogonal basis functions which are determined from the data. The basis functions are chosen to be different from each other and to account for as much variance as possible.

18.3 Sea Level in European Seas as Seen by Satellite Altimetry

About a decade later, the altimeter time series already covered a longer time period and the statistics were expected to be more reliable. This time, my interest extended to the whole region of the European seas. The sea level features in Fig. 39 are presented based on processing Envisat[64] (tracks in red) and Jason (tracks in blue) satellite altimetry data. Envisat orbits the Earth in about 101 min with a repeat cycle of 35 days. The corresponding values for Jason-2 are 112.4 min and 9.9156 days. The longer repeat cycle of Envisat allows for a greater density of tracks.

The winter sea level anomaly, defined as the difference between the December–January–February mean and the annual mean, reveals a number of specific manifestations of sea level patterns (Fig. 39). Winter anomalies are positive in shelf seas (the Baltic Sea and the northwest European shelf) and negative in deep basins. The

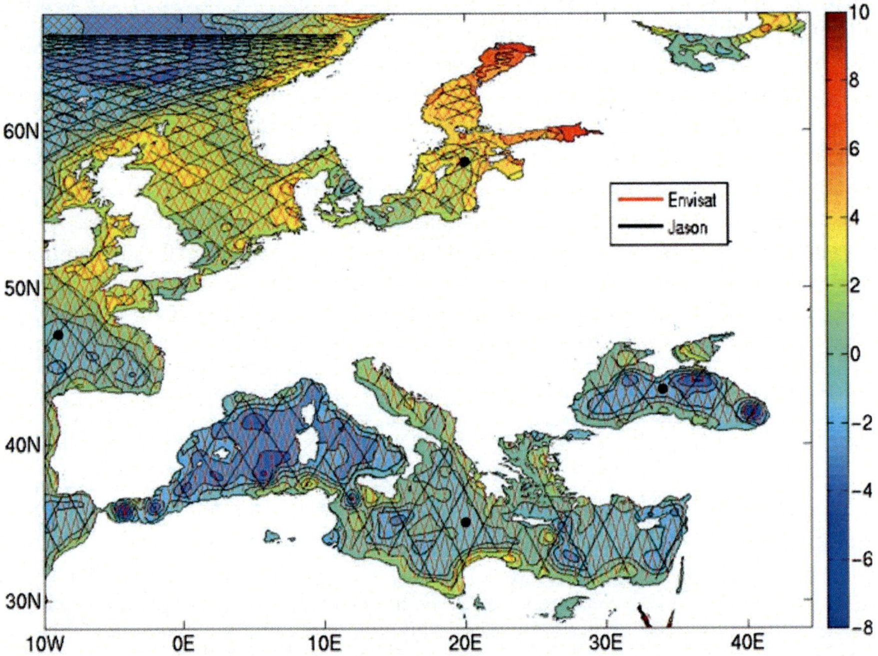

Fig. 39 Difference between the mean for winter months (December–February) and the annual mean sea level anomaly from altimeter data averaged from 1993 to 2011. Satellite tracks over the European seas are overimposed. From Stanev and Xi (2013)

[64] Envisat (Environmental Satellite) is a European Space Agency (ESA) environmental satellite that was launched in 2002 and operated until 2012. Its main tasks were to continuously monitor the climate, the ocean, and the land surface. It was the largest Earth observation satellite ever put into orbit by ESA.

largest difference between the winter anomalies in the northwest European shelf and the Atlantic reaches a maximum of about 10 cm in the continental slope region. A similar magnitude of difference exists between the winter sea level anomaly in the coastal and deep regions of the Mediterranean and the Black Sea. Since the sea surface pressure gradient, which determines the geostrophic transport, is very well recorded in the altimeter data, it can be concluded that satellite observations suggest a seasonal intensification of circulation in winter in the Mediterranean and the Black Sea (see Sect. 18.2 and also Stanev et al. 2000). Figure 40 allows to detect an essential difference between deep and shallow seas. The patterns of anomalies in the two estuarine basins, the Baltic Sea (shallow) and the Black Sea (deep), are too different, while there is a clear similarity between the patterns of anomalies in the deep seas, the Black Sea and the Mediterranean.

European semi-enclosed seas show very different and specific temporal variability in the sea level compared to the adjacent ocean (Fig. 40). This reflects the specific manifestations of the freshwater budget, surface heat fluxes, and control by the connecting straits on the exchange between the basins. The main difference between the temporal variation of sea level in the three semi-enclosed basins, the Black Sea, the Baltic Sea, and the Mediterranean Sea, and that in the Atlantic, where

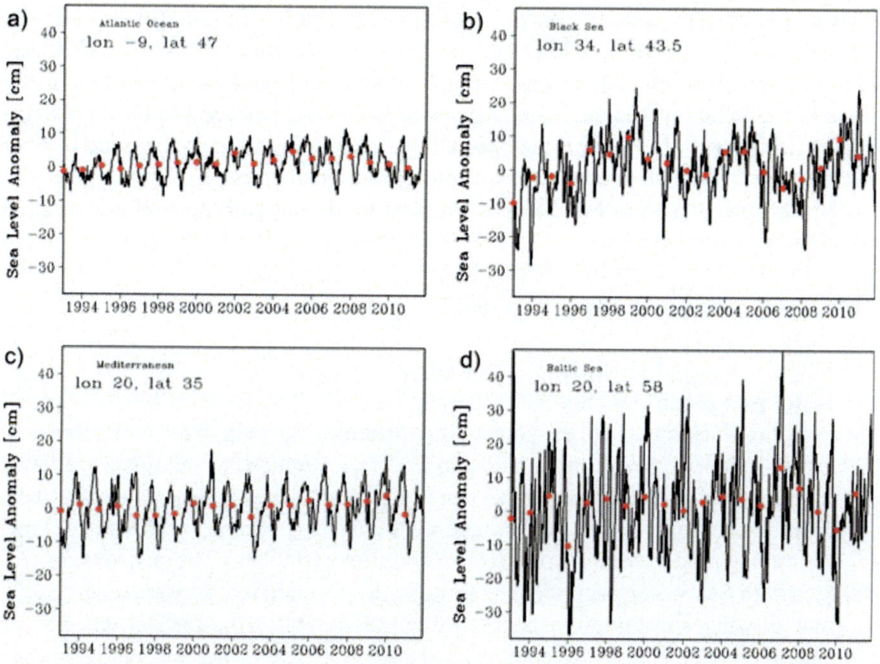

Fig. 40 Temporal change in sea level in 1993–2011 from satellite altimetry in the Atlantic Ocean (47° N, − 9° E), Black Sea (43.5° N, 34° E), Mediterranean Sea (35° N, 20° E), and Baltic Sea (58° N, 20° E). For the positions see the black dots in Fig. 39. The red dots in the graphs illustrate the annual mean sea level. From Stanev and Xi (2013)

the Gulf of Biscay site is used for comparison, is that in the ocean the magnitude of the temporal variation of 4.2 cm is low compared to the Mediterranean (6.3 cm), the Black Sea (8.4 cm), and the Baltic (14.1 cm). While the Atlantic and Mediterranean Sea show a pronounced annual cycle, the temporal variability in the Black and Baltic Seas is characterized by a strong intraseasonal periodicity (between a few months and one year) and interannual periodicity (Fig. 40). The predominant seasonal variability of the sea level in the Mediterranean Sea (the site presented here is in the Ionian Basin, see Fig. 39) is indicative of the role of heat and water fluxes at the sea surface, the latter also having very clear seasonal variability. In the two estuarine basins, the temporal variability is less "monochromatic", resulting from variations in transport in the straits caused by atmospheric influences (e.g., wind) as well as from less periodic river runoff. The strait control is a key factor that determines sea level variations in the two estuarine basins.

With satellite altimetry, a "revolution in oceanography" was underway (Le Traon 2013). I will add, and it is still ongoing.

19 Ventilation of Black Sea Waters

Looking back, I realize how winding my scientific path has been. On various occasions, I had very productive contacts with chemists, biologists, and geologists. Some of these contacts were very long and persistent, as the case was with J. Murray. In 1988, he organized the American-Turkish expedition in the Black Sea and discovered the suboxic layer at the boundary between the oxic surface and the anoxic deep water (Murray et al. 1989). He subsequently became involved in the issue of ocean acidification,[65] which is now a major field of marine chemistry.

Jim Murray was interested in my work on modeling currents and tracers in the Black Sea and the ventilation of its waters. He wrote to me: "We have some interesting data from measurements in the Black Sea. Are you interested in visiting us for twenty days and working with us on these data?".

The Penetration of Marmara Sea Water into the Deep Layers of the Black Sea

Prof. Jan Backhaus from the University of Hamburg provided me with an easy-to-use reduced-gravity numerical model,[66] with which my graduate student Julian Simeonov had studied the penetration of Marmara Sea water into the deep layers of the Black Sea (Simeonov et al. 1997; Stanev et al. 2001). We set up the model with a horizontal resolution of 600 m. The penetration of Marmara Sea water occurs in the form of the so-called gravity current, i.e., a flow of salty and therefore denser water which, following the bottom, sinks and reaches ever greater depths. The key significance of this current is that it supplies the deeper layers of the sea with oxygen. Therefore, the numerical model was coupled with a simple chemical model simulating the oxidation

[65] Acidification has very serious consequences for ocean ecosystems. One of these is that it damages coral reefs. A continued increase of carbon dioxide uptake would result in a level of ocean acidity higher than at any other time over the last 130 million years.

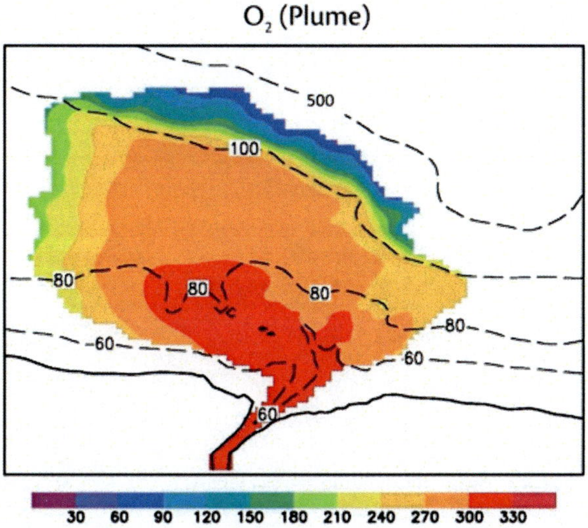

Fig. 41 Distribution of O_2 (molar units in the color bar) in the Bosporus plume. The plotted area extends between 28.8° E and 29.51° E and 41.18° N and 41.6° N. Depth contours down to 100 m are plotted with dashed lines with a contour interval of 20 m, the last dashed line is the isobath of 500 m. The figure is from Stanev (2005) and is made available under the terms of the Creative Commons Attribution 4.0 International License (https://creativecommons.org/licenses/by/4.0/)

of H_2S by O_2. Figure 41 illustrates the pattern of the plume as seen in the distribution of concentrations of O_2.

Despite my busy schedule during the semester, I visited J. Murray in April 2000. We were interested in the exchange regime between surface and deep water. Determining it is difficult because the various characteristics we measure are largely uncertain; however, passive tracers prove useful. One such tracer is freon (chlorofluorocarbon, CFC), which is used in refrigeration, for example. The function describing its rise in the last century is shown in Fig. 42. CFCs destroy the ozone layer, therefore the signatory countries to the Montreal Protocol of 1987 committed to reducing and eventually eliminating the emission of CFC. However, over the time of release into the atmosphere, CFCs have left clear traces in the ocean, too. Measurements of the distributions of two chlorofluorocarbons, CCl3F (CFC-11) and CCl2F2 (CFC-12), made during the June/July 1988 cruise of the research vessel "Knorr" in the Black Sea were available to J. Murray and John Bullister, the latter from the Pacific Marine Environmental Laboratory of NOAA (National Oceanic and Atmospheric Administration). According to their previous analysis (Lee et al. 2002), the situation in the Black Sea was very different from the ocean, as well from the one in the neighboring Marmara Sea (Fig. 43).

Figure 43 provides a key to understanding the main differences between the Black Sea and the Sea of Marmara in terms of ventilation of their deep waters. In both seas,

[66] Reduced gravity is the changed acceleration of gravity acting on one fluid in contact with a fluid of different density due to buoyancy force. Many reduced gravity models consist of two layers with different densities; the deep layer is at rest below the active upper ocean layer. In the model presented here, it was the deep layer, which was the active one.

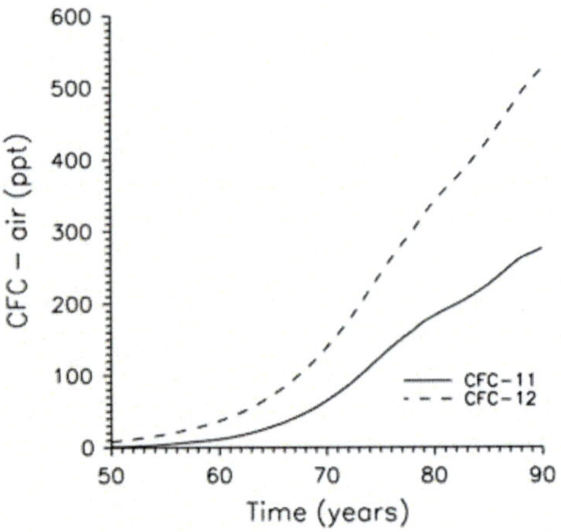

Fig. 42 Basin mean atmospheric CFC-11 and CFC-12 concentrations. All atmospheric CFC concentrations are given in units of mole fraction CFC in air (parts-per-trillion or ppt). From Stanev et al. (2004), with permission from Elsevier

salinity increases with increasing depth (Fig. 43a), but the vertical changes are much stronger in the Sea of Marmara. The layers below 50 m in this basin are filled with Mediterranean water, so salinities are much higher there (note the two different salinity axes in Fig. 43a for the two individual basins). The same data are presented in Fig. 43c in density coordinates (σ_t). The Black Sea water is in the lower left part and the Sea of Marmara water in the upper right part of the plot. The dashed line is the mixing curve between the Marmara and Black Sea waters. It is remarkable that while the Marmara Sea water is very homogeneous at depths below 50 m, the density–salinity relationship in the Black Sea shows a hockey stick-like line, due to the large influence of temperature on the density in the upper layers.

The panels on the right show the vertical distribution of CFC-12. In the upper 100-m layer (Fig. 43b), CFC-12 has comparable values in the two basins. Below this depth, the curve for the Black Sea quickly tends to zero, while the deeper layers in the Sea of Marmara are rich in CFC-12. This means that the Black Sea is poorly ventilated in its deep layers, while the opposite is true for the Sea of Marmara.

My goal was to develop a model for the whole Black Sea, i.e., not so fine in space, and at the same time to reproduce the gravity current decently. However, gravity currents are very small in size; hence, the model resolution needs to be very fine in order to resolve them (see Fig. 41). It seems, at first sight, that one cannot easily reconcile the two conditions, a basin-wide model and a fine resolution. This is where the ideas of parameterizing processes with scales smaller than what the model grid can resolve are of use. These small-scale processes need to be described in a simplified form on the basis of physical concepts known from theory, and at the same time they have to reproduce adequately what happens in nature. Such parameterization was developed by Stanev et al. (2004), and it was particularly important to test results from the numerical model against measurement data. Figure 44 illustrates

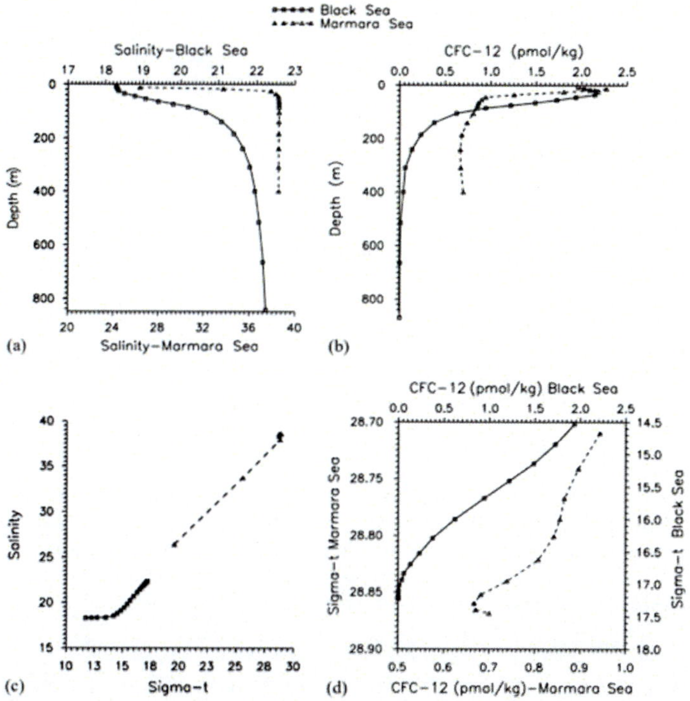

Fig. 43 Vertical profiles from the 1988 Knorr station 14, (—) and Knorr station 1 (- - -), see Lee et al. (2002). **a** Salinity versus depth, **b** CFC-12 concentration versus depth, **c** salinity versus density, **d** CFC-12 concentration versus density. Surface values at densities less than 14.0 (for Sta. 14) and less than 28.7 (for Sta. 1) are not included in the graph. All dissolved CFC concentrations are given in units of picomoles per kg seawater (pmol kg^{-1}). From Stanev et al. (2004), with permission from Elsevier

that the new parameterization does an excellent job of representing the ventilation of deep Black Sea water (model results and observations are plotted by solid line and triangular symbols, respectively). Furthermore, by excluding or including different CFC-12 sources in the model set up, one can estimate their individual contributions. Obviously, most of CFC-12 enters the sea from the atmosphere. The inflow from the Bosporus Strait only slightly increases the concentration of CFC-12 in the layers between 100 and 200 m, this is the interval where most of Bosporus intrusions occur. This explains indirectly the low oxygenation capabilities associated with the gravity currents.

Numerical simulations using parametrizations of the gravity currents confirmed the results from observations showing that the concentration of CFC-12 at 400 m in 1988 was less than 1% of that in the surface layers. As we have seen from the measurements, this is not at all the case in the ocean, nor in the neighboring Sea of Marmara. This unique characteristic of the Black Sea is explained by the fact that vertical mixing in highly stratified fluids is extremely weak. It turned out that the

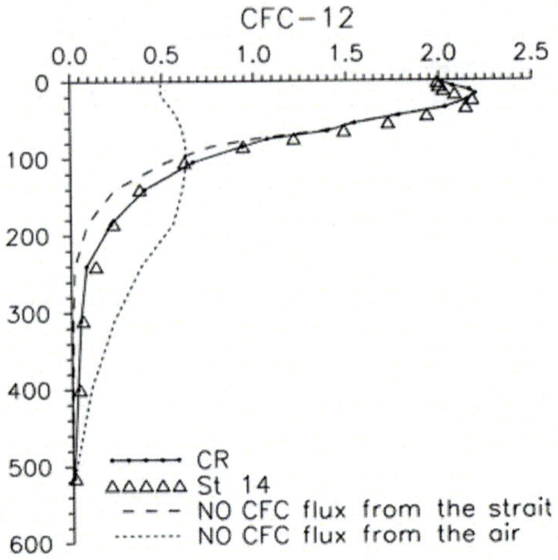

Fig. 44 CFC-12 (in pmol/kg) simulated with a 1D model where convection is included as parameterization. The different lines and experiments are indicated as follows: (—) Control Run (CR), (– – –) experiment with no tracer input from the Bosporus, (- - -) experiment with no tracer flux from the atmosphere. The CFC-12 observations from Knorr station 14, Lee et al. (2002) are shown with triangles. From Stanev et al. (2004), with permission from Elsevier

numerical model not only reproduced the observations well, but the large amount of model data also made it possible to elucidate the exchange and mixing pathways of the waters (Stanev et al. 2004).

The visit in Seattle gave me the opportunity to make personal contacts with Peter Rhines,[67] from whom I have learned in the past a lot, mainly about eddies in the ocean or so-called macroturbulence. To the theoretical aspects of his work in the 1970s and 1980s, P. Rhines added laboratory modeling. With a youthful fervor, he showed me around his lab. It was a huge hall equipped with unique instrumentation where he developed laboratory models and applied sophisticated measurement techniques to study the rotation of the fluids of the Earth and the planets. To illustrate fundamental way of thinking of Peter Rhines, I will quote something I found on his web page: "Chaos, singular perturbation theory, fractals, and solitons are all examples in which natural, environmental science provided the root discovery that grew into a major branch of physics or mathematics. Thus, the intellectual flow is not always from the "mother sciences" to the "environmental sciences" but has gone in both directions". The role of geophysical hydrodynamics could not be better and more clearly appreciated.

[67] Peter Rhines is professor at the University of Washington, a member of the U.S. National Academy of Sciences, fellow of the American Academy of Arts and Sciences and of many other important scientific societies and committees.

20 The End of the Millennium

In parallel with the growing number of scientific and organizational activities I have described, my career at the University of Sofia was also developing. In 1986, I became an associate professor, and in 1996, a professor of physical oceanography. What changed my professional trajectory was an announcement I came across during my visit in Hamburg in 1999. The University of Oldenburg, a town in northern Germany, 40–50 km from Bremen, was looking to recruit a research fellow to the newly established Physical Oceanography group of Professor Jörg-Olaf Wolf. I had a big dilemma whether to apply. There were many arguments against, but there were also some for. My work in Sofia was going well, but not quite as I wanted it to. V. Roussenov left for England in 1997. Before him, two good collaborators of S. Panchev, Vasil Andreev and D. Syrakov, moved to the BAS. After several years of working at the Department of Meteorology and Geophysics, J. Staneva also moved to the BAS. The golden years of the Department had passed. Nationally, in the face of disappearing political tutelage, the once established coordination between scientific institutions, though with its many shortcomings, had been replaced by complete freedom of action, opening up opportunities for unfair competition.

I observed that more than half of my students found good employment in the West, mostly in the USA. This was easily explainable. At that time there were strong teaching schools at the University of Sofia. I realize now that 50 years ago, the physics education we received was comparable to that of the best world universities. This became one of the prerequisites for many colleagues to continue their scientific work outside Bulgaria after 1989. On December 12, 1999, I submitted my application for the vacant position, and shortly there after J.-O. Wolff wrote to me that I had been selected. I accepted, but decided not to break with either my scientific or teaching activities in Sofia. For more than 20 years, I remained split between Bulgaria and Germany.

References

Aubrey DG (1991) International workshop on the Black Sea Varna, Bulgaria, 30 Sept–4 Oct 1991. Intergovernmental Oceanographic Commission workshop report no. 86, p 31

Bekov V, Petrova-Klisarova D (2013) Eighty years Marine Biological Station with aquarium in Varna. Bulg J Agric Sci 19(Supplement 1):1–5

Beuvier J, Sevault F, Herrmann M, Kontoyiannis H, Ludwig W, Rixen M, Stanev E, Béranger K, Somot S (2010) Modeling the Mediterranean Sea interannual variability during 1961–2000: focus on the Eastern Mediterranean Transient. J Geophys Res Oceans 115:C08017

Bryan K (1969) A numerical method for the study of the circulation of the World Ocean. J Comput Phys 4:347–376

Clementi E, Aydogdu A, Goglio AC, Pistoia J, Escudier R, Drudi M, Grandi A, Mariani A, Lyubartsev V, Lecci R, Cretí S, Coppini G, Masina S, Pinardi N (2021) Mediterranean Sea physical analysis and forecast (CMEMS MED-Currents 2016–2019) (Version 1). Copernicus Monitoring Environment Marine Service (CMEMS). http://doi.org/10.25423/CMCC/MEDSEA_ANALYSIS_FORECAST_PHY_006_013_EAS4

Cummings J, Bertino L, Brasseur P, Fukumori I, Kamachi M, Martin MJ, Mogensen K, Oke P, Testut C-E, Verron J, Weaver A (2009) Ocean data assimilation systems for GODAE. Oceanography 22:96–109. https://doi.org/10.5670/oceanog.2009.69

Donlon C, Casey K, Robinson I, Gentemann C, Reynolds R, Barton I, Arino O, Stark J, Rayner N, Le Borgne P, Poulter D, Vazquez-Cuervo J, Armstrong E, Beggs H, Llewellyn-Jones D, Minnett P, Merchant C, Evans R (2009) The GODAE high-resolution sea surface temperature pilot project. Oceanography 22:34–45. https://doi.org/10.5670/oceanog.2009.64

Frins I (2019) Ehrung für Prof. em. Dr. Jürgen Sündermann. https://blogs.helmholtz.de/kuestenforschung/2019/02/22/ehrung-fuer-prof-em-dr-juergen-suendermann/

Groom S, Sathyendranath S, Ban Y, Bernard S, Brewin R, Brotas V, Brockmann C, Chauhan P, Choi JK, Chuprin A, Ciavatta S, Cipollini P. Donlon C, Franz B, He X, Hirata T, Jackson T, Kampel M, Krasemann H, Lavender S, Pardo-Martinez S, Mélin F, Platt T, Santoleri R, Skakala J, Schaeffer B, Smith M, Steinmetz F, Valente A, Wang M (2019) Satellite ocean color: current status and future perspective. Front Mar Sci 6:485. http://doi.org/10.3389/fmars.2019.00485

International Altimetry Team (2021) Altimetry for the future: building on 25 years of progress. Adv Space Res 68(2):319–363. https://doi.org/10.1016/j.asr.2021.01.022

Josey SA (2003) Changes in the heat and freshwater forcing of the eastern Mediterranean and their influence on deep water formation. J Geophys Res 108(C7):3237. https://doi.org/10.1029/2003JC001778

Klein B, Roether W, Manca BB, Bregant D, Beitzel V, Kovacevic V, Luchetta A (1999) The large deep water transient in the eastern Mediterranean. Deep Sea Res Part I 46:371–414

Konovalov SK, Murray JW, Luther GW III (2005) Basic processes of Black Sea biogeochemistry. Oceanography 18:24–35

Korotaev GK, Saenko OA, Koblinsky CJ (2001) Satellite altimetry observations of the Black Sea level. J Geophys Res Oceans 106:917–933

Lancelot C, Martin J-M, Panin N, Zaitsev Y (2002a) The North Western Black Sea: a pilot site to understand the complex interactions between human activities and the coastal environment. Estuar Coast Shelf Sci 54:279–283

Lancelot C, Staneva J, van Eeckhout D, Beckers JM, Stanev E (2002b) Modeling the Danube-influenced north-western continental shelf of the Black Sea. II: ecosystem response to changes in nutrient delivery by the Danube River after its damming in 1972. Estuar Coast Shelf Sci 54:473–499

Lass HU, Mohrholz V (2003) On the dynamics and mixing of inflowing salt-water in the Arkona Sea. J Geophys Res Oceans 108:3042

Le Traon PY (2013) From satellite altimetry to Argo and operational oceanography: three revolutions in oceanography. Ocean Sci 9:901–915

Lee BS, Bullister JL, Murray JW, Sonnerup RE (2002) Anthropogenic chlorofluorocarbons in the Black Sea and the Sea of Marmara. Deep Sea Res I 49:895–913

Levitus S (1982) Climatological atlas of the world ocean. NOAA/ERL GFDL professional paper 13, Princeton, NJ, p 173 (NTIS PB83-184093)

Madec G, NEMO System Team (2022) NEMO ocean engine. Scientific notes of Climate Modeling Center (27), Institut Pierre-Simon Laplace (IPSL). ISSN 1288-1619. http://doi.org/10.5281/zenodo.6334656

Meier HEM (2001) On the parameterization of mixing in three-dimensional Baltic Sea models. J Geophys Res Oceans 106:30997–31016

Millot C (1999) Circulation in the western Mediterranean Sea. J Mar Syst 20:423–442

Mohrholz V, Dutz J, Kraus G (2006) The impact of exceptionally warm summer inflow events on the environmental conditions in the Bornholm Basin. J Mar Syst 60:285–301

Munk WH (1950) On the wind-driven ocean circulation. J Meteorol 7:79–93

Murray JW, Jannasch HW, Honjo S, Anderson RF, Reeburgh WS, Top Z, Friederich GE, Codispoti LA, Izdar E (1989) Unexpected changes in the oxic/anoxic interface in the Black Sea. Nature 338:411–413

Oguz T, Aubrey DG Latun VS Demirov, E, Koveshnikov L, Sur HI, Diaconu V, Besiktepe S, Duman M, Limeburner R, Eremeev V (1994) Mesoscale circulation and thermohaline structure of the Black Sea observed during HydroBlack '91. Deep Sea Res Part I Oceanogr Res Pap 41(4):603–628

Özsoy E, Mikaelyan A (eds) (1997) Sensitivity to change: Black Sea, Baltic Sea and North Sea. In: Proceedings of the NATO advanced research workshop on sensitivity of North Sea, Baltic Sea and Black Sea to anthropogenic and climatic changes, Varna, Bulgaria, 14–18 Nov 1995. NATO security through science series: sub-series C: environmental security, vol 27, p 516

Özsoy E, Ünlüata Ü (1997) Oceanography of the Black Sea: a review of some recent results. Earth Sci Rev 42(4):231–272

Peneva E, Stanev E, Belokopytov V, Le Traon P-Y (2001) Water transport in the Bosphorus Straits estimated from hydro-meteorological and altimeter data: seasonal to decadal variability. J Mar Syst 31:21–33

Peneva E, Stanev E, Ciliberti S, Lima L, Aydogdu A, Marinova V, Valcheva N (2021) Interannual variations of the Black Sea Rim Current. J Oper Oceanogr s53–s59

Petersen W, Colijn F, Dunning J, Hydes D, Kaitala S, Kontoyiannis H, Lavin A, Lips I, Howarth J, Ridderinkhof H, Pfeiffer K, Soerensen K (2006) European Ferry Box project: from online oceanographic measurements to environmental information. In: Dahlin H, Flemming NC, Marchand P, Petersson SE (eds) European operational oceanography: present and future. Proceedings of the 4th international conference on EuroGOOS. Brest (F), 06.-09.06.2005. Office for Official Publ. of the EUR, Luxembourg, pp 551–560. ISBN: 92-894-9788-2

Pinardi N, Korres G, Lascaratos A, Roussenov V, Stanev E (1997) Numerical simulation of the interannual variability of the Mediterranean Sea upper ocean circulation. Geophys Res Lett 24:425–428

Pinardi N, Navarra A (1993) Baroclinic wind adjustment processes in the Mediterranean Sea. Deep Sea Res Part II 40:1299–1326

Rachev NH, Stanev EV (1997) Eddy processes in semi-enclosed seas. A case study for the Black Sea. J Phys Oceanogr 27:1581–1601

Ribal A, Young IR (2019) 33 years of globally calibrated wave height and wind speed data based on altimeter observations. Sci Data 6(77). http://doi.org/10.1038/s41597-019-0083-9

Robinson AR (ed) (1983) Eddies in marine science. Springer, Berlin, p 609

Roether W, Klein B, Manca BB, Theocharis A, Kioroglou S (2007) Transient Eastern Mediterranean Deep Waters in response to the massive dense-water output of the Aegean Sea in the 1990s. Prog Oceanogr 74:540–571

Roether W, Manca BB, Klein B, Bregant D, Georgopoulos D, Beitzel V, Kovacevic V, Luchetta A (1996) Recent changes in Eastern Mediterranean Deep Waters. Science 271:333–335

Roussenov V, Stanev E, Artale V, Pinardi N (1995) A seasonal model of the Mediterranean Sea. J Geophys Res Oceans 100:13515–13538

Semtner AJ (1974) An oceanic general circulation mode with bottom topography. Numerical simulation of weather and climate. Technical report № 9, University of California, Los Angeles, p 99

Simeonov J, Stanev E, Backhaus J, Jungclaus J, Roussenov V (1997) Heat and salt intrusions in the pycnocline from sinking plumes. Test case for the entrainment in the Black Sea. In: Özsoy E, Mikaelyan A (eds) Sensitivity to change: Black Sea, Baltic Sea and North Sea. Kluwer Academic, Dordrecht, pp 417–438

Stanev EV (1986) Numerical modeling of the Black Sea dynamics and hydrogen sulphide and oxygen distribution in the Black Sea. Proceedings, Report on the chemistry of the sea water, XXXIII. International meeting on the chemical and physical oceanography of the Black Sea, Goteborg, June 2–4, 1986, p 423

Stanev EV (1989a) Numerical modeling of the circulation and hydrogen sulphide and oxygen distribution in the Black Sea. Deep Sea Res 36:1053–1065

Stanev EV (1989b) On the response of the Black Sea eddy field to seasonal forcing. In: Nihoul JCJ, Jamart BM (eds) Mesoscale/synoptic coherent structures in geophysical turbulence. Elsevier, Amsterdam, pp 423–433
Stanev EV (1990) On the mechanisms of the Black Sea circulation. Earth Sci Rev 28:285–319
Stanev EV (1992) Numerical experiment on the spreading of Mediterranean water in the North Atlantic. Deep Sea Res 39:1747–1766
Stanev EV (1994) Assimilation of sea surface temperature data in a numerical ocean circulation model. A study of the water mass formation. In: Brasseur PB, Nihoul JCJ (eds) Data assimilation: tools for modeling the ocean in a global change perspective. NATO ASI series, I, vol 19, pp 33–57
Stanev EV (2005) Understanding Black Sea dynamics: overview of recent numerical modeling. Oceanography 18:56–75
Stanev EV, Beckers JM (1999a) Barotropic and baroclinic oscillations in strongly stratified ocean basins: numerical study of the Black Sea. J Mar Syst 19(1–3):65–112
Stanev EV, Beckers JM (1999b) Numerical simulations of seasonal and interannual variability of the Black Sea thermohaline circulation. J Mar Syst 22:241–267
Stanev EV, Bowman MJ, Peneva EL, Staneva JV (2003) Control of Black Sea intermediate water mass formation by dynamics and topography: comparisons of numerical simulations, survey and satellite data. J Mar Res 61:59–99
Stanev EV, Buesseler KO, Staneva JV, Livingston HD (1999c) A comparison of modelled and measured Chernobyl 90Sr distribution in the Black Sea. J Environ Radioact 43(2):187–203
Stanev EV, Chtirkova B (2021) Interannual change in mode waters: case of the Black Sea. J Geophys Res Oceans 126:e2020JC016429. http://doi.org/10.1029/2020JC016429
Stanev EV, Friedrich HJ (1991) On the assimilation of climatological data by means of numerical circulation models, exemplified for the Mediterranean Sea. Oceanol Acta 14:97–114
Stanev EV, Friedrich HJ, Botev SV (1989) On the seasonal response of intermediate and deep water to surface forcing in the Mediterranean Sea. Oceanol Acta 12:141–149
Stanev EV, Le Traon PY, Peneva EL (2000) Sea level variations and their dependency on meteorological and hydrological forcing: analysis of altimeter and surface data for the Black Sea. J Geophys Res Oceans 105(C7):17203–17216
Stanev EV, Ricker M (2020) Interactions between barotropic tides and mesoscale processes in deep ocean and shelf regions. Ocean Dyn 70:713–728
Stanev EV, Roussenov V (1985) A numerical study of the seasonal variability of sea currents. In: Izrael Yu (ed) Integrated global ocean monitoring. Proceedings of the I international symposium, USSR, Tallinn, Oct 2–10, 1983, Leningrad, Gidrometeoizdat, pp 120–129 (In Russian: Станев Е, Русенов В. Численное исследование сезонной изменчивости морских течений)
Stanev E, Schulz-Stellenfleth J (2014) Methods of data assimilation. Die Küste 81:133–152
Stanev EV, Simeonov JA, Peneva EL (2001) Ventilation of Black Sea pycnocline by the Mediterranean plume. J Mar Syst 31:77–97
Stanev EV, Staneva J, Bullister JL, Murray JW (2004) Ventilation of the Black Sea pycnocline. Parameterization of convection, numerical simulations and validations against observed chlorofluorocarbon data. Deep Sea Res 51:2137–2169
Stanev EV, Staneva JV, Roussenov VM (1997) On the Black Sea water mass formation. Model sensitivity study to atmospheric forcing and parameterization of physical processes. J Mar Syst 13:245–272
Stanev EV, Xi L (2013) European semi-enclosed seas: basic physical processes and their numerical modeling. In: Soomere T, Quak E (eds) Preventive methods for coastal protection. Springer, Heidelberg. http://doi.org/10.1007/978-3-319-00440-2_5
Staneva JV, Buesseler KO, Stanev EV, Livingston HD (1999) The application of radiotracers to study Black Sea circulation: validation of numerical simulations against observed weapon testing and Chernobyl ^{137}Cs data. J Geophys Res Oceans 104(C5):11099–11114
Staneva JV, Dietrich D, Stanev EV, Bowman M (2001) Rim current and coastal eddy mechanisms in an eddy-resolving Black Sea general circulation model. J Mar Syst 3:137–157

Staneva JV, Stanev EV (1998) Oceanic response to atmospheric forcing derived from different climatic data sets. Intercomparison study for the Black Sea. Oceanol Acta 21:393–417
Stigebrandt A (1987) A model for the vertical circulation of the Baltic deep water. J Phys Oceanogr 17:1772–1785
Stommel HM (1957) A survey of ocean current theory. Deep Sea Res 4(3):149–184
Theocharis A, Klein B, Nittis K, Roether W (2002) Evolution and status of the Eastern Mediterranean Transient (1997–1999). J Mar Syst 33–34:91–116
Ünlüata Ü, Aubrey DG, Belberov Z, Bologa A, Eremeev V, Vinogradov M (1993) International program investigates the Black Sea. EOS 74(36):401–416
Ünlüata Ü, Oğuz T, Latif MA, Özsoy E (1990) On the physical oceanography of the Turkish straits. In: The physical oceanography of sea straits. Springer, Dordrecht, pp 25–60, 881 p
Yankov A et al (2006) History populated with people. Gutenberg Press, Sofia, 780 p (In Bulgarian: Историята населена с хора. Българското общество през втората половина на XX век)
Zhang YJ, Stanev EV, Grashorn S (2016) Unstructured-grid model for the North Sea and Baltic Sea: validation against observations. Ocean Model 97:91–108

From Regional to Coastal Oceanography

This part is dedicated to research and development in the European coastal zone, which is of great importance, as coastal regions are among the most productive, rich in terms of ecosystems, resources, and services, but also among the most affected and exploited areas on the planet. The physical specificity of the coastal environment is presented, focusing on the large role of friction in coastal regions and the small spatial and temporal scales of some key processes. Attention is paid to tidal dynamics and concrete illustrations based on continuous observations and numerical modeling in the Wadden Sea. The role of tidal dynamics in erosion, deposition, and sediment distribution is explained.

Strong impetus for the R&D in coastal areas has been given by the European COastal-shelf sea OPerational observing and forecasting system (ECOOP) project. It aimed to integrate existing coastal and regional marine observations into a pan-European observing system and to integrate existing coastal and regional marine forecasting systems into a pan-European forecasting system. The objectives, structures, and achievements of this initiative are presented, along with several national coastal observing and forecasting systems. The German Coastal Observing System for Northern and Arctic Seas (COSYNA) is being examined in detail. This system put marine monitoring on a qualitatively new footing and started to provide the necessary data coverage of the area in the long term. This was achieved by an optimal use of existing data streams and numerical models for their assimilation. Radar remote sensing was to play a particular role. Scientific products of COSYNA included hourly electronic maps of surface currents, salinity, temperature, and waves.

The project "Towards a Joint European Research Infrastructure network for Coastal Observatories (JERICO)" is a natural step in the process of consolidation of observing systems in Europe. This led to the development and operationalization of a pan-European coastal measurement framework (European Research Infrastructure Consortium, ERIC), which ensures continuous and sustainable provision of high-quality environmental data and information products in the European coastal seas. Activities in the frame of the "Policy-oriented marine Environment Research for the Southern European Seas (PERSEUS)" project presented a response of the scientific

E. Stanev, *Trajectories in Oceanography*, Springer Oceanography,
https://doi.org/10.1007/978-3-031-33720-8_3

community to the Europe's Integrated Maritime Policy, which sets a target of Good Environmental Status in EU marine waters. This constituted an initial step to management of the marine environment and the establishment of clear environmental targets and monitoring programs.

1 Institut Für Chemie Und Biologie Des Meeres

So far, I was trying to follow the chronology of events. However, various activities spanned over years, or others took place in parallel, leading me to present some of them one by one, which is why I will go back in time in several places to start again with something new.

The territorial principle plays a role in the development of oceanography in Germany. The aim is to saturate the Northern provinces bordering the North Sea and the Baltic Sea evenly with oceanographic institutions. Thus, in 1987, the Conference of Science Ministers and Senators of Northern Germany adopted a Coordination Concept for Marine Research and Technology. The Wilhelmshaven-Oldenburg region was recommended to carry out research on the biogeochemistry of shallow seas, and the Institute for Chemistry and Biology of the Sea (Institut für Chemie und Biologie des Meeres, ICBM) was established at the University of Oldenburg for this purpose.

Today, ICBM is an interdisciplinary research institute where fundamental and applied questions of the marine environment and ecology are investigated. By bringing together the research expertise and scientific potential from the fields of chemistry, biology, physics, and modeling, the institute aims to understand the functioning of marine systems. The main focus is on the cycles of transformation of marine matter and energy fluxes in bottom sediments, as well as in the water column. This research is also of great importance at a global scale, because coastal regions are among the most productive, but also among the most heavily impacted and exploited areas of the planet. They are rich in terms of resources, ecosystems, and services (Borgese 1998), gradually evolving into regions of industrial and infrastructure development. Coastal areas account for 40–50% of the global population. Over 70% of the population in the East Asian Seas live in the coastal zone—covering 100 km.

In the late 1990s, the need for ICBM to also develop the physical component of research and training was realized. The newly established working group of J.-O. Wolff had to fill this gap. It was intended to deal mainly with numerical modeling of shallow seas, e.g., the North Sea, and in particular the German Bight region and its southern parts (Fig. 1). You may ask why a distinction should be made between shallow and deep seas. This is because the processes in the two environments are very different, and these differences arise mainly from the large role of friction in coastal regions and the small scales of some basic processes there (see Sect. 4.2 in chapter "In Search of a Destination" and Dyke 2016).

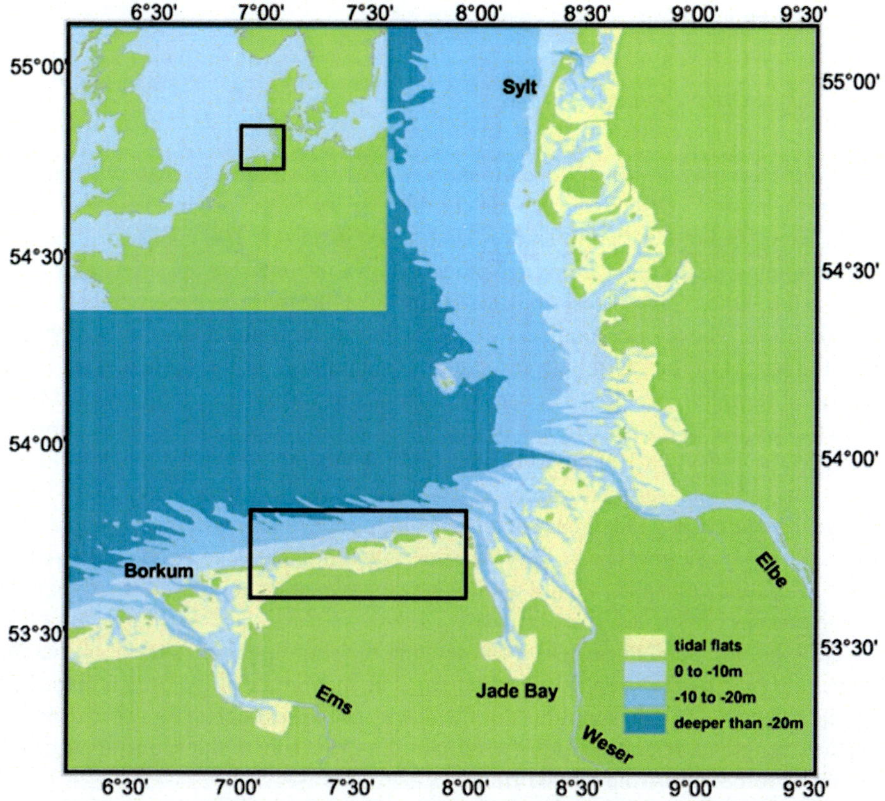

Fig. 1 North Sea (top-left inset), German Bight (in the square frame on the inset and in the big map), and the East Frisian Wadden Sea (in the rectangle frame). The plotted depths are negative. The yellow color gives approximately the areas which could become dry during low tide. In the adjacent area (lightest blue), the depth is less than 10 m (see legend), but this area cannot become dry. From Stanev et al. (2003b)

In his career, before moving to Oldenburg, J.-O. Wolff had worked mainly on currents in the Southern Ocean; the smallest basin I had worked on so far was the Burgas Bay. None of us have worked at scales (and with horizontal resolution) as fine as those needed to properly address coastal ocean dynamics. We had information that Hans Burchard from the University of Hamburg (now a professor at the Baltic Sea Research Institute, Das Leibniz-Institut für Ostseeforschung, IOW, in Warnemünde) was developing a model for the physical processes in a shallow sea. We invited him, and he visited us soon after.

2 Where the Sea, Land, and Air Meet

2.1 Dynamic Coasts

I guess if I ask the question "Where do the sea, land and air meet?", many will answer "At the seashore". Of course, they do. In the interior of the ocean, the atmosphere is far from the bottom (land), and on land, there is no sea. It remains for the three environments, so different from each other, to come in contact at the shore. However, what is the shore? Can we describe it with a line just like those on geographical maps? Furthermore, who tells geographers what that line is? It is known from fractal[1] theory that the length of the coast between point *A* and point *B* is not defined. This is the so-called shoreline paradox. The length of the shore varies depending on what level of detail we use to measure it. Let us take the coastline of the UK as an example. If we measure it using a 100 km long measuring device, the length of the coastline is 2800 km. If the measuring device is 50 km long, the length becomes 3400 km. These are mathematical or geodetical problems. Things get even more complicated, as they often do, if we ask physicists. They will then say that there is no fixed coastline that the coastline is constantly changing and that its dynamics are part of the dynamics of the ocean.

Walking on a sandy beach, one can see that the shoreline is changing, or that the waves are making the sea advance or retreat. The magnitude of the back and forth of the shore depends on the amplitude of the waves, the latter depending approximately on the strength and duration of the wind. Similar is the role of tides, one result of the action of forces of astronomical origin.

2.2 The History of Tidal Research

Isaac Newton was the first to explain tides as the product of the combined effects of the gravitational forces of the Moon and the Sun on the one hand and the rotation of the Earth on the other. Most people know that there are tides because the Moon attracts water masses. They also know that the Earth makes one revolution a day on its axis. Why, then, do tides happen almost everywhere in the ocean twice a day? Isn't it only once a day that a point on the Earth's surface points to the Moon? However, it is not just the moon that matters. In addition to the gravitational pull from the Moon, the water masses are accelerated by a centrifugal force due to the Earth's rotation. On the Moon's side, the two forces act in the same direction and add up. On the opposite side, the centrifugal force is opposed by the gravitational force, the two forces weaken each other, and the combined force points in the opposite direction to the Moon. For this reason, the second tide during the day is usually weaker.

[1] In mathematics, a fractal is a geometric shape containing detailed structure at arbitrarily small scales.

Furthermore, the positions of the Sun and Moon change continuously. With the Sun and Moon roughly aligned (when there is a new or full Moon), the gravitational pull on the sea is the greatest and the tides are on average slightly larger than usual. These tides are known as spring tides. When the Sun and Moon are at right angles to each other (at *quadrature)*, there is least gravitational pull, and the tide is at its lowest. These tides are known as neap tides and occur 7 days after the spring tides.

Newton regarded the tidal problem as static insofar as he did not raise the question of the dynamics of the ocean under the action of tidal forces. Leonard Euler[2] was the first to realize that the horizontal component of the tidal force is much more important to the motion of sea masses than the vertical component. In other words, the sum of the force of lunar attraction and the centrifugal force makes a certain angle with the direction of the gravity acceleration. The gravity force is much greater than this sum force, so the astronomical force is negligible in the vertical direction. However, its horizontal component is comparable or superior to the other forces that set the ocean in motion, such as that the force due to wind acting on the sea surface.

In the early nineteenth century, Pierre-Simon Laplace,[3]instead of adding up all forces acting on an individual body, like in the Solar System, introduced the concept of "field". He defined a property, for instance acceleration, acting in every point of space and formulated a fundamental system of equations relating the horizontal flow in the ocean to the position of its surface. This represents the first major step toward a modern dynamical theory of tides. In the second half of the nineteenth century, William Thomson[4] proved theoretically the existence of coastal waves (tidal waves propagate along coasts) known as Kelvin waves. These waves represent one of the most important elements of the tidal dynamics of the oceans.

2.3 *Tidal Dynamics*

Tides carry water masses from one place to another and then bring them back. As a result, the sea level falls in some places and rises in others. That is, tides are not just a rising and lowering of the ocean level, but rather a horizontal propagation of waves across the ocean surface. Because of their large size, we do not see them, but they are essentially similar to the waves caused by a stone falling on the surface of

[2] Leonhard Euler (1707–1783) was a Swiss mathematician, physicist, astronomer, geographer, logician, and engineer.

[3] Pierre-Simon (Marquis de) Laplace (1749–1827) was a French mathematician, physicist, and astronomer. Among other things, he was concerned with differential equations, a branch of mathematics of high importance for the development of tidal theory.

[4] William Thomson, first Baron Kelvin or Lord Kelvin for short, (1824–1907) was a British physicist in the fields of electricity and thermodynamics. Thomson is known also for his theoretical work in the field of tidal dynamics and for the development of instruments measuring tides. He was the president of the Royal Society from 1890 to 1895.

a pond. The difference is "only" that the length of tidal waves is much greater than the depth of the ocean. This is not the case with the perturbation in a pond created by the stone.

You will ask why there are no tides in the Black Sea or in other small and almost enclosed ocean basins. Not that there are none, but their amplitude is so small (in the order of 1–3 cm) that only measuring instruments can catch them. The tides in the Black Sea have a very small amplitude because of the small size of the basin. Due to this, the horizontal forces that cause tides are almost the same over the entire sea, i.e., there are almost no horizontal differences in tidal forces that cause significant tides. Thus, no meaningful tide can occur because water cannot flow freely from or to another place with an opposing tide on the other side of the world. To that, one has to add that narrow straits connecting semi-enclosed basins with the open ocean efficiently filter out the tidal influences from the open ocean (Stanev et al. 2015a).

Tides receive their energy from astronomical forces, and some of this energy is dissipated, i.e., is converted into heat. This conversion of energy occurs in a cascade, from large scales to small scales, with marine turbulence and friction at the ocean bottom and along the shores eventually taking the brunt of the dissipation. Tidal dissipation can be viewed as a process whereby the Earth-Moon pair loses some of the energy associated with their mutual rotational motion. This changes the torque of the Moon, resulting in a gradual increase in the distance between the Earth and its satellite. The separation of the two bodies from each other is about 3.8 cm per year. The length of the Earth's day is also increasing: by about 2 h in the last 600 million years.

In many coastal areas, the tides are semidiurnal—two nearly equal high tides each day. There are also coasts where the tides are diurnal—one high tide each day

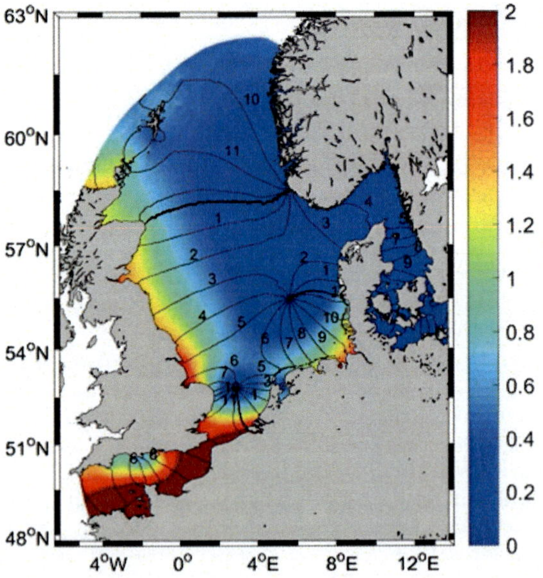

Fig. 2 Tidal characteristics in the North Sea. The amplitude of the tides (A_{M2}) with a semidiurnal frequency (twice a day) is shown with different colors. The black lines illustrate their phase—for example, each successive line indicates the time of high water onset. The points where these lines converge to a single point are called amphidromic points. Courtesy of B. Jacob

or mixed—two tides a day of different heights. There are also tidal motions with longer or shorter periods. The height of the tides in different areas of the Earth varies between centimeters to more than 10 m.

The amplitude of Kelvin waves is at its maximum at the coast; therefore, the coastal ocean plays a particular role in the process of dissipation of tidal energy. Actually, the coast and the ocean floor shape the way tides propagate. In most areas of the Northern Hemisphere, these (tidal) waves propagate such that, when viewed in the direction of wave propagation, the shore remains to the right. This is also the case in the North Sea, where Kelvin waves circulate in a counterclockwise direction (Fig. 2).

2.4 *The Wadden Sea*

There are coastal areas which are very flat; in certain areas, the elevation above the mean sea level is 1 m, for example. If there the height of the tide (high water) is 2 m, the tidal wave will flood this shallow bottom and continue to spread, inundating the whole area lower than 2 m. At low tide, the entire area with an elevation greater than − 2 m will lose its water cover (see Fig. 3). This example shows that the area with an elevation of − 2 to 2 m will be periodically flooded and drained. These areas are known as Wadden Sea (Watten Meer), or as tidal flats[5] (Fig. 4). Such flats form off the coasts of the Netherlands, Germany, and Denmark. These coastal areas regularly dry out twice a day at low tide and are flooded at high tide. They extend over distances of many kilometers and are usually fronted by so-called barrier islands, which protect the tidal flats from the destructive effects of the open sea. Individual islands are separated by tidal channels (Fig. 4). Due to the high tides, huge volumes of water are transported along the tidal channels from the open sea to the tidal flats and back. These channels can be traced deep into the tidal flats as large rivers into which smaller rivers flow, facilitating the efficient drainage of water at low tide. Currents in tidal channels often exceed 1 m s^{-1}, being highly erosive to the seabed. Fine bottom sediments are subject to the greatest erosion; therefore, the bottom in the tidal channels is usually sandy.

Together with sediments brought in from land by river flows, fine sediment fractions eroded from areas with strong currents deposit onto the tidal flats, making the latter particularly smooth (Figs. 5 and 6). Due to the frequent reversal of tidal currents, the sediment on the bottom does not have time to compact[6] sufficiently, so it is in the form of a soft, sticky mud, and the Watten Meer is in fact a mud sea. Tidal flats are also known as mudflats. During low water, the distance from the shore to the barrier islands can be walked, but it is advisable to have a guide who knows the area and the habits of the local tide.

[5] Sometimes the word Watt is used instead of Watten Meer, which is also known as marine tidal flats.

[6] Compaction is the process by which sediments turn into solid material.

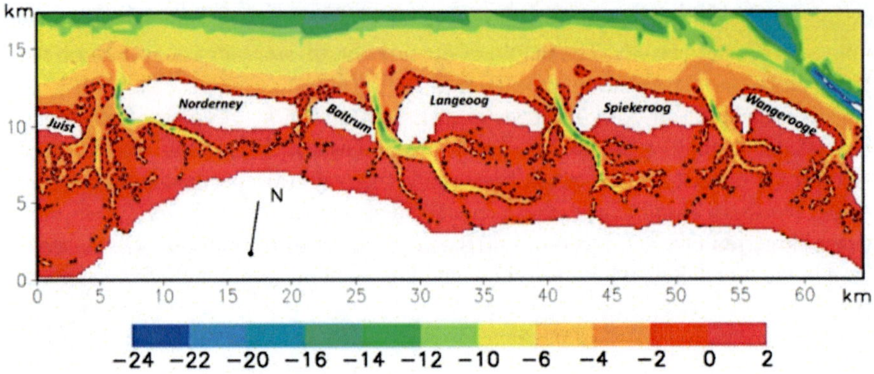

Fig. 3 East Frisian tidal basins. Bottom depth is in m. The same area (Wadden Sea) is indicated by a rectangle in Fig. 1. Areas shallower than 2 m (red) are periodically drained at low tide. From Stanev et al. (2003b)

Fig. 4 German coasts in the North Sea and the estuaries of Ems-Dollart, Weser and Elbe. Wadden Sea is seen as green-gray area between the ocean (blue) and land (green). From Stanev et al. (2006)

Fig. 5 This is what the Wadden Sea looks like. The bottom-left inset is the barrier island of Spiekeroog, and the top-right inset shows the sticky mud. From Stanev (2022)

Similar coastal areas are found in many parts of the world. In the tropics, they are overgrown with mangroves (tidal forests). Mudflats are thought to be among the most productive natural ecosystems on Earth and play an important role in global BGC cycles. They are not only a preferred habitat for many bird species, but also a breeding ground for a wide variety of marine organisms (Fig. 7). All this, because sea, land and air come together. In these areas between ocean and land, both shore and bottom are changing over time. The impact of these changes on marine systems is not yet fully understood, let alone the predictability of the changes taking place within them.

3 The New Turbid Waters

At the time of my arrival in Oldenburg, the research group[7] "Biogeochemistry of the Tidal Flats" (BioGeoChemie des Watts) was being finalized. It was launched in 2001, combining the efforts of scientists from ICBM and the Marine Physics

[7] Research groups (Forschungsgruppen) are organizational forms of cooperation and funding science, which bring together scientists from different fields, universities, and other academic institutes.

Fig. 6 Another view of the tidal flats during low tide. They are grooved by numerous tidal channels, along which water flows out to sea. The barrier island can be seen in the distance. Courtesy of J. Staneva

Working Group (both at the University of Oldenburg), Senckenberg Research Institute (Senckenberg am Meer), and the Terramare Research Centre (incorporated in 2008 into ICBM), both in Wihlhelmshaven, as well as the Max-Planck Institute for Marine Microbiology in Bremen. The research group, called "Watt", was led by Prof. Jürgen Rulköter, the director of ICBM.

The area of interest included part of the most extensive tidal flats in the world located in front of the German North Sea coast. Climate change, sea level rise and the ever-increasing importance of coastal areas for populations require accessing and predicting future evolution of this unique marine environment. One issue is how its natural indicators are changing and how to optimally counteract undesirable changes in the short and long term. Hence, the aim of the Watt research group was to further understand the various physical and BGC processes occurring in the tidal flats. This includes the influence of the tides as well as the weather, but also the changing world of microorganisms. There was a need to understand the balance between processes stimulating or inhibiting the growth of biological components of the marine environment. Solving these tasks required novel observations and the development of predictive models.

Before the work of the Watt research group began, it was known that microorganisms in the water column and in bottom sediments ensure the healthy functioning of

Fig. 7 Roughness of the otherwise very smooth mud surface gives an evidence of active processes occurring on the periodically drying bottom. Courtesy of J. Staneva

the tidal flat system by decomposing the remains of dead plankton and other biological substrates and returning the resulting products to the nutrient cycle. The most important role is played by aerobic bacteria[8] in the water column and upper sediment layers, the latter often only a few millimeters thick and still containing oxygen. Below this layer is the anoxic zone, where another type of bacteria, anaerobic bacteria, takes control of the BGC processes. The anaerobic zone is almost always covered by the oxic (aerobic) zone. However, sometimes the boundary between two zones is disrupted because of anomalous environmental conditions, anoxic materials surface and large black areas appear on the bottom, as was the case after the severe winter of 1995–1996. This event led to the death of large numbers of mussels and worms due to lack of oxygen. To this day, the role of bacteria that drive the processes of decomposition of biological material remains largely unclear.

The wide range of scientific fields required the Watt research group to be interdisciplinary (Rullkötter 2005). It therefore included geochemists, sedimentologists, marine physicists, microbiologists, and mathematical modelers. The construction of the Continuous Measurement Station in 2002 (Fig. 8) near the southwestern tip of the island of Spiekeroog (see Fig. 3 for the position of this island) is a particular success

[8] Aerobic (anaerobic) bacteria are bacteria living in oxic (anoxic) environments.

Fig. 8 Station for continuous measurement of air and seawater parameters near the southwestern tip of Spiekeroog Island. From Baschek et al. (2017)

of the research group. This station, operating under the maintenance and supervision of ICBM, continues to provide a wealth of information on salinity, temperature, and current velocity as well as various meteorological data. It later became part of the North Sea Observing Network and has proven to be reliable even in extreme weather conditions such as hurricanes or ice storms in winter.

The data obtained from the measurements of the Watt research group have made it possible to develop a detailed picture of the transport of sediment and organic material. This helped elucidate the microbiological activity in this specific ocean environment, which for 1000 years has been subject to adverse impacts from such activities as fishing, land reclamation, and dike building. Jürgen Rulkötter sees success above all in the close collaboration between the participating institutes: "We set up the group from the start in such a way that scientists and research institutions inspire each other". The researchers found high levels of microbial exchange in the sediments around Spiekeroog Island. It became clear that the sandy bottom represents an active bioreactor for remineralizing dead organic material. At the same time, a wide variety of microorganisms were found to neutralize toxic sulfur compounds on the surface of bottom sediments and may even use them to accelerate growth. New to science was the research on morphodynamics[9] in shallow water behind barrier islands carried out by the Senckenberg am Meer Institute.

The understanding of the influence of hydrodynamic conditions on various BGC processes, as well as quantifying the budgets and balances of various substances, was of paramount importance to the research group. The development of a mathematical ecosystem model for BGC processes in tidal flats, as well as sediment dynamics was central to all research parties. The challenge was to make our research applicable to other similar tidal areas in the ocean.

4 Dynamics of the German Bight

During the second week after my arrival in Oldenburg, H. Burchard visited us and gave me his model and the necessary descriptions. He had named it "Mudflat". It was professionally written and relatively easy to learn. The modern version of this model is known as the General Estuarine Transport Model (GETM, Burchard and Bolding 2002). I applied the computational code for the area of the East Frisian tidal flats, the main region of interest of the Watt research group (Fig. 3). For this, I had to prescribe reliable boundary condition data that is the physical state of the ocean at the open boundary. However, no appropriate measurement data existed (even to this day); therefore, I used the data provided to me by the German Hydrographic Institute (Bundesamts für Seeschifffahrt und Hydrographie, BSH). These data were produced by a model covering a larger area. You will ask, having a model operating over a large area, why do we need a model for a subarea of it. The problem is that the "big model" usually has a coarse resolution, which does not allow describing adequately the bathymetric features in the tidal flats, and the narrow tidal channels. The raster I used for this area was 200 m, whereas the BSH model had a raster of 1.8 km, i.e., about ten times coarser. Data from observations were also needed to validate the model. Some regional coastal services provided me with sea level data at several stations, and GKSS (Gesellschaft für Kernenergieverwertung in Schiffbau und Schifffahrt)[10] provided me with data on currents measured in the Akum Strait separating Langeoog and Spiekeroog Islands.

One year after I started working at ICBM, a critical mass of good results was collected that were worth publishing. The tidal wave asymmetry and its modulation in the coastal zone appeared very interesting (Stanev et al. 2003b). This asymmetry

[9] Morphodynamics deals with the study of the interaction between the seafloor and hydrodynamic processes (wind waves, tides, and currents) and their mutual adjustment, which includes sediment transport, erosion, and sediment deposition.

[10] GKSS is a research center where I was to continue my work years later and which I will talk about in next sections.

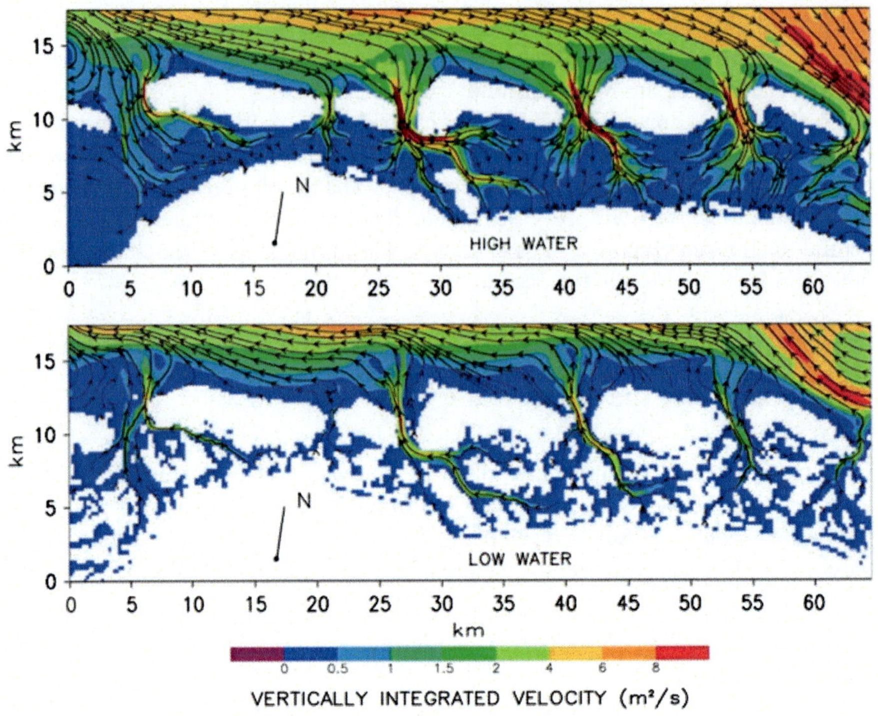

Fig. 9 Vertically integrated velocity ($m^2\ s^{-1}$) during high water (top) and low water (bottom) at spring tide. From Stanev et al. (2007), with permission from Elsevier

is revealed by the fact that the currents at high tide and low tide can be significantly different.[11] These differences occur because friction on the seabed tends to slow currents in shallow water. In high water, the water column is thicker and friction is usually smaller. Because of this, the tidal currents at high water may be faster than at low water. This asymmetry leads to different erosion and specific sediment distribution during the two phases of the tide. For this reason, a redistribution of sediments can occur that changes the topography of the bottom. This turned out to be the first publication on numerical modeling of three-dimensional ocean processes in the region of East Frisian tidal flats and their time evolution.

The regime of filing tidal flats with water from the open ocean (Fig. 9) depends on the bottom slope. This results in another type of asymmetry of the tidal signal (Fig. 10), with the tidal transition time from flood to ebb shorter than the remaining part of the tidal period (Stanev et al. 2003a). This asymmetry can change over time, providing a mechanism, which controls the sediment accumulation regime in tidal basins. This issue was a major focus of interest for the Watt research group.

[11] I remind that characteristics of simple periodic oscillations are the same during the phases of minimum and maximum, but just mirrored (symmetric).

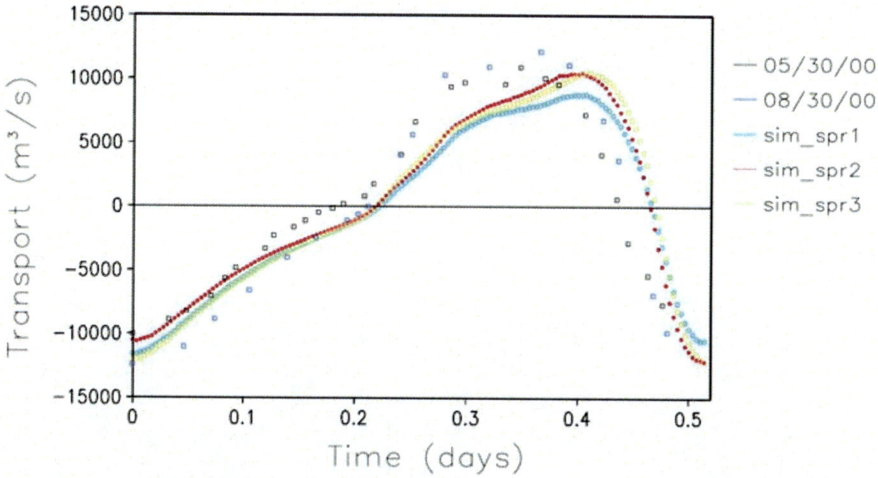

Fig. 10 Simulated (colored lines) and observed (dots) transports during spring tide across the main channel (Accumer Ee) connecting the Langeoog basin with the open ocean. Positive values correspond to transport toward tidal flats. The three colored lines correspond to three consecutive tidal periods. Observations include data during two different campaigns in 2000 (between 18 and 27 sections per tidal cycle). See legend for the dates. From Stanev et al. (2003b)

In the area that we studied, the East Frisian Wadden Sea, strong sea storms are known to cause redistribution of sediment and sand, as well as changes to the shoreline. For example, the devastating storm in the North Sea on the night of 11–12 October 1634 caused in the northern part of the tidal flats the accumulation of a layer of marine sand several meters thick (Fruergaard et al. 2013). The importance of extreme events for the dynamics of Wadden Sea motivated us to develop the model further so that it could reproduce the process of sediment transport and the influence of wind waves on that transport. In the deep ocean, this influence is negligible, but when the depth of the sea becomes comparable to the wavelength, wind waves become the major erosive factor. When approaching the coast, the steepness of the waves increases until the instability becomes so great that waves break. At this point, a large amount of energy is released, the water column becomes fully mixed, and sediment is eroded from the bottom, carried by the currents, and spreading over large distances. The individual roles of tides and wind waves as drivers for sediment erosion and deposition can be seen in Fig. 11 illustrating the spatial distribution of friction velocity. This characteristic measures the intensity of turbulence in the bed layer, which is indicative of erosion conditions. Results from numerical model show that tides have a strong erosive role in the tidal inlets (Fig. 11a), while the wind waves show greater erosive impact in the shallow tidal flats (Fig. 11b).

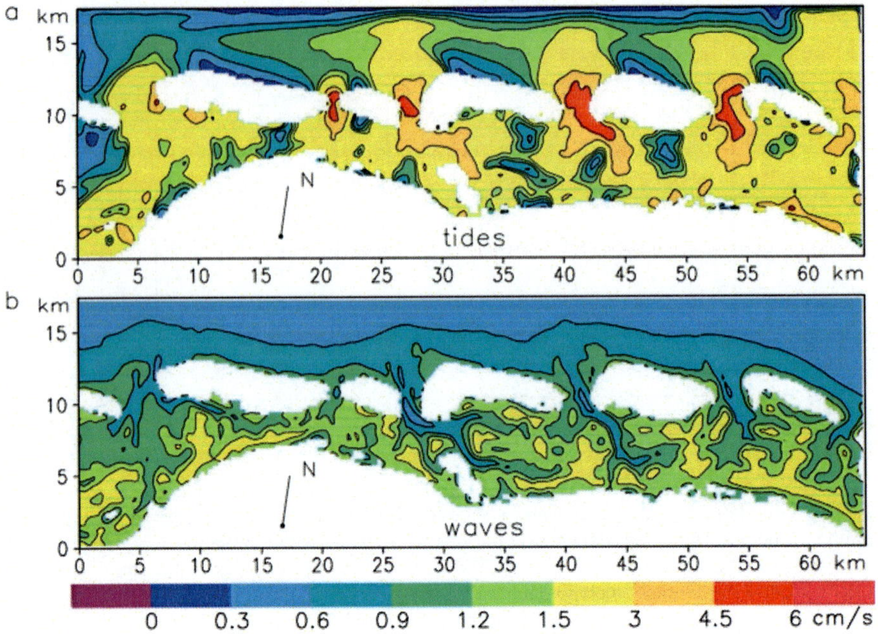

Fig. 11 Friction velocity due to tidal forcing (**a**) and forcing by wind waves (**b**). From Stanev et al. (2006)

Acoustic Doppler Current Profiler (ADCP)

The principle of the measurements with ADCP uses the Doppler Effect, which is that the sound wave has a higher frequency when the source is moving toward you than when it is moving away. Recall that the sound you hear from a car coming toward you is higher in pitch than when it is moving away. ADCP works by transmitting "ping" pulses at a constant frequency. These pulses are so weak that humans and even dolphins cannot hear them. The sound waves are reflected by particles suspended in the moving water and return to the instrument. Because of the Doppler Effect, sound waves reflected from particles floating away have a slightly reduced frequency; the opposite is true when particles are moving toward the instrument. In order to accurately determine the direction and magnitude of the flow, several emitters pointing in different directions are used. The power of the reflected signal can also be used to determine the sediment concentration in the water.

The Senckenberg am Meer Institute in Wilhelmshaven was one of the first in Germany and in the world to measure flow velocity using ADCP. Scientists, under the direction of its then director Burghard Flemming, used the instrument to determine how fast the water is moving at different levels in the water column (in our case the vertical discretization was 25 cm). Flemming's colleague Alexander Bartholomä had put the ADCP instrument on a motorboat and had been continuously crossing the strait between the islands of Spiekeroog and Langeoog for more than twenty-four hours on almost the same route. In these early days of ADCP use, the data were far from perfect. One reason was that the rocking of the boat degraded their quality. Nevertheless, invaluable measurement material was accumulated on the structure of currents in space and time (Fig. 12). It was a challenge to compare this data from direct measurements and our numerical models. This exercise gave us confidence that the model predictions were correct. The combined efforts of experimenters and theoreticians proved successful. The results that elucidate the temporal and spatial variability of tides in tidal inlets turned out to be the first for the region and one of the few in the world at that time (Stanev et al. 2007). Today we use ADCP extensively in our practice, and the data seem perfect compared to the once "noisy" first measurements.

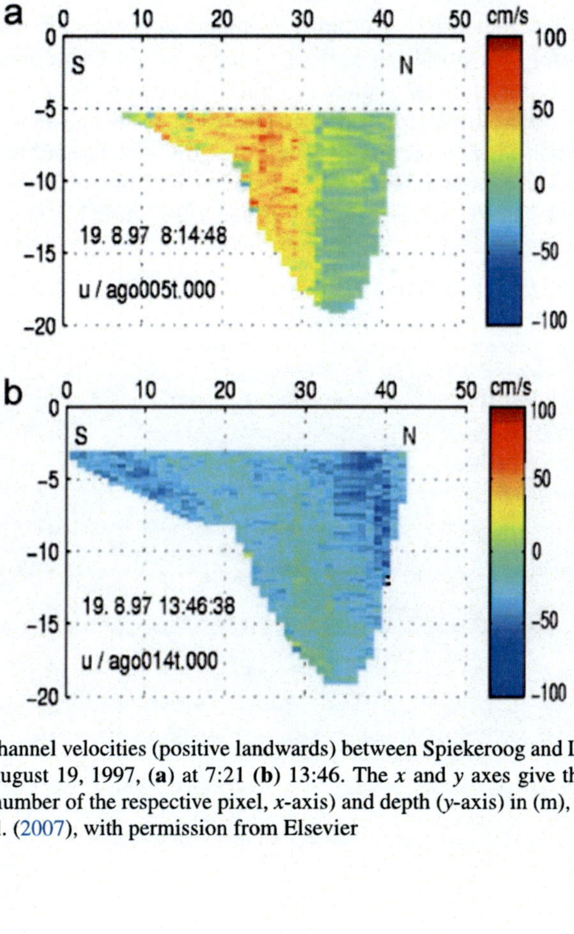

Fig. 12 Along-channel velocities (positive landwards) between Spiekeroog and Langeoog Islands (see Fig. 3) on August 19, 1997, (**a**) at 7:21 (**b**) 13:46. The *x* and *y* axes give the position along the transect (the number of the respective pixel, *x*-axis) and depth (*y*-axis) in (m), correspondingly. From Stanev et al. (2007), with permission from Elsevier

Surprises in the Coastal Area Never End: Subterranean Estuaries

About ten years after the time I am describing, from February 21 to 23, 2011, I visited Nordernei, one of the biggest barrier islands (Fig. 3). The occasion was a workshop I was conducting with my research department from the institute, where I continued my work later. During one of the breaks, we went to the beach to take a group photo. It was low tide, and the seashore was frozen (Fig. 13). The island is about five kilometers from the mainland, and there is no river mouth nearby. It is normal for seawater, which in this area has a salinity of about 30, not to freeze. However, there is a large volume of freshwater beneath the island (this is also the situation for many areas on the mainland coast). When the sea level is high, the bottom pressure increases due to the thicker water column, and the freshwater recedes into the island area. At low tide, the bottom dries out, there is no longer anything to "push" the groundwater beneath the island, and it flows out. This intermittent process affects the balance of chemical and biological substances—something that my colleague, professor of micro-biogeochemistry of the ICBM Hans Brumsack was very interested in (Waska et al. 2019). It turns out that these coastal systems, interfacing land and ocean, "work" like subterranean estuaries. In these estuaries, meteoric[12] groundwater mixes with incoming seawater in a coastal aquifer[13] before discharging into the adjacent water column. Subterranean aquifers can increase concentrations of some ionic constituents such as copper (Cu) and iron (Fe), for example, due to long residence times[14] and reductive dissolution of mineral phases along the groundwater pathway. These Cu and Fe enriched waters trace the mixing pathway of freshwater with seawater, which provides useful information needed to understand interactions between coasts and the open sea.

Fig. 13 Ocean is also freezing (near the shore). Courtesy of Gayer

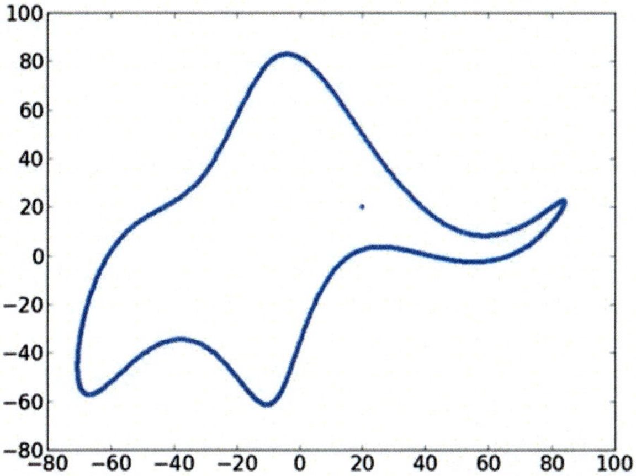

Fig. 14 This is what an elephant looks like drawn with a mathematical formula that has four free parameters. Courtesy of K. Wahle. The mathematical program that does the drawing is freely available at https://www.johndcook.com/blog/2011/06/21/how-to-fit-an-elephant/. Last accessed on 02.02.2023

The work in the Watt research group also involved numerical modeling of biological processes. One of the authors of the most complex ecological model for ocean systems at that time was Prof. Wolfgang Ebenhöh of the ICBM. The European Regional Seas Ecosystem Model (ERSEM, Baretta et al. 1995), created with his participation, describes in detail the marine biogeochemistry and especially the lower trophic levels[15] of the marine food chain. Its advantage is the ability to work with complex ecosystems both in the water column and on the bottom of the sea. I was faced with the dilemma of either coupling the physical model I was using with the ecosystem model, or to provide the simulated with GETM physical data to Wolfgang's research group and have them use that data to drive the biological system. I had enough other work at the time; thus, Wolfgang's collaborators explored this domain. Another reason for my decision was that I distrusted the large number of model parameters in ERSEM because I was not sure if their values were optimally applicable to any marine area. In an article on the great physicist Enrico Fermi, Freeman

[12] Meteoric water is the water from atmospheric precipitations.

[13] Aquifer: an underground layer of water-bearing permeable rock, or unconsolidated materials.

[14] The residence time of a fluid parcel is the total time the parcel has spent in a control volume (e.g., lake, ocean).

[15] Lower trophic levels mean that the ecosystem model is mainly applied for soluble matter, phyto- and zooplankton, but not for fish and other large marine organisms. More general models are the so-called end-to-end models, which combine physicochemical oceanographic descriptors and organisms ranging from microbes to higher-trophic-level organisms, including humans, in a single modeling framework (Rose et al. 2010). Another class of ecosystem models is the so-called individual-based models (Daewel et al. 2008). Individual-based models are based on the explicit representation of individual organisms.

Dyson [16] wrote: "In desperation I asked Fermi whether he was not impressed by the agreement between our calculated numbers and his measured numbers. He replied, "How many arbitrary parameters did you use for your calculations?" I thought for a moment about our cut-off procedures and said, "Four." He said, "I remember my friend Johnny von Neumann used to say, with four parameters I can fit an elephant, and with five I can make him wiggle his trunk."" (Fig. 14). There are at least a hundred parameters in ERSEM, but there are no parameters in nature; they were created by modelers to roughly describe something that we cannot represent in full in models. I personally liked the simple models where you understand what is going on.

5 The Black Sea Flood

In the late 1990s, William Ryan and Walter Pitman launched the Black Sea Flood hypothesis (Ryan and Pitman 1998). They assumed that 7150 years BC a huge flow of Mediterranean Sea water invaded the then Black Sea freshwater lake. This had disastrous consequences for the population that lived on the shores. I use the term "the Black Sea freshwater lake" because, in remote geological epochs, there were periods when the salinity of the Black Sea was very low; the sea was, in effect, a freshwater lake. These drastic changes are due to changes in the ice cover of the Northern Hemisphere during the ice ages when huge amounts of water were taken from the ocean, and deposited on land, and the level of the World Ocean was about 100–150 m lower than it is today. After the warming of the climate, the ocean level gradually rose and the connection with the Black Sea was restored (Georgievski and Stanev 2006). The salty Marmara Sea waters invaded the Black Sea and, being very dense, sunk to the bottom. A sharp boundary was created between the saline bottom waters and the less saline surface waters, which gradually rose over several thousand years until it reached its present equilibrium position.

The dominant processes in the Black Sea are controlled primarily by the freshwater from the rivers, their mixing with the Black Sea water, and a two-way flow through the Bosporus to and from the Marmara Sea (Fig. 2 in chapter "Further into the Ocean"). When the hydrological regime in the Black Sea catchment area changes, the water balance is disturbed and sea level can undergo significant variations. Let us consider the following thought experiment, in which we have two basins filled with water of different densities separated by a barrier. If the water levels in the two basin are the same, at the same depth in the basin with denser water, the pressure is higher. If we connect the two basins with a pipe at some depth below the surface (Fig. 15), a flow will be created from the higher density basin to the lower density basin. The flow will stop when the rise in level in the lower density basin creates a pressure difference that is able to compensate for the pressure force due to the

[16] F. Dyson was an English-American theoretical physicist and mathematician. The citation that follows is from Dyson (2004), p. 297.

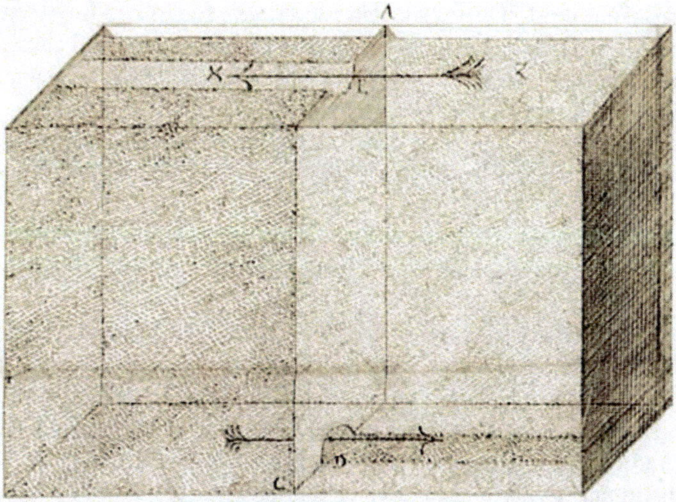

Fig. 15 Schematic representation of Luigi Ferdinando Marsigli's physical model for a two-layer flow published in 1681. The model consists of a box separated by a baffle (A) with openings at the top and bottom containing water of different densities in the two compartments. The denser water (darker colored) is directed to the opposite compartment through the bottom opening, causing the lower density water (lighter colored) to move in the opposite direction through the top opening. The figure is from Soffientino and Pilson (2005) and is made available under the terms of the Creative Commons Attribution 4.0 International License (https://creativecommons.org/licenses/by/4.0/)

density differences (the sea level in the low-density ocean areas is higher than in the high-density ones). Let us now remove the barrier between two basins. Then, two mutually compensating flows will be created. This thought experiment has been known since the time of Luigi Marsigli (1658–1730), the Italian nobleman, scholar, eminent natural scientist, military engineer, and diplomat. He was the one, who first proved the existence of a deep current in the Bosporus, directed from the Sea of Marmara to the Black Sea (Fig. 15).

Suppose now that for some reason, the flow of freshwater into the Black Sea increases. This will cause the surface current to increase, which in turn will tend to suppress the bottom current.[17] More and more salt will be exported from the Black Sea, and less and less salt will come in. As a result, salinity will continuously decrease. It may happen that the bottom current is completely blocked. This will turn the Bosporus into a river through which the Black Sea water flows and with it the salts. In such conditions, the sea will continuously decrease its salinity and so until it becomes a freshwater lake. Such processes took place during the ice ages.

How does the flood in the Black Sea happen? The process of glaciation of the Earth is in effect a displacement of water (a liquid state) out of the ocean by its transformation into a gaseous state (evaporation from the surface of the ocean), its

[17] Similar conditions to what is explained in the text could happen if the flow of freshwater remains the same, but the ocean level sinks.

removal from the atmosphere as snow (a solid state), and its deposition on continents (the ice caps). The displacement of water from the sea to the land causes the ocean level to fall. The level of the Black Sea kept pace with the ocean level until it reached the sill of the Bosporus. Due to the different freshwater balance in the Black Sea area compared to the average one in the ocean, the sea level in this basin fell faster than the ocean level and was much lower. In the reverse global climate trend, associated with rising ocean level (upper-left panel in Fig. 16), there was a point when, according to some researchers, the ocean level first reached the Bosporus sill. They assume that at that point the level of the Black Sea was more than a hundred meters below the ocean level (bottom-left panel in Fig. 16).

Now we can present Ryan and Pitman's hypothesis (Ryan et al. 1997). The rising Mediterranean Sea level pours over the stone sill of the Bosporus. 100,000 km^2 of land are flooded, mainly in the shallow areas in the northwestern Black Sea. According to these researchers, 50 km^3 of water is pouring out every day, which is two hundred times the flow of the Niagara Falls, and this over the course of a year. The "catastrophic scenario" of reconnection between the Black Sea and the ocean (bottom-left panel in Fig. 16) assumes that the flood deeply affected prehistoric

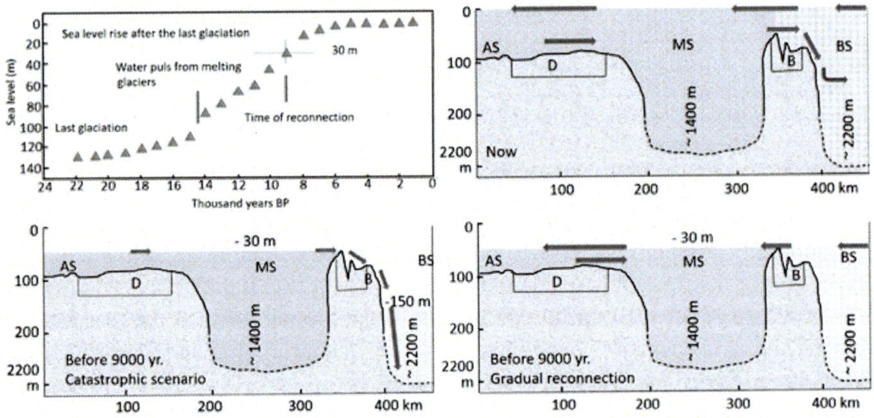

Fig. 16 Two scenarios of the reconnection between the Black Sea and the Mediterranean. The top-left panel shows the rise in ocean level over the last 25 millennia, while the one on the top-right shows the current state of water exchange between the Black Sea and the Mediterranean (see also Fig. 2 in chapter "Further into the Ocean"). The bottom diagrams represent the catastrophic scenario (left) and the gradual reconnection scenario (right). The different lengths of the arrows (representing currents) are meant to schematically visualize the different intensities of exchange in the straits. The bottom relief in the Sea of Marmara and the straits is shown schematically. The leftmost and rightmost parts of the diagrams represent only the adjacent parts of the Black Sea and the Aegean Sea. The depth (vertical coordinate) is represented differently in the upper 200 m layer and below it. The bathymetry in the deep part is therefore drawn with a dotted line. The following abbreviations are used: AS—Aegean Sea, MS—Marmara Sea, BS—Black Sea, D—Dardanelles, B—Bosporus. The different shading reflects the different water characteristics of the different sea basins. Replotted from Stanev (2022)

settlements and that these events are the basis of the Bible's legends about Noah and the flood.

As is often the case in science, there is an alternative hypothesis that contrasts the "catastrophic scenario" which assumes a gradual reconnection between the two seas (bottom right panel in Fig. 16). According to this hypothesis, 10–11 thousand years BC, the level of the Black Sea first rose above the sill in the Bosporus Strait and Black Sea waters began to spill into the Sea of Marmara. Deposits of sapropels[18] at this time, reported in the eastern Mediterranean, support this assumption. The outflow from the Black Sea continued for about two thousand years, until the ocean level also reached the Bosporus sill. Another thousand years followed, when the sea level in the Mediterranean was still rising, but the "Black Sea River" flew unidirectionally at all depths in the strait. The theory says that when the thickness of the water column in the strait increases above a certain critical value, this can happen as a result of the global rise of the ocean level, bidirectional flows occur. These flows are such that currents to the Sea of Marmara are at the surface and a countercurrent, directed to the Black Sea, flows in the bottom layers (remember the thought experiment of Marsigli). The authors of the gradual reconnection theory (Aksu et al. 2002) argue that this happened eight thousand years BC when the salty bottom waters entering the Black Sea from the strait contributed to the gradual increase in salinity in the Black Sea.

The current state of the exchange is shown in the top-right panel in Fig. 16, which depicts the two-layer current in the Bosporus. It also illustrates that much of the water sinking from the Bosporus does not reach the deep layers of the water column.

Shortly before I left for Germany, Gilles Lericolais from the Institut français de recherche pour l'exploitation de la mer (IFREMER) approached me. He wanted to organize a consortium of scientists with an interest in Black Sea paleo-evolution. The project name was "Assessment of the Black Sea sedimentary system since the last glacial extreme" (ASSEMBLAGE). The European Commission approved our submission, and the project started on February 1, 2003. My participation was on behalf of ICBM. The other Bulgarian participants were Petko Dimitrov from IO-BAS and J. Staneva from the part of NIMH.[19]

The rationale of ASSEMBLAGE was based on some unique features of the Black Sea, namely the presence of a particularly wide shelf in the north-west sector, the hydrological balance in its catchment area, and the high degree of sedimentation that creates favorable conditions for paleoclimate studies. Investigations of the sedimentary system of the northwestern Black Sea, through the continental shelf, to the deep sea were also envisaged. Complex geomorphological, geophysical, stratigraphic, and

---

[18] Sapropel (from the Greek sapros and pelos, rot and mud or clay, respectively) is a term used in marine geology to describe dark-colored sediments rich in organic matter. The term sapropel events may also refer to cyclic oceanic anoxic events, in particular those affecting the Mediterranean Sea with a periodicity of about 21 000 years.

[19] After my move to Oldenburg, a position was opened at the Alfred-Wegener-Institut (AWI) in Bremerhaven for a numerical ocean modeler. Joanna Staneva applied and was given this position. She started working in AWI at the beginning of 2001. In 2003, she started working at ICBM.

sedimentological studies were planned to be carried out to help decipher past climatic and environmental changes.

My work focused on clarifying the chronology of events related to sea level change. Together with my Ph.D. student Goran Georgievski, I had to reconstruct the history of the Black Sea level using available data and theoretical concepts about the outflow of waters through the Bosporus. In doing so, we had to "play out" several alternative scenarios of climate change and associated precipitation and evaporation regimes over land and sea. The hydrological data we used came from available modern data and paleo-observations.

The Eurasian Paleo-Cascade

Hypotheses about the disconnection and reconnection between the Black Sea and the ocean would be incomplete if we did not mention the paleo-cascade connecting the Aral, Caspian, and Black Seas (Fig. 17). During the severe glaciations of the Earth in the last 100 kyr,[20] the northern parts of Asia and Europe were covered with glaciers that cut off the path of the great Siberian rivers to the north, and they turned their way southward toward the Aral Sea. The Scandinavian glacier was covering the Baltic Sea and made it impossible for waters to flow out toward the Atlantic. According to some paleo-observations, there have been two such river diversion events in the last hundred thousand years. During the first, about 90–80 kyr BP,[21] the Ob and Yenisei rivers reached the Aral Sea and from there headed toward the Caspian Sea and further through the Kumo-Manich Depression toward the Black Sea (Fig. 17). The next event took place around 18–17 kyr BC. At that time, the outflow of water through the Baltic Sea was blocked by the Scandinavian glacier. The meltwater continued southward to the Volga River and thence through the Caspian Sea and the Kumo-Manicha Depression to the Black Sea and the Bosporus, eventually forming the so-called Cascade of the Eurasian Basins (Georgievski and Stanev 2006; Chepalyga, 2007). Later studies of oxygen isotopes (Badertscher et al. 2011) have shown that there have been at least seven periods during the last 670 kyr when Caspian waters have penetrated the Black Sea.

Data from old geological periods have large errors; hence, Georgievski and Stanev (2006) have dealt only with the last 20 kyr. Experiments to determine the sensitivity of the Black Sea to climate conditions showed that the climatic conditions in its catchment area would have to be extremely dry, similar to today's climate in the Caspian Sea area, in order for the level of the Black Sea to drop to about 100 m below the level of the Bosporus sill. The estimates of sea level changes were consistent with estimates based on shoreline evolution, showing that in a dry climate scenario, the sea level reached the Bosporus sill about 8.8–7.4 kyr BP.

[20] Kyr, thousand years.

[21] BP, before present time.

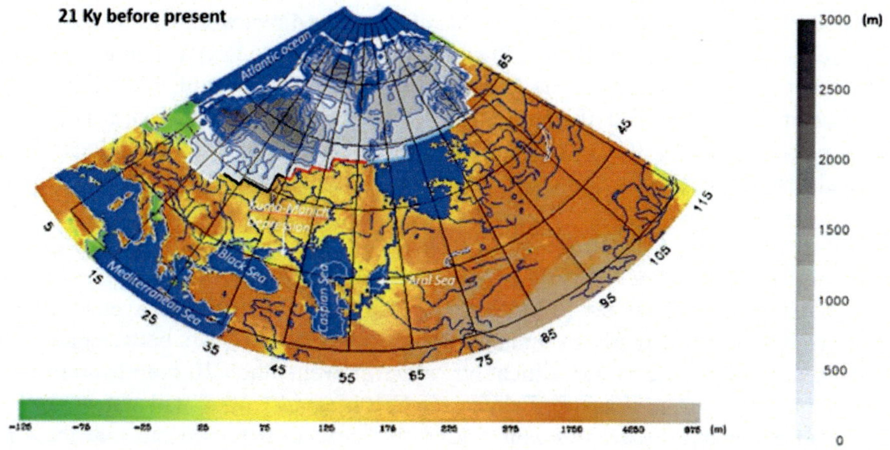

Fig. 17 Paleo-cascade connecting the Aral, Caspian, and Black Seas. The area north of the solid broken line is covered by glaciers. The gray bar on the right gives a measure of the height of the glaciers. The bar below the figure shows the height of the land not covered by glaciers in meters. Thin white lines show the modern coasts. Courtesy of G. Georgievski

IFREMER oceanographic surveys conducted in the northwestern Black Sea in 1998 and 2002 gave indications that the level of the Black Sea Lake had risen sharply on the continental shelf, which presented clear evidence of the onset of 7150 BC sea conditions. Interpretation of the data leads to the conclusion that well-preserved "drowned" beaches, sand dunes and soils demonstrate a rapid inundation. The researchers also found eroded areas of land surface well below the sill of the modern Bosporus, which they interpret as exposure of these depths to weathering (i.e., the sea level at the time of reconnection was low). The insufficient quality and conclusiveness of the data cannot yet put an end to the dispute about the catastrophic or gradual way in which the reconnection between the Black Sea and Mediterranean occurred.

6 Again at the John Stennis Space Center

A new scientific "star" was rising in the Black Sea, and his name was Birol Kara. He was working at the NRL on the interaction of the atmosphere and the ocean as well as on the development of the part of the US Navy model covering the Black Sea. Apparently, colleagues from the NRL had information about the quality of our model for the Black Sea and were not aware of why some metrics, such as the CIL characteristics, could not achieve this quality in their model. They invited me and I visited them from 8 to 29 May 2004.

My host was Harley Hurlbur from the Ocean Dynamics and Forecasting Division. He is a world-leading expert in ocean modeling and the use of satellite altimetry data to understand the dynamics of equatorial currents. My direct work was with B. Kara, as well as with Joseph Metzger, with whose help I started to learn running the model that they were using named the HYbrid Coordinate Ocean Model (HYCOM). At the heart of this model is the notion that in a density-stratified ocean, it is not the geometric depths, which are the natural vertical coordinates, but the depths of the equal density levels. Thus, in areas of strong stratification, the model uses more levels, which is very reasonable. Furthermore, different coordinate systems can be used in different layers of the ocean, i.e., HYCOM allowed hybrid coordinates.

Birol Kara and I had to find an optimal approach to improve the quality of model prediction of the CIL in the Black Sea. I remind the reader that this layer maintains relatively low temperatures in summer because turbulent diffusion between layers is reduced by strong density stratification, which prevents the CIL core from mixing with the warm surface water. In winter, the CIL is rapidly filled with a new portion of cold water due to strong cooling from the surface. It turned out that the CIL simulated in HYCOM was very weak in summer, because the mixing of waters was so strong that the CIL almost disappeared. I was familiar with this problem. The solution I found for my models was to use mixing schemes, which sufficiently isolated the layers and prevented strong mixing with waters above and below the CIL. As mentioned already, vertical mixing in numerical models is not computed in an ideal way because the processes governing mixing (e.g., breaking internal waves[22]) occur over such small distances that they cannot be properly described in numerical models. For this, different parameterizations, known as mixing schemes, are used. In HYCOM, there were three such parametrizations, and B. Kara performed tests with all of them to find the optimal one. It turned out that all three schemes gave almost identical (unsatisfactory) results for the CIL. My explanation of this failure was that transferring data to the new coordinates at each step in HYCOM, because water density changes continuously, involves small inaccuracies in the calculations, which is equivalent to adding a weak mixing. This does not mean that the model is bad, but that the processes behind the formation of vertical stratification in the Black Sea are so specific that they require very careful and special treatment.

The NRL had enormous computing resources, and B. Kara could afford to use very fine resolution of 3.2 km. To use the Black Sea HYCOM in Germany, I had to configure it with coarser resolution. I had the great pleasure of working with J. Metzger, a very thoughtful person, on this. With him, we went all the way in configuring the model for the three-basins system: the Azov Sea, the Black Sea, and the Marmara Sea (Stanev 2005), coarsening the grid size used by Kara et al. (2008) to 6.4 km. Here, the advantage of using density coordinates was very telling, because in the Bosporus area the density of the water changes dramatically.

Figure 18 shows two snapshots of simulated sea level: one in winter (top panel) and one in summer (bottom panel). The dynamics in winter are characterized by stronger

[22] Internal waves are gravity waves in the interior of a stratified fluid. They are the undersea analogue of surface gravity waves (wind waves), but have much higher amplitudes than surface waves. Their speed of propagation depends on the strength of the density stratification. Depending upon distribution of density, internal waves can propagate vertically as well as horizontally through the fluid.

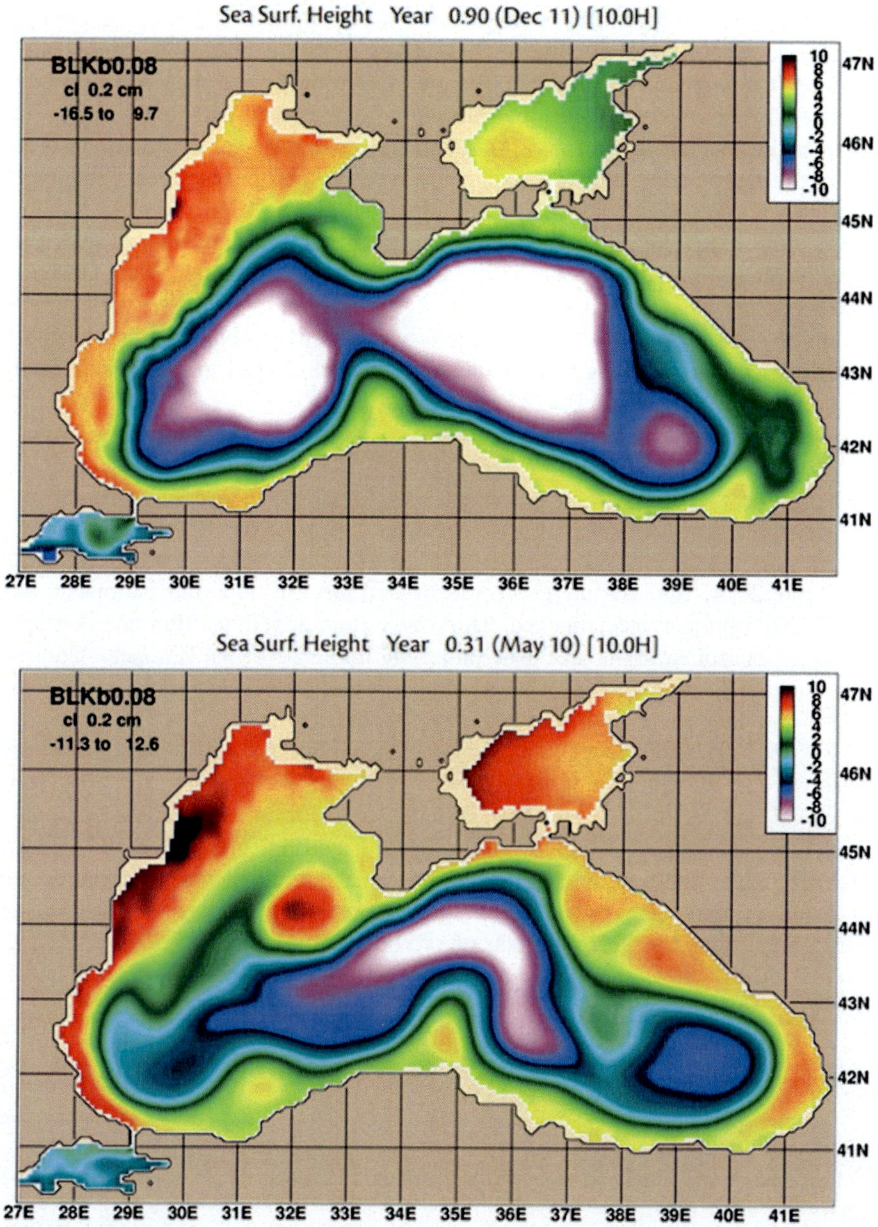

Fig. 18 This figure shows the first simulations for the system of three cascading basins, from the Azov Sea to the Marmara Sea. It was a joint effort of H. Hurlburt, J. Metzger, B. Kara, A. Wallcraft (all at the NRL) and the author. The HYCOM was set up for this region with a resolution of ~ 6 km. Figure is from Stanev (2005)

sea level gradients following approximately the continental slope, i.e., geostrophic surface currents are stronger during this season. In summer, the anticyclonic eddies between the Rim Current and the coast are more pronounced; the Crimean Eddy is a typical representative of quasi-permanent coastal eddies (compare with Fig. 6 in chapter "Further into the Ocean"). Although the horizontal resolution of model is very coarse for reproducing realistic exchange between basins, the results are encouraging and motivate the development of downscaling[23] strategies. More advanced numerical simulations extending the cascade down to the Mediterranean Sea and with much finer resolution in the straits are discussed in Sect. 7.2 in chapter "New Observing and Modeling Systems and New Scientific Applications".

Byrol Kara, who turned out to be a very productive young scientist, died four years after my visit. Along with the personal tragedy, I think this is a great loss for numerical modeling of the Black Sea.

7 ECOOP

In April 2004, the Marine Environment and Security for the European Area (MERSEA) project was launched. This was a time of reformatting oceanographic research, consolidating it (not least under the influence of the European Commission), and in the longer term, an orientation of research toward operational problems, and toward the provision of services for different users.

MERSEA (Desaubies 2009) aimed to develop a European system for operational observations and forecasting of ocean physics, biogeochemistry, and ecosystems at global and regional scales. This system was to become the ocean component of the future European Global Monitoring for Environment and Security initiative (GMES, Rider and Stel 2003). The core activity is the collection, validation, and assimilation into ocean circulation models of data from satellite observations and direct measurements. The focus was on providing information products on past and present ocean states, as well as predictions of future ocean states, having uniform and good coverage in time and space. The applications developed were expected to address the needs of intermediate and end users,[24] whether from the public or private sector. It was envisaged that the research would improve the safety and efficiency

[23] Dynamical downscaling is an approach that transfers information at large scales to subareas, e.g., regional domains (Kourafalou et al. 2015).

[24] Users of ocean information are of a different nature. There are those who are involved as beta-testers (beta testing is the final round of testing before releasing a product to a wide audience) in the verification of the service and its continuous improvement (Hernandez et al. 2015). Others simply use the data, but there are also users who can develop specific downstream products and services from the data provided and distribute them freely or sell them. The latter are known as intermediate users.

of maritime transport and operations; sustainable exploitation and management[25] of ocean resources (offshore oil and gas industry, fisheries); mitigating the effects of environmental hazards (oil spills, harmful algal blooms[26]); studies on ocean climate variability and seasonal climate forecasts; as well as national security.

Developing such a system requires an unprecedented level of collaboration and coordination of multidisciplinary research, as well as a range of operational activities. The project brought together the resources and expertise of 36 European institutes in the areas of satellite data processing, direct ocean observing systems, modeling of the ocean and its ecosystems, and the production of marine forecasts. MERSEA included the global ocean as well as four regional components: the Arctic, the Baltic-North Sea pair, the northeast Atlantic, and the Mediterranean. The main variables to be provided free of charge to users were temperature, salinity, and current velocity fields. Activities were also developed to include BGC and ecosystem parameters.

Several of the MERSEA participants took the initiative to create a similar consortium, but for coastal areas. The Black Sea was not included in MERSEA, and I assume that this was one of the reasons why three Black Sea scientists, G. Korotaev, T. Oguz and I, were invited in the development phase of ECOOP (European COastal-shelf Sea OPerational observing and forecasting system). A call for proposals was soon announced, and a Steering Group was formed, which over several meetings built the skeleton of the ECOOP project. In March 2005, a consensus meeting was organized in the Turkish town of Erdemli, on the Asia Minor coast of the Mediterranean, north of Cyprus, with numerous scientists expressing interest and showing their expertise. A consolidation of European experts in the field appeared to be in place.

A group of approximately 12 people, led by the project coordinator Erik Buch from the Danish Meteorological Institute (DMI), finalized the project submission. The overall objective of ECOOP became the consolidation, integration, and further development of existing operational observing and forecasting systems for European coastal and regional seas, into an integrated pan-European system aimed at detecting climate and environmental changes and predicting their evolution. Project activities were aimed at the timely production of quality-assured forecasts providing marine information services (including data, information products,[27] knowledge, and scientific advice) and facilitating decision-making. This speaks of a big similarity between MERSEA and ECOOP. The major difference is that consolidating activities in different coastal areas, many of them with a number of specificities, under the governance of different jurisdictions and local regulations, with different degrees of scientific maturity and willingness to undertake scientific commitments, becomes particularly difficult. The path of ECOOP to MERSEA will keep us busy for a long time to come because, although the distance between the two initiatives has

[25] According to Kullenberg (2020), sustainability science aims to illuminate the interactions between nature and society and, in so doing, provides evidence that management interventions can improve ecosystem services and human well-being.

[26] See Kudela et al. (2017) about oceanography of harmful algal blooms.

[27] Aznar et al. (2022) define "product" as a usable set of data (or one or more datasets) with its descriptive information (metadata).

been much shortened in the last fifteen years, the convergence of coastal operational oceanography with regional and global operational oceanography is still a work in progress to this day.

Objectives of ECOOP

The challenge was great, and it was foreseen to:

- Integrate existing coastal and regional marine observations into a pan-European observing system; and
- Integrate existing coastal and regional marine forecasting systems into a pan-European forecasting system and assimilate pan-European observation database into the system.

A number of more specific actions were also envisaged to secure the above objectives, such as assessing the quality of existing observing and forecasting systems, developing new observing and forecasting technologies, establishing a pan-European marine information system, and building capacity for education and training.

From Germany, BSH, AWI, GKSS, ICBM, and IOW participated, whereas I was acting as the representative of the University of Sofia. A total budget of € 11,238,655 was foreseen, of which € 6,990,251 came from the European Union; the rest was provided by the partner institutes. The twelve work packages of ECOOP were structured into four groups of activities with their respective leaders: observing systems (Kai Sötje, from the BSH), forecasting systems (Pierre Bahurel, at that time, from the French Naval Hydrographic and Oceanographic Service, SHOM), information systems (Martin Holt, from the Meteorological Office, UK) and technology transfer, and education and training (N. Pinardi). I led work package (WP) 7 "Synergy between coastal forecasting and newly available data and methodologies (a step toward next generation forecasting systems)" in the group of forecasting systems.

With the decrease of the characteristic scales of the motions in the ocean, for example from tens or hundreds of kilometers in the Gulf Stream region to sizes on the order of hundreds of meters in the coastal region, the balance between the acting forces changes substantially. The very nature of the motions changes too. These changes were not very clear and required extensive research before the forecast models became part of an operational package.

Coastal oceanography required the maximum use of high-quality data with high information content. Such are the data provided by ADCPs, ferries, fine-resolution satellite imagery (around 200 m), and autonomous underwater vehicle (AUV[28]) robots. Numerical modeling at very fine resolution necessitated the use of new models

[28] AUV is a self-propelled, unmanned, untethered, underwater vehicle capable of carrying out simple activities with little or no human supervision.

and advanced data processing techniques. Sediment transport is also an important issue that distinguishes coastal regions from the World Ocean. Models for such transport are needed in order to improve the ability to predict sediment deposition on or erosion from the bottom due to anthropogenic influences, such as regulatory actions in estuaries, as well as climate change. The predictability timescales for coastal ocean processes are very short and small uncertainties in models, or in the use of data, can dramatically reduce quality of predictions. We also needed to provide an objective assessment and understanding of the causes of errors in regional and coastal forecasting systems. As I read today what we promised nearly twenty years ago, I am amazed at the maturity of consortium then, on the one hand, and its optimism on the other. Most of the tasks listed above were just getting started. Years of hard work followed, but despite this ECOOP became a milestone in the development of European coastal operational oceanography.

Structure of Works in ECOOP

1. Observational System

   WP1: Quantitative assessment of observation systems: gaps and quality
   WP2: Integration of observing networks (pan-EU baseline system)
   WP3: Optimal synergy between remote sensing—in situ[29]

2. Forecasting System

   WP4: Modeling and forecasting system assessment: standards, quality, functions
   WP5: Integration of modeling and forecasting systems (baseline system)
   WP6: System development (downscaling, data assimilation, eco-modeling)
   WP7: Synergy between coastal forecasting and newly available data and methodologies (a step toward next generation forecasting systems)

3. Information system

   WP8: European Marine Information System of Systems (EuroMISS)
   WP9: European decision supporting system (EuroDESS)
   WP10: Hindcast[30] and scenario studies

4. Technology transfer, education, and training

   WP11: International cooperation, technology transfer
   WP12: Education and training

[29] In situ is a term used to indicate that data are measured directly rather than, for example, by satellites or aircraft measuring from a distance.

Synergy Between Coastal Forecasting and Newly Available Data and Methodologies (A Step Toward Next Generation Forecasting Systems)

Twenty European institutes participated in this WP and activities were structured in three parts. The overall leader of the *observing systems* part was Giuseppe Manzella from the ENEA.[31] I, along with the overall leadership of the work package, led the *coastal models* part. This part of the WP was to provide the scientific basis needed for high-resolution coastal modeling[32] and to produce demonstrations of improved forecast models. One such step was developed under the leadership of Jens Schroeter from AWI, who, together with English and Dutch scientists, demonstrated the effect of using unstructured computational grids[33] in coastal modeling. Joanna Staneva led the first steps in the assimilation of new types of data specific to coastal areas. Agustín Sánchez-Arquila from the Catalonia University of Technology, Barcelona, led the research in the third part of the work package on *sediment dynamics and modeling*. Here the Dutch experience in the person of Han Winterwerp from WL Delft Hydraulics (now Deltares) was of great help. One of the best research achievements was in the field of marine storm forecasting (Pérez et al. 2012) in the shelf area under the leadership of Enrique Alvarez Fanjul from the Spanish agency PdE.[34]

8 Redistribution of the Earth's Masses

8.1 Gravity Field of the Earth

The current estimate for the mass of the Earth is $M_{\oplus} = 5.9722 \times 10^{24}$ kg, and the respective average density is 5515 kg m^{-3}, which varies between 2700 kg m^{-3} in the upper crust and ~ 13,000 kg m^{-3} in the inner core. The Earth's core accounts for 15% of Earth's volume, the mantle for 84% of the volume, while the crust accounts for less than 1%. The mass of the Earth's oceans is 1.35×10^{21} kg, which is

30 A posteriori analysis (reanalysis) of the state of a dynamic system using a numerical model. In this analysis, observations are available for the entire time period and can be integrated into the calculation of the hindcast at any point in time.

31 The ENEA is the National Agency of Italy for New Technologies, Energy and Sustainable Economic Development.

32 Fringer et al (2019) provide an overview of the current state-of-the-art of coastal modeling.

33 Grids that are made of irregularly shaped elements (e.g., triangles of different size and shape) rather than identical rectangles (see Fig. 3 in chapter "In Search of a Destination").

34 Puertos del Estado is a public entity in Spain that has overall responsibility for the state port system.

Fig. 19 Geoid 2011, the Earth's gravitational field (vertically enhanced). This image is known as the "Potsdam gravity potato". Courtesy of the GFZ (German Research Center for Geosciences)

Fig. 20 Twin satellites from the US-German GRACE flying in orbit as a tandem. Illustration courtesy of NASA/JPL-Caltech

approximately 2.3 × 10^{-4} of the total mass of the planet. The atmosphere has a mass of about 5.15 × 10^{18} kg. While the mass in the solid Earth almost does not change, the annual range in the atmosphere due to variations in the water vapor is ~ 1.2 to 1.5 × 10^{15} kg. It is known that there is a transport of air masses across the equator due to the opposite seasonal variations of the atmospheric state in the northern and southern hemisphere. This transport changes the angular momentum of the planet. Its measurable effect results in a change of the length of the day of ~ 0.8 ms. Water masses in the ocean also undergo substantial redistributions and present an important element of the conservation of the planetary momentum of rotation.

The changes in the distribution of Earth's masses together with the change in the geometry of the surface of the continents and the ocean cause changes in the gravitational field. Earth satellites are able to determine changes in the gravitational field of the Earth with great precision (Fig. 19). Thus, the gravitational field becomes an indicator of the redistribution of mass in the atmosphere, at the surface of Earth, in its crust, and in its mantle. It is of fundamental importance for science and for many practical activities to have a good knowledge of the changes in the distribution of masses. This can enable the determination of their transport and evolution, changes in the total amount of water on the continents, the balance of ice, regional sea level variations, changes in masses in the interior of Earth, etc. In this context, there are a number of unresolved issues. One of the most fundamental one is the interrelationship between the diverse geophysical processes leading to mass redistribution. Hence, it is necessary to unite the efforts from different disciplines, seeking the active participation of geodesists. In order to meet these challenges, the program "Mass Transport and Mass Distribution in the Earth System" was established. It was elevated to the rank of "priority program" within the activities of the German Research Foundation (Deutsche Forschungsgemeinschaft, DFG).

GRACE

The Gravity Recovery And Climate Experiment (GRACE) is a US-German mission based on a pair of minisatellites recording the change in distance between them (Satellite-to-Satellite Tracking). Each of the satellites carries a GPS receiver and high-precision accelerometers, allowing precise orbit determination. Fluctuations in the strength of the gravitational field are reflected in the distance between the tandem satellites (Fig. 20). From the latter information, changes in the distribution of mass in the ocean, atmosphere, and solid ground and in the storage of water, snow, and ice on land can be derived.

8.2 STREMP and Hydrological Modeling

As it turned out, without having planned for it in advance, I dived into a scientific issue that became a hot topic. The reason for this is that the Black Sea and the

Mediterranean Sea are an interconnected pair, the mass of both parts of which can be determined by satellites. In these semi-enclosed basins, errors in balance estimates derived from regional models of precipitation, evaporation, river runoff, and currents in the Bosporus and Gibraltar Strait are relatively well known. Thus, hydrological data can be used to validate mass transport and determine mass change estimated from satellite data and numerical models. A key requirement of the DFG priority program was that satellite measurements should be linked to the numerical environmental modeling framework.

In 2005, together with Matthias Becker (professor at the Technical University of Darmstadt) and Jürgen Kusche (then associate professor at the Technical University of Delft, now professor at the University of Bonn, and subsequently head of the whole priority program), we joined forces and developed a project called "Spatial and temporal resolution limits for regional mass transport and mass distribution in Mediterranean and Black Sea catchments (STREMP)". The main theme of the project was the coupling of continental hydrology with oceanography in coastal oceans and the refinement of the water balance in the straits.

Hydrological Models. Basic Information

Hydrological models aim to quantify real hydrologic features, such as surface water (lakes and rivers), soil water, wetland, groundwater, estuary, and estimate the response to meteorological forcing (e.g., precipitation, evaporation) at local scales, as well as globally. Precipitation (rain, snow) can be measured at a point (gauges) or estimated over an area (satellites, radars, or atmospheric models). Evaporation can be computed using specific physical models. While evaporation is a pure physical process, transpiration refers to the transfer of liquid water from plants into the atmosphere as vapor, thus we speak about evapotranspiration. Balance equations (continuity equations) are at the heart of hydrological modeling. The change S in the amount of water in one element must equal precipitation P minus evapotranspiration ET, groundwater G, and streamflow Q: $S = P - ET - G - Q$.

The streamflow Q is the water that, when it reaches the land surface, flows over the land surface. It is composed of runoff and infiltration. The infiltration is the water that flows into the land surface. It is not easy to decide how much water is drained and how much is infiltrated. The infiltration capacity of the soil decreases as water enters the soil, which is governed by the properties of the soil, as well as by the characteristics of rain (amount and intensity). Furthermore, the infiltration capacity is spatially variable, which necessitates developing appropriate parameterizations and their calibrations against known characteristics of soil and region. Hence, knowing precipitation, evapotranspiration, and infiltration, one can compute using specific physical models (parameterizations) the amount of water over the earth's surface and within the soil layers. The former represents a fast response to precipitation, the latter, the slow response.

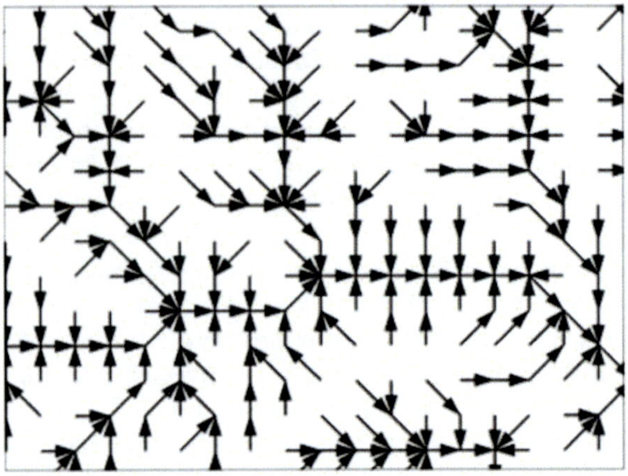

Fig. 21 Drainage network extracted for watershed defined by a digital elevation map. From Vieux (2008)

The next question is about the routing, or what happens to the surface and soil layers runoff, that is how flow is routed until it reaches a river channel. A number of parameters determine the routing, such as the distance and the slope between where the precipitation occurred and the channel, the soil saturated hydraulic conductivity, or the roughness of land. Adjusting these parameters (calibrating them) makes a simulation in a model more or less accurate, that is, the model produces more or less accurate output, streamflow, and soil moisture.

Although atmosphere and ocean sciences, from one side, and land hydrospheric science, from another, deal with equations describing fluid dynamics, there are substantial differences between hydrological models and the models for the atmosphere and ocean. Hydrological motion is much stronger constrained by friction, and regional specificities play a much larger role than in the atmosphere and ocean. Therefore, hydrological models are less generic and need much more calibration.

Hydrological Models. Networks of Reservoirs and Outflows

The model area of interest in hydrological models is considered as consisting of individual model cells. The processes in each of them are described in the box "Hydrological models. Basic information". These cells are connected with the neighboring cells building a complex network of reservoirs, which predicts the outflow from each reservoir (Fig. 21). This outflow is then used as the input to the next cell in the series. Continuity and storage-discharge equations are used to describe outflow from each reservoir (the change of water amount in the cell equals the difference between what goes in and out).

The hydrology and water use model WaterGAP (Water—Global Assessment and Prognosis) assesses the global freshwater system with a specific focus on the interactions between humans and the terrestrial part of the global water cycle. It computes both water resources (surface runoff, groundwater recharge, river discharge, water storage in the soil, groundwater, and surface water bodies) and water use (surface water and groundwater abstractions as well as return flows in the water use sectors irrigation, livestock, households, thermal power plants, and manufacturing). This model has been developed at the University of Kassel (Center for Environmental Systems Research). Its fine-resolution version (5′, which is ~7 km × 7 km), WaterGAP3, operates on the global scale but for each continent separately. In the study of aus der Beek et al. (2012), a special landmask has been developed, as the study region includes parts of three continents (Europe, Africa, and Asia, Fig. 22). For each grid cell, WaterGAP3 takes into account spatially distributed physiographic information about elevation, slope, hydrogeology, land cover, and soil properties, as well as location and extent of lakes, wetlands, and reservoirs. The upstream/downstream relationship among the grid cells is defined by a global drainage direction map (DDM) which indicates the drainage direction of surface water (Lehner et al. 2008). WaterGAP3 was used in STREMP to simulate the characteristic behavior of the terrestrial water cycle in order to estimate total water storage (TWS) for each grid cell and river basin (Fig. 22). Herein, TWS is defined as the sum of all potential water storages: rivers, groundwater, soil water, canopy water, snow, lakes, and wetlands. WaterGAP3 was applied to correct estimates of seawater mass derived from the GRACE gravity mission and to account for freshwater inflow into the Black Sea and Mediterranean. The performance of the WaterGAP3 regional solution has been evaluated by comparing the seawater mass derived from GRACE, corrected for the leakage of continental hydrology, to an independent estimate derived from satellite altimetry.

8.3 *Mediterranean and Black Sea*

My part of the cooperative research aimed at clarifying the synergy between satellite data and numerical modeling of mass transport in the Mediterranean and Black Sea. To this aim, satellite altimeter data were also used, which helped determine the sea level height relative to a geocentric reference system. Particular attention had to be paid to the coarse grid on which the GRACE data were provided (about 500 km resolution) and the issues arising in this respect regarding their usability in small regional seas. The next important contribution was to account for the steric effect.[35] For this, temperature and salinity data, which were provided from the model of

[35] Ocean volume can change due to changes in density (expansion and contraction of water masses). The dependence of density on temperature and salinity and the temporal and spatial variability of temperature and salinity is responsible for the halosteric (salinity) and thermosteric (temperature) effect. This effect can be measured by the respective sea level changes.

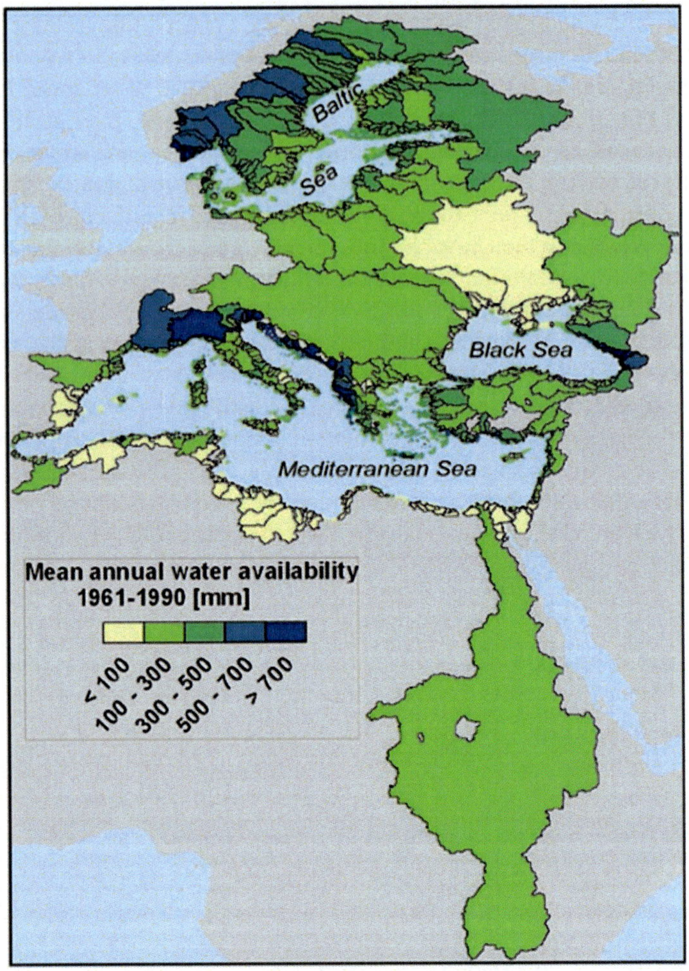

Fig. 22 European semi-enclosed seas and their catchment basins. The colors give the annual mean water availability for 1961–1990. From Stanev and Xi (2013). Courtesy of Tim aus der Beek, Center for Environmental Systems Research, University of Kassel.

Grayek et al. (2010), were used. The steric effect for the Mediterranean Sea was estimated using the output from the numerical model of Tonani et al. (2008).

Two time series[36] were constructed for the two basins for the interval between August 2002 and July 2008 representing the basin mean mass signal in terms of equivalent water height. The first series is obtained from steric-corrected altimetry[37]

[36] A series of data points indexed in time order.

[37] Altimeter data in which the sea level is corrected considering the steric effect.

while the other is deduced from GRACE data corrected for the contamination by continental hydrology (Fig. 23). These time series show a good agreement in terms of annual and interannual signals. In the Mediterranean Sea, we obtain the best agreement using a steric correction from the regional oceanographic model of Tonani et al. (2008) and a continental hydrological leakage correction derived from the global continental hydrological model. In the Black Sea, the time series are less consistent, with lower accuracy of the GRACE-derived estimate (keep in mind the stronger amplitude in the Black Sea (Fig. 23b) caused by the large river runoff). Based on mass conservation, the mass-induced sea level variations, river runoff, and precipitation minus evaporation are combined to derive the strait flows between the Black Sea and the Mediterranean and between the Mediterranean and the Atlantic Ocean. At the Gibraltar Strait, the net inflow varies annually with an amplitude of 52 $\pm$ 10 $\times$ 10^{-3} Sv peaking end of September (1 Sv = 10^6 m^{-3} s^{-1}). The inflow through the Bosporus strait displays an annual amplitude of 13 $\pm$ 3 $\times$ 10^{-3} Sv peaking in the middle of March. Additionally, an increase of the Gibraltar net inflow (3.4 $\pm$ 0.8 $\times$ 10^{-3} Sv yr^{-1}) was detected. One important result in STREMP was that the regional hydrological model is capable of realistically quantifying water mass variation in the region.

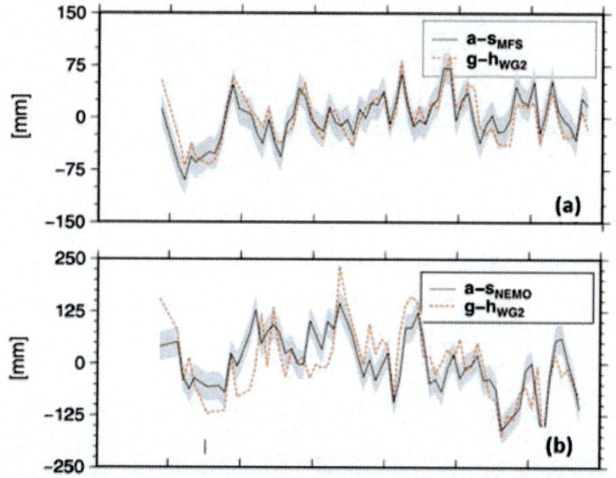

Fig. 23 Basin average of water mass variation expressed in equivalent water height in (**a**) the Mediterranean and **b** the Black Sea. Water mass change is from steric-corrected altimetry (solid line) and from hydrology-corrected GRACE gravity solutions (dashed line), where g is the rescaled mass change computed as GRACE-load + GAD-GIA as described in Fenoglio-Marc et al. (2012). GAD is monthly average of the removed non-tidal effects restricted over the ocean, GIA is the Glacial Isostatic Adjustment. Error bounds correspond to the RMS differences, and unknown systematic errors may remain. The figure is from Fenoglio-Marc et al. (2012), with permission from Elsevier

9 Southern European Seas on the Road to a Good Environmental Status

9.1 SESAME

The project "Southern European Seas: Assessing and Modeling Ecosystem changes (SESAME)" aimed to assess and predict changes in the Mediterranean and Black Sea ecosystems. This project was a natural extension of POEM and the NATO-TU-Black Sea program. What was new was that the pair of basins was approached as coupled (interacting) climatic ecosystems. Oceanographic expeditions were planned in the two seas, and mathematical models were to be validated against new and other available data. Central to this project was the issue of optimal use of data to provide qualitative assessments and predictions of the marine environment. SESAME took an innovative step toward merging economic and natural sciences to study changes in the Western and eastern Mediterranean and the Black Sea over a period of 50 years in the past to 50 years in the future.

SESAME was a big project as 47 institutes were involved, totaling EUR 14,794,318, of which the European Union covered € 9,999,121. The coordinator was Vangelis Papathanassiou from the Hellenic Centre for Marine Research (HCMR), Greece. From Bulgaria, IO-BAS, and the University of Sofia participated. I participated as a representative of ICBM. SESAME succeeded in creating a platform for dissemination of results to a wide public and has also been instrumental in strengthening international cooperation in the Mediterranean and Black Sea region through the participation of research organizations from European Union Member States, Associated Countries, Associated Candidate Countries, as well as non-Mediterranean countries and countries from former Soviet Republics. The project started on November 1, 2006, and continued until April 30, 2011. In Sofia, a computer literate student, Rostislav Kandilarov, joined E. Peneva to help with the ECOOP and SESAME work.

Given my planned work in the STREMP and SESAME projects, I decided to check the performance of the model known as NEMO. This model was gaining a recognition as a state of the art modeling framework for research activities and forecasting services in ocean and climate sciences, and I wanted to know whether it was superior to the MOM. At that time, the latter had been developed to its fourth version, MOM4. I was working with this model in Bulgaria, and it would be easiest for me to continue with it in the new conditions. The HYCOM, which I had recently become familiar with, was also a candidate. Nevertheless, I set about learning NEMO.

9.2 Marine Strategy Framework Directive

The Marine Strategy Framework Directive (MSFD[38]) is the environmental component of Europe's Integrated Maritime Policy adopted in June 2008. It sets a target of Good Environmental Status to be achieved in EU marine waters by 2020. The Directive defines Good Environmental Status (GES[39]) as "The environmental status of marine waters where these provide ecologically diverse and dynamic oceans and seas which are clean, healthy and productive". GES means that the different uses of the marine resources are conducted at a sustainable level, ensuring their continuity for future generations.

In addition, GES means that:

- Ecosystems, including their hydro-morphological (i.e., the structure and evolution of the water resources), physical, and chemical conditions, are fully functioning and resilient to human-induced environmental change.
- The decline of biodiversity caused by human activities is prevented, and biodiversity is protected.
- Human activities introducing substances and energy into the marine environment do not cause pollution effects. Noise from human activities is compatible with the marine environment and its ecosystems.

To help Member States interpret what GES means in practice, the directive sets out, eleven qualitative descriptors which describe what the environment will look like when GES has been achieved.

MSFD Descriptors

Annex 1 of the Directive 2008/56/EC of the European Parliament and of the Council of June 17, 2008, specified the following descriptors

Descriptor 1. Biodiversity. The quality and occurrence of habitats and the distribution and abundance of species are in line with prevailing physiographic, geographic, and climatic conditions.

Descriptor 2. Non-indigenous species. Non-indigenous species introduced by human activities are at levels that do not adversely alter the ecosystems.

Descriptor 3. Commercial species. Populations of all commercially exploited fish and shellfish are within safe biological limits, exhibiting a population age and size distribution that is indicative of a healthy stock.

Descriptor 4. Marine food webs. All elements of the marine food webs, to the extent that they are known, occur at normal abundance, diversity, and levels capable of ensuring the long-term abundance of the species and the retention of their full reproductive capacity.

---

[38] Full title: Directive 2008/56/EC of the European Parliament and of the Council of 17 June 2008.

[39] https://ec.europa.eu/environment/marine/good-environmental-status/index_en.htm, last accessed on 03.02.2023.

Descriptor 5. Human-induced eutrophication. Human-induced eutrophication is minimized, especially adverse effects thereof, such as losses in biodiversity, ecosystem degradation, harmful algae blooms, and oxygen deficiency in bottom waters.

Descriptor 6. Sea floor integrity. Sea floor integrity is at a level that ensures that the structure and functions of the ecosystems are safeguarded, and benthic ecosystems, in particular, are not adversely affected.

Descriptor 7. Hydrographical conditions. Permanent alteration of hydrographical conditions does not adversely affect marine ecosystem.

Descriptor 8. Contaminants. Contaminants are at a level not giving rise to pollution effects.

Descriptor 9. Contaminants in seafood: They are below safe levels.

Descriptor 10. Marine litter. Properties and quantities of marine litter do not cause harm to the coastal and marine environment.

Descriptor 11. Energy and underwater noise. Introduction of energy, including underwater noise, is at levels that do not adversely affect the marine environment.

The MSFD mandates an ecosystem approach to management of the marine environment. This approach cuts across sectors, organizations, and policies. The marine strategies to be developed by each Member State must contain a detailed assessment of the state of the environment, a definition of GES at regional level, and the establishment of clear environmental targets and monitoring programs. Building capacity of stakeholders and policymakers is essential to implementing the MSFD. These guidelines necessitate using the social, economic, and environmental science base and methods to tackle key management tasks and help decision-makers prepare spatial plans. The ecosystem-based management helps direct policy to minimize the environmental impact of human activities.

9.3 PERSEUS

The research project "Policy-oriented marine Environment Research for the Southern European Seas (PERSEUS)" timely responded to the MSFD. It assesses the impact of human activity and natural pressures on the Mediterranean and Black Seas by merging natural and socioeconomic sciences to predict the long-term effects of these pressures on marine ecosystems. PERSEUS aimed to design an effective and innovative research governance framework, to provide the basis for policymakers to turn back the tide on marine life degradation. This project, which was effectively a continuation of the SESAME project, included 53 European participants and had an overall budget € 16,994,500 with a European Union contribution of € 12,973,123. It started in January 2012 and ended in December 2015.

The overall scientific objectives of PERSEUS were to identify the interacting patterns of natural and human-derived pressures on the Mediterranean and Black Seas, assess their impact on marine ecosystems, and, using the objectives and principles of the Marine Strategy Framework Directive as a vehicle, to design an effective and innovative research governance framework based on sound scientific knowledge. Well-coordinated scientific research and socioeconomic analysis were foreseen to be applied at a wide-ranging scales, from a basin to coastal scales. The new knowledge gained aimed at advancing our understanding on the selection and application of the appropriate descriptors and indicators of the MSFD. In order to achieve these goals, new tools were developed to evaluate the current environmental status, by way of combining monitoring and modeling capabilities and existing observational systems.

Basic achievements of PERSEUS are the development of a scenario-based framework of adaptive policies and management schemes.[40] The project results can be considered as a promotion of the principles and objectives outlined in the MSFD for the Mediterranean and Black Sea. Of great value is the PERSEUS oceanographic information management system, which provides user-friendly and fast access to physical, geochemical, and biological data of the ecosystems of the Mediterranean and Black Sea. The database includes both data collected by PERSEUS and previous European projects such as SESAME, MyOcean (Sect. 3 in chapter "Operational Oceanography and Future Scientific and Societal Challenges"), SeaDataNet,[41] and the European Marine Observation and Data Network (EMODnet[42]). About 98% of the data is freely available to any user after online registration.

10 GKSS

At the end of 2005, I came across an advertisement that the Institute for Coastal Studies at the GKSS Research Centre near Hamburg was looking for a scientist to lead the Data Analysis and Assimilation Department. More specifically, expertise in sea currents and waves with a focus on the development of operationally applicable methods for coastal observation and modeling was sought. The procedure was a joint one with the University of Oldenburg, and the selected candidate was to become a professor at ICBM. I applied for the position and on February 26, 2007,

---

[40] https://open.metu.edu.tr/bitstream/handle/11511/58248/CORDIS_287600_Report.pdf. last accessed on 03.02.2023.

[41] SeaDataNet is a pan-European infrastructure to ease the access to marine data measured by the countries bordering the European seas.

[42] EMODnet is a network of organisations supported by the EU's Integrated Maritime Policy. It is comprised of more than 150 organizations, which gather marine data, metadata, and data products and make them more easily accessible for a wider range of users. EMODnet consists of seven subportals: bathymetry, geology, physics, chemistry, biology, seabed habitats, and human activities (Míguez et al. 2019).

the Minister of Science and Culture of Lower Saxony formally invited me[43] to take up the professorship. I was to work in the sector of the institute headed by Prof. Hans von Storch.

Hans von Storch is a scientist with broad interests in coastal climate, its possible future changes, statistical climatology, and the social aspects of climate science. His popularity is largely due to the popularity of climate change issues. However, there are many people dealing with these issues, some of them in a much more fundamental way, and yet they are less popular. In my view, the popularity of H. von Storch has a lot to do with his clear opinion and his power to impose it, as well as his talents for communication and leadership.

The Helmholtz Association of Research Centers (Hermann von Helmholtz-Gemeinschaft Deutscher Forschungszentren e.V., HGF) comprises at least five research centers studying different aspects of processes on our planet. The prerequisites for conducting large-scale research on Earth are excellent in HGF because the individual centers have a unique variety of observational devices and infrastructures. GKSS, which is one of these centers, was found in 1956 as a company for the utilization of nuclear energy in shipbuilding and shipping. The first directions of research were marked by the construction of a research reactor and work on the use of nuclear energy in civil shipping. On October 11, 1968, the nuclear cargo ship Otto Hahn[44] was launched. It was the world's third civilian ship powered by a nuclear reactor, after the Soviet icebreaker Lenin and the American Savannah. Because many ports did not allow nuclear-powered ships, GKSS oriented its research work toward nuclear reactor safety. Later, environmental research was added.

Over several decades, the subject of research has changed so much that the name of the GKSS research center had to be changed. On November 1, 2010, it was renamed to the Helmholtz-Zentrum Geesthacht Center for Material and Coastal Research (HZG). The new name was intended to show the link with the Association of German Research Centers Helmholtz and the town of Geesthacht, where the institute has its main headquarters. In 2021, the name of the institute was changed again, this time, as Helmholtz-Zentrum hereon GmbH. The name Hereon is a combination of the terms "Helmholtz", "resilience", and "innovation".

[43] The practice in Germany is that the invitation to take up a professorship comes from the relevant provincial minister.

[44] The ship is named after one of the founders of radiation chemistry. Otto Hahn (1879 - 1968) has been nominated 22 times for the Nobel Prize in Chemistry and 16 times for the Nobel Prize in Physics. In 1944, he received the Nobel Prize in Chemistry. He was the founder and the president of the Max Planck Society from 1948 to 1960. He is considered by many to be a model of scientific excellence and personal integrity, which is why he is regarded as one of the most influential and most respected citizens of postwar West Germany.

The Early Years of GKSS Environmental Research

In the early 1970s, work began on developing offshore and underwater technologies. In the late 1990s, work began on environmental and climate research, particularly in coastal areas. This led to the establishment of the Institute of Coastal Research. Priority was given to investigating natural hazards such as storms or tsunamis, as well as to research on anthropogenic climate change and sea level rise. Studies on the Elbe and Weser estuaries and their relationship to coastal processes were also being stepped up. Further, research on atmospheric processes was also initiated, as well as the development of the MEdium Resolution Imaging Spectrometer (MERIS), which was installed on Envisat (see Sect. 18 in chapter "Further into the Ocean"). Direct measurements by GKSS were used to calibrate the Envisat data, which contributed to the reliability of estimates of the distribution of BGC properties in the ocean based on data from space.

When I moved to GKSS, about 800 employees were working there. The Institute of Coastal Research consisted of two sectors: "System analysis and modeling" under the direction of H. von Storch and "Operational Systems" under the direction of Prof. Franciscus Colijn, a marine biologist. The department I was to head was known as "Data Analysis and Assimilation" (Auswerteverfahren und Datenassimilation, KSD[45]). Soon after me, J. Staneva was employed as a researcher at GKSS.

11 COSYNA and Partners Gave Birth to JERICO

11.1 COSYNA

Before I started at GKSS, I was involved in the preparation of the project "Coastal Observing System for Northern and Arctic Seas" (COSYNA, Baschek et al. 2017). This was a German investment program aimed at establishing a long-term observing network for the southern coastal regions of the North Sea, to be supported by pre-operational models providing data for scientific and management purposes. In short, COSYNA consisted of two basic parts: observations and forecasting.

The institute's sector "Operational Systems" was primarily concerned with oceanographic measurements. The scope of work of "Operational Systems" included observational platforms based on Earth satellites, FerryBoxes, radars as well as

[45] The abbreviation KSD comes from KS (KüstenSysteme—coastal systems, the short name of the sector of the institute headed by H. von Storch) and "D", Daten—data, the short name of my department.

continuously operating data stations. Taken together, these data represented a wealth of information, the usefulness of which was not yet known. It was also planned to integrate the BSH and ICBM permanent stations into a common observing system, as well as to install new sensor arrays on offshore platforms, on commercial vessels and on autonomous underwater floating robots. The acquisition of such data requires enormous resources, including skilled personnel. Since the spring of 2007, I have been in intensive communication with one of F. Colijn's coworkers, Dr. Rolf Ritmüller. From 2000 to 2015, he was the head of the department "Coastal Oceanographic Measurement Systems". Rolf Ritmüller worked on drafting the observational part of COSYNA proposal, and I worked on drafting the forecasting part. The experience was already there. Even before J. Staneva and I came to GKSS, we had developed, with a few other colleagues, a model for the southern part of the North Sea (German Bight, Staneva et al. 2009). There remained work to be done on the optimal use in numerical modeling of the data from observations in developing the scientific basis for coastal forecasts.

COSYNA started as a pilot investment project at a cost of €2.5 million. Nine scientific institutes were involved, with coordination taken over by GKSS in the person of F. Colijn. He was very communicative and carried out the overall management with good tact and kindness. COSYNA had to put marine monitoring on a qualitatively new footing and provide the necessary data coverage of the area in the long term. This was to become the basis for both the development of short-term

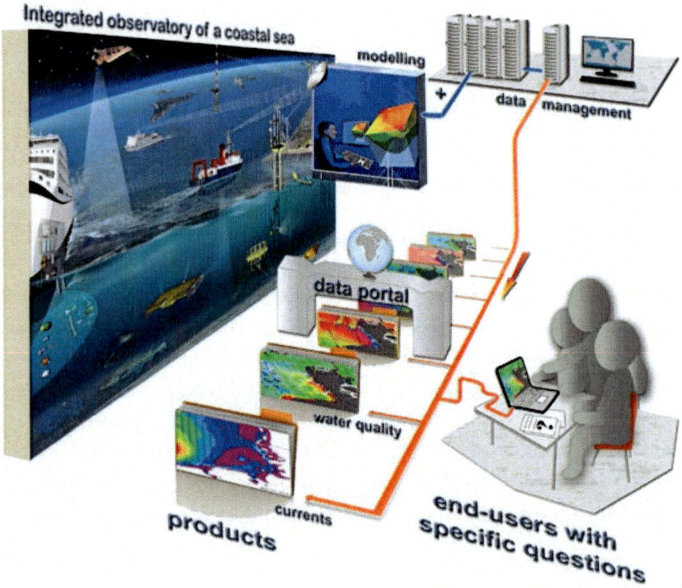

Fig. 24 Concept of COSYNA implies a massive use of measurement data, their assimilation into numerical models, and the provision of operational products to interested users. From Eschenbach (2017)

forecasts and scenarios identifying trends in the changing marine environment. The synergism between observations and models was underpinned by the idea of data analysis and assimilation (Fig. 24). It was also intended that COSYNA, in its early development phase, would support the establishment of the Northwest European Observing and Forecasting System.

More Details About COSYNA

The interest in COSYNA was focused on the physical parameters of the environment, as well as on sedimentary, geochemical, and biological parameters that were to be monitored at a high temporal resolution throughout the water column. COSYNA can be considered as a response of scientists to needs of environmental management and was intended to enrich our understanding of the processes associated with ocean warming[46] and acidification of waters. Bringing together the efforts of several institutes, COSYNA created a centralized management of scientific and practical activities. This provided a basis for future research projects and support for BSH in their operational practice in the coastal regions. As one of the most complete observing systems, located in one of the most intensively used coastal areas in the world, the North Sea, COSYNA has created a model for other parts of the World Ocean. In doing so, it has helped to strengthen the role of Germany in the international development of marine monitoring and long-term observing and forecasting strategies.

The development of COSYNA occurred in two phases: a conceptual phase and an implementation phase. The first phase involved the formation of an international council consisting of representatives of the participating institutes, foreign institutes, and representatives of local authorities, agencies, and companies. At the end of the first phase, formal agreements were concluded with all partners, including those dealing with rights, obligations, and financial issues. The completion of the first phase was evaluated very positively and additional funding of around € 9 million ensured that the actual work of the project was carried out in the period 2010–2014.

The main goal of KSD was to improve the quality of the existing physical and BGC models for the North Sea, in particular the German Bight, and demonstrate that models are in a position to adequately reproduce the relevant processes. Of great importance was the optimal use of existing data streams and numerical methods for their assimilation. Radar remote sensing was to play a particular role. These (radar) measurements are based on the backscattering of electromagnetic signals from the ocean surface. The roughness of the ocean surface depends on the characteristics of the sea waves, and these are determined by the winds blowing on the sea surface

[46] The global mean sea surface temperature increased by 0.67 °C over the last century. In the North Atlantic, warming has penetrated well below 1,000 m. The mean temperature in the 0–2000 m layer of the World Ocean increased by 0.09 °C from 1955 to 2010. Over the past 25 years, warming of above 0.96 °C has been observed in the North Sea and Black Sea.

Fig. 25 This is what a group of WERA antennas deployed along the coast look like. They are used to determine the velocity of ocean surface currents. Courtesy of W. Ziemer

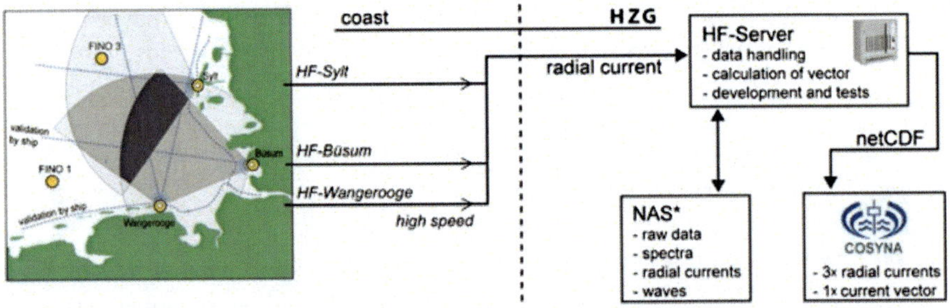

Fig. 26 HF WERA system covering the German Bight. (left) Positions of the three COSYNA HF radar stations: Wangerooge, Büsum, and Sylt. (right) Processing and data flow scheme. Radial current[47] components are transferred to the HZG WERA server. Here, the current vector field is calculated from the radial components and presented on the COSYNA data portal. The radar data are stored on a Network-Attached Storage (NAS) system. From Stanev et al. (2015b), © American Meteorological Society. Used with permission

as well as the currents. Hence, the analysis of the scattered radar signal gives us information about the currents and waves at the sea surface. The radars we used emit waves in the high-frequency (HF) range (3–30 MHz), which is why they are called HF radars. They cover distances greater than 150 km from the shore.

There are two types of HF radars: the German WERA (WavE RAdar) system and the American CODAR (Coastal Ocean Dynamics Applications Radar). The latter is more compact, but the WERA (Fig. 25) that we used gave very high-quality information. In Europe, there were a few short-lived attempts to use WERA. For

[47] Radial current or radial speed is the speed with which the water moves away from the observer (HF-radar).

the first time, COSYNA carried out a multiyear operation of three stations along the German coast (Fig. 26), which made available, every 20 min, i.e., in near real time, the maps of surface currents.

Assimilation of HF radar measurements, which was based on Spatio-Temporal Optimal Interpolation (STOI, Stanev et al. 2015b), allowed the provision of short-term surface current forecasts with hourly updates. There have been other attempts, mostly in the USA, to assimilate HF radar data, but from which the high-frequency oscillations were clipped. COSYNA demonstrated that coastal forecasts could be improved also in the high-frequency range by assimilating unfiltered HF radar data (Fig. 27). The most important takeaway from this work was that measurement data in a relatively small area of German Bay contributed to improved forecasts throughout the whole model domain where we conducted our studies (much bigger than the observed area).

The positive outcome of the assimilation of HF radar data in numerical models is in line with the later research of Sandery and Sakov (2017) who showed decreased ocean predictive skill as model horizontal resolution increases. This is explained by the fact that the observational data are usually available on coarser grids compared to the model data. It occurred that by using HF radar data, characterized by high spatial and temporal resolution, this problem was substantially reduced.

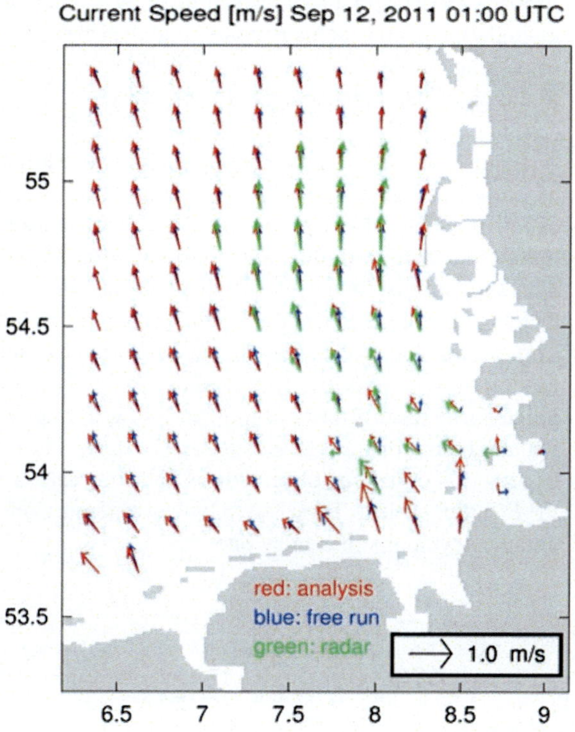

Fig. 27 Comparison of the analysis (red) and the free run (blue). Green arrows are radar observations. From Stanev et al. (2014)

This success of using HF radar data brought COSYNA to achieving one of its main goals, namely the development of pre-operational products.[48] For example, updated hourly charts and 12-h ocean current forecasts were made available by COSYNA as a scientific product, pointing the way to improving current routinely used observational and modeling techniques. Subsequently, further pre-operational products were developed such as hourly electronic maps of salinity, temperature, and waves.

11.2 European Coastal Operational Activities on the Road to JERICO

COSYNA was not alone. A pilot coastal observatory was established in Liverpool Bay (East Irish Sea) in 2002 with an expected lifetime of at least 10 years (Howarth et al. 2006). John Howard and Roger Proctor of the Proudman Institute in Liverpool led the work on the measurement part (John) and the modeling (Roger). Together with other colleagues from the UK, they set out to solve some of the same problems that we set out to solve in COSYNA. The East Irish Sea, and in particular Liverpool Bay, had a great similarity to the German Bight. The measurement systems were also similar—in Liverpool Bay, they used WERA. There were also initial attempts of data assimilation of coastal measurements at the Liverpool Bay Pilot Coastal Observatory, something similar to what we were doing in COSYNA. It was unfortunate that, due to the high maintenance costs of the radar in the Liverpool Bay, its operation was canceled.

English colleagues had received national funding to organize a study of the status of European coastal ocean observing systems. One form of work was to organize expert meetings to raise awareness, interaction, and cooperation between those developing and using coastal observing and forecasting systems worldwide. The first workshop on this topic was organized in the UK, and the second one was held from October 22 to 24, 2007, at the Hotel Fermentor, Mallorca. The local host was Joaquín Tintore, the initiator of the Balearic Islands Coastal Ocean Observing and Forecasting System (SOCIB, Tintore et al. 2013). The venue of the conference was close to the north-easternmost point of Mallorca, Fermentor Peninsula, a beautiful natural feature. I mention this second meeting because it brought together the main European experts on coastal sea observations and forecasts, as well as some key members of the ECOOP project. They had to look at emerging observing and forecasting systems and the development of new ones.[49]

48 Operational products are provided on a regular basis (twenty-four/seven hours) by the authorized agencies. COSYNA is not authorized for such provision; its pre-operational products are scientific products bearing most of the technical characteristics of the operational products, or even more advanced.

49 COSYNA was not just an emerging, but an established system.

The French national project PREVIMER (Lecornu et al. 2008; Lecornu and De Roeck 2009) was established in parallel with COSYNA. It was aimed at developing a system capable of providing real-time information on the current status and forecasts of physical and environmental conditions in the English Channel, Bay of Biscay and northwestern Mediterranean regions. PREVIMER was built on four components: (1) measurement systems (including the development of new instruments), (2) numerical models (including models for swell[50]), (3) a data center, and (4) a website to disseminate the information to users. The actual project phase lasted from 2008 to 2013 under the coordination of Jacques Legrand and Fabrice Lecornu from IFREMER. The major aim of PREVIMER was to demonstrate that it could function as an operational coastal forecasting system, providing synoptic observations and 4- to 6-day forecasts.

The development of the four coastal systems I mentioned above (COSYNA, Liverpool Bay Pilot Coastal Observatory, SOCIB, and PREVIMER) kept pace with the scientific developments in ECOOP. One major activity in the in ECOOP WP7 was on observing system design. The main issue was how to optimize the observing activity so that the measurement data would be as informative as possible for the area being observed, in other words, where, when, how, and what to measure. Jacques Legrand led these specific works in ECOOP. We had already had several personal meetings with him. One of them was at IFREMER HQ in Paris on March 26–27, 2007, when he hosted the partners meeting of ECOOP work package 7 (see Sect. 7).

There was a natural process of consolidation of observing systems in Europe. In 2008, J. Legrand visited the GKSS together with Patrick Farcy from IFREMER. We discussed our next engagements in future actions. Soon Patrick took over the leadership of the project "Towards a Joint European Research Infrastructure network for Coastal Observatories (JERICO)". This project represents the response of the scientific community to the realization that there is considerable heterogeneity across Europe in terms of the technological design of monitoring systems, the parameters measured, maintenance and quality control practices, sensor quality, and data exchange standards. So far, the expansion of coastal observatories has been driven by intra-institutional or national interests and has been mainly carried out through short-term research projects. Increasing the coherence of dispersed coastal infrastructures has been a major challenge for the research community, along with the need to develop and operationalize a pan-European coastal measurement framework (European Research Infrastructure Consortium, ERIC[51]).

[50] A sea swell is a series of surface gravity waves generated by distant weather systems raging hundreds of miles out to sea. Swells represent more regular or smooth waves, often with longer periods unlike the choppier, more confused wind waves within the generating area.

[51] This is a specific legal form that facilitates the creation and operation of research infrastructures of European interest.

Aims of JERICO

JERICO (Farcy et al. 2019) establishes a robust and transparent organization for the timely, continuous, and sustainable provision of high-quality environmental data and information products related to the marine environment in European coastal seas. Such a research infrastructure is innovative in the sense that for the first time, there would be a common European organization to consolidate and harmonize fragmented coastal observation activities. In other words, JERICO was proposing a pan-European approach to creating a network of coastal marine observatories, integrating observational infrastructure and technology. It also aimed to conduct new collaborative research to identify novel and strategic technologies to be deployed in the next generation of European coastal observatories. JERICO is thus positioning itself as a leader in international and global efforts in climate change research, operational observation, and coastal ocean forecasting. This project addressed some of the needs of environmental research and the general public by providing the opportunity for open access to advanced and innovative facilities, resources, data, and services that foster international scientific collaboration.

JERICO is one of the projects under the European Strategy Forum on Research Infrastructures (ESFRI) mission, which aimed to support coherent strategies and policy approaches on research infrastructures in Europe. The strategy of ESFRI is to overcome the limits of existing fragmentation and to provide Europe with state of the art research infrastructures that respond to rapidly evolving scientific trends and technologies. The first phase of JERICO ran from 2011 to 2015. Dr. Willi Petersen was the responsible GKSS representative in the project. KSD's participation was relatively smaller because the focus of JERICO was on observations and KSD was involved in numerical modeling and optimization of measurement strategies based on proxy (model) data (see Sect. 4.4 in chapter "New Observing and Modeling Systems and New Scientific Applications").

References

Aksu AE, Mudie J, Kaminski MA, Abrajano T, Yasar D (2002) Persistent Holocene outflow from the Black Sea to the Eastern Mediterranean contradicts Noah's flood hypothesis. GSA Today 12:4–10

aus der Beek T, Menzel L, Rietbroek R, Fenoglio-Marc L, Grayek S, Becker M, Kusche J, Stanev EV (2012) Modeling the water resources of the Black and Mediterranean Sea river basins and their impact on regional mass changes. J Geodyn 59–60:157–167, https://doi.org/10.1016/j.jog.2011.11.011

Aznar R, Ciliberti S, Crosnier L, Drevillon M, Drillet, Pérez Gómez B, Reppucci A, Sudheer J, Sotillo M, Tonani M, Vinaychandranand PN, Zhong A (2022) Architecture of ocean monitoring and forecasting systems. In: Fanjul, EA, Ciliberti S, Bahurel P (eds) Implementing operational ocean monitoring and forecasting systems; IOC-UNESCO, GOOS-275, pp 30–76

Badertscher S, Fleitmann D, Cheng H, Edwards RL, Göktürk OM, Zumbühl A, Leuenberger M, Tüysüz O (2011) Pleistocene water intrusions from the Mediterranean and Caspian seas into the Black Sea. Nature Geosci 4:236–239

Baretta JW, Ebenhöh W, Ruardij P (1995) The European regional seas ecosystem model, a complex marine ecosystem model. Neth J Sea Res 33:233–246. https://doi.org/10.1016/0077-7579(95)90047-0

Baschek B, Schroeder F, Brix H et al (2017) The coastal observing system for northern and Arctic Seas (COSYNA). Ocean Sci 13:379–410

Borgese EM (1998) The Oceanic circle—governing the seas as a global resource. United Nations University Press, Tokyo, p 240

Burchard H, Bolding K (2002) GETM—a general estuarine transport model. Scientific documentation, no EUR 20253 EN, European Commission, printed in Italy, p 157

Chepalyga AL (2007) The late glacial great flood in the Ponto-Caspian basin. In: Yanko-Hombach V, Gilbert AS, Panin N et al (eds) The Black Sea flood question: changes in coastline, climate and human settlement. Springer, Dordrecht, pp 118–148

Daewel U, Peck MA, Schrum C, St. John MA (2008) How best to include the effects of climate-driven forcing on prey fields in larval fish individual-based models. J Plankton Res 30:1–5

Desaubies Y (2009) The MERSEA project. Ocean Sci 5(2):1–3

Dyke P (2016) Modeling coastal and marine processes, 2nd edn. Imperial College Press. https://doi.org/10.1142/p1028

Dyson F (2004) A meeting with Enrico Fermi. Nature 427(297). https://doi.org/10.1038/427297a

Eschenbach CA (2017) Bridging the gap between observational oceanography and users. Ocean Sci 13:161–173. https://doi.org/10.5194/os-13-161-2017

Farcy P, Durand D, Charria G, Painting SJ, Tamminen T, Collingridge K, Grémare AJ, Delauney L, Puillat I (2019) Toward a European coastal observing network to provide better answers to science and to societal challenges; the JERICO research infrastructure. Front Mar Sci 6:529. https://doi.org/10.3389/fmars.2019.00529

Fenoglio-Marc L, Rietbroek R, Grayek S, Becker M, Kusche J, Stanev E (2012) Water mass variation in the mediterranean and black seas. J Geodyn 59–60:168–182

Fringer OB, Dawson CN, He R, Ralston DK, Zhang YJ (2019) The future of coastal and estuarine modeling: findings from a workshop. Ocean Model 143:101458. https://www.sciencedirect.com/science/article/pii/S1463500319301015

Fruergaard M, Andersen T, Johannessen P, Nielsen LH, Pejrup M (2013) Major coastal impact induced by a 1000-year storm event. Sci Rep 3:1051

Georgievski G, Stanev E (2006) Paleoevolution of Euroasian watersheds: water transport through the Bosporus Straits as an indicator of the Lateglacial-Holocene transition. Clim Dyn 26(6):631–644

Grayek S, Stanev EV, Kandilarov R (2010) On the response of Black Sea level to external forcing: altimeter data and numerical modeling. Ocean Dyn 60:123–140

Hernandez F, Blockley E, Brassington GB, Davidson F, Divakaran P, Drévillon M, Ishizaki S, Garcia-Sotillo M, Hogan PJ, Lagemaa P, Levier B, Martin M, Mehra A, Mooers C, Ferry N, Ryan A, Regnier C, Sellar A, Smith GC, Sofianos S, Spindler T, Volpe G, Wilkin J, Zaron ED, Zhang A (2015) Recent progress in performance evaluations and near real-time assessment of operational ocean products. J Oper Oceanogr 8(sup2):221–238. https://doi.org/10.1080/1755876X.2015.1050282

Howarth MJ, Proctor R, Knight P Smithson MJ, Mills DK (2006) The liverpool bay coastal observatory towards the goals. In: OCEANS 2006 conference proceedings. Richardson, TX, USA, IEEE, pp 381–386

Kara AB, Wallcraft AJ, Hurlburt HE, Stanev EV (2008) Air–sea fluxes and river discharges in the Black Sea with a focus on the Danube and Bosphorus. J Mar Syst 74:74–95

Kourafalou VH, De Mey P, Staneva J, Ayoub N, Barth A, Chao Y, Cirano M, Fiechter J, Herzfeld M, Kurapov A, Moore AM, Oddo P, Pullen J, Van der Westhuysen A, Weisberg R (2015) Coastal Ocean forecasting: science foundation and user benefits. J Oper Oceanogr 8(sup1):s147–s167. https://doi.org/10.1080/1755876X.2015.1022348

Kudela RM, Berdalet E, Enevoldsen H, Pitcher G, Raine R, Urban E (2017) GEOHAB–the global ecology and oceanography of harmful algal blooms program: motivation, goals, and legacy. Oceanography 30(1):12–21. https://doi.org/10.5670/oceanog.2017.106

Kullenberg, G (2020) Ocean science and international cooperation: historical and personal recollections. Intergovernmental Oceanographic Commission, ISBN : 978-92-3-100487-2, 299 pp.

Lecornu F, De Roeck Y-H (2009) PREVIMER—coastal observations and forecasts. La Houille Blanche 95(1):60–63. https://doi.org/10.1051/lhb:2009006

Lecornu F, Paillet J, Ravenel H (2008) PREVIMER—coastal observations and forecasts 2 years making-up and future perspectives. SeaTechWeek 2008, coastal operational oceanography–brest 13 14 October 2008. https://citeseerx.ist.psu.edu/viewdoc/download?doi=10.1.1.601.5872&rep=rep1&type=pdf

Lehner B, Verdin K, Jarvis A (2008) New global hydrography derived from spaceborne elevation data. EOS Trans AGU 89(10):93–94. https://doi.org/10.1029/2008EO100001

Míguez BM, Novellino A, Vinci M, Claus S, Calewaert JB, Vallius H et al (2019) The European marine observation and data network (EMODnet): visions and roles of the gateway to marine data in Europe. Front Mar Sci 6:313. https://doi.org/10.3389/fmars.2019.00313

Pérez B, Brouwer R, Beckers J, Paradis D, Balseiro C, Lyons K, Cure M, Sotillo MG, Hackett B, Verlaan M, Alvarez-Fanjul E (2012) ENSURF: multi-model sea level forecast implementation and validation results for the IBIROOS and Western Mediterranean regions. Ocean Sci 8:211–226. https://doi.org/10.5194/os-8-211-2012

Rider P, Stel JH (2003) An introduction to the global monitoring for environment and security (GMES) initiative. Oceanogr Ser 69:655–666

Rose KA, Icarus Allen J, Artioli Y, Barange M, Balckford J, Carlotti F, Cropp R, Daewel U, Edwards K, Flynn K, Hill SL, Ris Lambers H, Huse G, Mackinson S, Megrey B, Moll A, Rivkin R, Salihoglu B, Schrum C, Shannon L, Shin Y, Smith L, Smith C, Soldoro C, St John M, Zhou M (2010) End-to-end models for the analysis of marine ecosystems: challenges, issues, and next steps. Mar Coast Fish Dyn Manage Ecosyst Sci 2:115–130. https://doi.org/10.1577/C09-059

Rullkötter J (2005) BioGeoChemie des watts, EINBLICKEN r. 4 1 (Frühjahr)

Ryan W, Pitman W (1998) Noah's flood: the new scientific discoveries about the event that0 changed history. Touchstone, New York, p 249

Ryan W, Pitman W, Major C, Shimkus K, Moskalenko V, Jones GA, Dimitrov P, Goriir N, Sakin M, Yüce H (1997) An abrupt drowning of the Black Sea shelf (PDF). Mar Geol 138(1–2):119–126

Sandery PA, Sakov P (2017) Ocean forecasting of mesoscale features can deteriorate by increasing model resolution towards the submesoscale. Nature Commun 8:1–8

Soffientino B, Pilson MEQ (2005) The Bosporus Strait: a special place in the history of oceanography. Oceanography 18(2):16–23. https://doi.org/10.5670/oceanog.2005.38

Stanev EV (2005) Understanding Black Sea dynamics: overview of recent numerical modeling. Oceanography 18:56–75

Stanev E (2022) A sea of people and numbers. St. Kliment Ohridski University Press, 744 pp (In Bulgarian: Станев, Е. „Море от хора и числа", Издателство на Софийския университет Св. Кл. Охридски)

Stanev EV, Flemming BW, Bartholomä A, Staneva JV, Wolff J-O (2007) Vertical circulation in shallow tidal inlets and back-barrier basins. Cont Shelf Res 27:798–831

Stanev EV, Floeser G, Wolff J-O (2003a) First- and higher-order dynamical controls on water exchanges between tidal basins and the open ocean. A case study for the East Frisian Wadden Sea. Ocean Dyn 53:146–165

Stanev E, Schulz-Stellenfleth J (2014) Methods of Data Assimilation. Die Küste, 81:133–152

Stanev EV, Wolff J-O, Brink-Spalink G (2006) On the sensitivity of sedimentary system in the East Frisian Wadden Sea to sea level rise and magnitude of wind waves. Ocean Dyn 56(3–4):266–283. ISSN 1616-7341

Stanev EV, Wolff J-O, Burchard H, Bolding K, Flöser G (2003b) On the circulation in the East Frisian Wadden Sea: numerical modeling and data analysis. Ocean Dyn 53:27–51

Stanev EV, Xi L (2013) European Semi-enclosed Seas: Basic Physical Processes and Their Numerical Modeling. In: Soomere T, Quak E (ed) Preventive Methods for Coastal Protection. Springer, Heidelberg. https://doi.org/10.1007/978-3-319-00440-2_5

Stanev E, Xi L, Grashorn S (2015a) Physical processes in the transition zone between North Sea and Baltic Sea. Numerical simulations and observations. Ocean Modeling 93:56–74

Stanev EV, Ziemer F, Schulz-Stellenfleth J, Seemann J, Staneva J (2015b) Blending surface currents from HF-radar observations and numerical modeling: tidal hindcasts and forecasts. J Atmos Oceanic Tech 32:256–281

Staneva J, Stanev E, Wolff J, Badewien T, Reuter R, Flemming B, Bartholomä A, Bolding K (2009) Hydrodynamics and sediment dynamics in the German Bight. A focus on observations and numerical modeling in the East Frisian Wadden Sea. Cont Shelf Res 29(1):302–319. https://doi.org/10.1016/j.csr.2008.01.006

Tintore J et al (2013) SOCIB: the Balearic Islands coastal ocean observing and forecasting system responding to science, technology and society needs. Marine Technol Soc J 47(1):101–117

Tonani M, Pinardi N, Dobricic S, Pujol I, Fratianni C (2008) A high-resolution free-surface model of the Mediterranean Sea. Ocean Sci 4:1–14

Vieux B (2008) Distributed hydrologic modeling. In: Shekhar S, Xiong H (eds) Encyclopedia of GIS. Springer, Boston. https://doi.org/10.1007/978-0-387-35973-1_317

Waska H, Brumsack H-J, Massmann G, Koschinsky A, Schnetger B, Simon H, Dittmar T (2019) Inorganic and organic iron and copper species of the subterranean estuary: origins and fate. Geochim Cosmochim Acta 259:211–232

New Observing and Modeling Systems and New Scientific Applications

Argo floats revolutionize ocean research. They provide a valuable information about the composition of water masses and complement another set of valuable data, namely data from satellite altimetry. The principles of the work of Argo floats are presented. More details are provided about BioArgo, which made it possible to carry out continuous biogeochemical measurements in the ocean interior over large areas and long periods of time. Along with the evolution of technology and practices, the steps taken by the European research institutes, which laid the foundations for the development of the Euro-Argo research infrastructure, are described. It is demonstrated that Argo floats allowed for the development of a new view on the physical, chemical, and biological oceanography of the Black Sea and contributed to the revision of some of the key concepts dominating in the past.

The question of how to optimally use the measurement data to get the most complete picture of the oceanic systems is of important theoretical and practical importance. To answer this question, one needs to know the extent to which it is possible to reconstruct the state of the ocean from a limited number of measurements. To this end, mathematical methods are introduced to describe the state of an observed system based on insufficiently accurate and complete observations and to estimate the relative importance of the different sensors operating in the observing network. This issue naturally leads us to the so-called observing system experiments (OSEs). In these experiments, the value of individual data is estimated by including and/or excluding data in an ocean data assimilation system. Observing system simulation experiments (OSSEs) use synthetic observational data as a proxy for real observations. One useful practical outcome from this research was the estimation of a number of floats operating simultaneously and representing in an optimal way the physical state of the Black Sea. Research in this field is further extended by using artificial intelligence for the reconstruction of the spatial (basin-wide) and temporal changes of the sea level in the entire North Sea from limited information measured by tide gauge stations operating along its shores.

Reliable monitoring and forecasting of coastal zones necessitate a deeper understanding of the dominant processes and development of adequate computational

E. Stanev, *Trajectories in Oceanography*, Springer Oceanography,
https://doi.org/10.1007/978-3-031-33720-8_4

tools. Unstructured grid modeling appears optimal for the coastal zone, and the semi-implicit cross-scale hydroscience integrated system model (SCHISM) is presented in several estuarine applications. The performance of this model is also illustrated on the example of three straits, Sound,[1] Great Belt, and Little Belt, which connect the North Sea and the Baltic Sea. Another example is the cascade including the Azov Sea, the Black Sea, the Marmara Sea, and the Aegean Sea connected by three narrow straits (the Kerch, the Bosporus, and the Dardanelles). SCHISM appears to be a perfect tool to solve some practical problems, e.g., the sediment erosion and distribution influenced by wind parks, or the effects of the morphological evolution in the Wadden Sea. Applications of numerical modeling to studying marine litter spreading in the Black Sea and the North Sea are also included in this part, along with the numerical modeling of light climate in the North Sea affected by sediment.

1 Argo, Another Revolution in Oceanography

1.1 Profiling Floats

In the 1950s, the USA developed floats, which follow currents at a certain depth, using the principle of neutral buoyancy. This type of floats was proposed by the British oceanographer John Swallow. The idea is that a float immersed in seawater remains stable at a given depth, where its density is equal to that of seawater, if its compressibility is less than that of water. Suppose, for some reason, the float moves away from its equilibrium depth (pressure) to a level of greater pressure. It will not be able to contract as much as the surrounding water contracts and will end up lighter. According to a principle of Archimedes, the float will rise back to the point where its weight equals the weight of the displaced water. If it passes this depth of neutral buoyancy, as it moves upwards, it will appear denser and will sink back. Thus, the float cannot "escape" from the level at which it is prescribed to operate. If we want the float to return to the surface, we will have to make it less dense. This is achieved by changing its volume. For this reason, there is a rubber bladder at the bottom of the float that can reduce its volume to make the float sink or inflate it to make it rise (Fig. 1). Upon reaching the surface of the ocean, its location can be fixed by satellite. By knowing the location of two successive surfacings, the current velocity at the depth at which the float is programmed to operate can be approximately determined.

At the very beginning of the twenty-first century, it was realized that while surfacing (Fig. 2), floats could also measure the temperature and salinity of water (Davis et al. 2001). Soon, the idea of real-time global ocean observations using profiling floats was born (Roemmich et al. 2001). The name Argo (Array for Real-time Geostrophic Oceanography) was given to this global observational platform to emphasize the strong complementary relationship of the global fleet of profiling floats

[1] Øresund or Öresund is commonly known in English as the Sound.

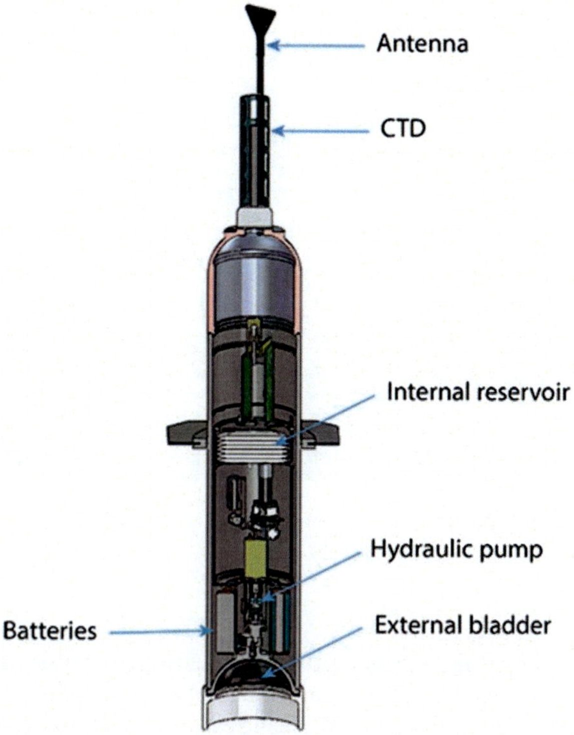

Fig. 1 To the explanation of what an Argo float is. Credit to Michael McClune and Argo Program at Scripps Institution of Oceanography

to the mission of the Jason satellites (Fig. 37 in chapter "Further into the Ocean"). Jason is known to be the leader of the Argonauts.[2]

The sea level provides information about the distribution of water masses. This level is lower in areas of denser water and higher in areas of less dense water (Fig. 15 in chapter "From regional to Coastal Oceanography" and the respective text in Sect. 5 in the same chapter). The altimeters on board the satellites used to determine the position of the sea level guarantee a high accuracy of sea level observations (see Sect. 18 in chapter "Further into the Ocean"). Unfortunately, this precise information about the ocean is incomplete because sea level can have the same position for waters with different vertical stratification. Argo provided the opportunity to overcome this problem. The global fleet of profiling floats measures depending on how the individual instruments are programmed (e.g., a profiling every 10 days or different resolution in the vertical direction, Fig. 2), temperature, and salinity down to 2000 m in all oceans.[3] P.-Y. Le Traon, who was part of the global team that developed the Argo program, says that this extraordinary achievement represents a second revolution in oceanography (Le Traon 2013).

[2] The Argonauts were a group of heroes in Greek mythology who accompanied Jason on his quest for the Golden Fleece to Colchis. Their name comes from their ship, the Argo.

[3] Recently, floats which can work much deeper than 2000 m have also been developed.

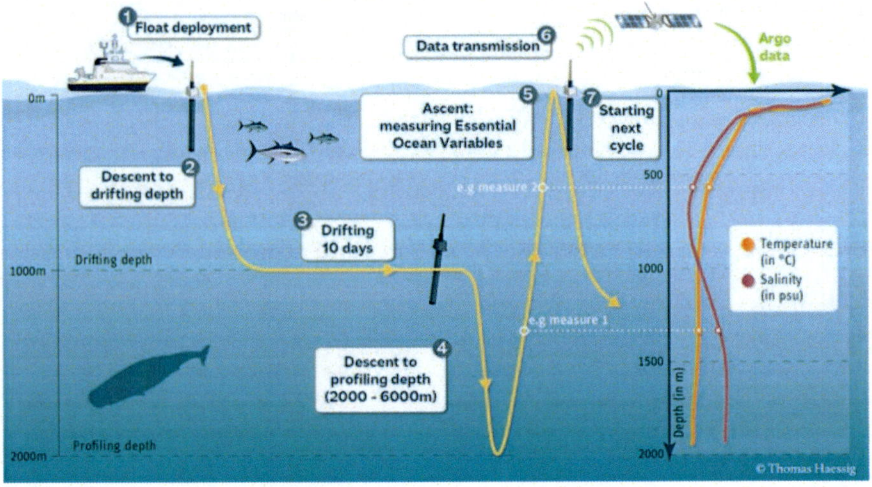

Fig. 2 Illustration of the regime of work of profiling floats. Credit to Argo Program at Scripps Institution of Oceanography

Today, about 4000 Argo floats are in operation in the global oceans simultaneously (Fig. 3), launched by 30 countries. Analysis of the data quickly revealed their uniqueness, such as their role in elucidating processes in areas of deep convection, improving estimates of ocean heat content, and changes in salinity that suggest amplification of the Earth's global hydrological cycle. Many of the studies made joint use of Argo and altimeter data. This was the best demonstration that understanding ocean processes can best be achieved by exploiting the strengths of both observing systems. In the following years, the basis of European operational oceanography was laid on the assimilation of altimetry and Argo data.

1.2 Euro-Argo

Pierre-Yves Le Traon took the initiative, together with colleagues from several European research institutes, to lay the foundations for the development of the Euro-Argo research infrastructure. This infrastructure is part of the ESFRI roadmap and aimed at organizing and pooling European contributions to global Argo.[4] The initial goal of Euro-Argo was to make the European contribution approximately a quarter of the global dataset, which meant Europe contributing around 250 floats per year. The specific European interest required more intensive measurements in the northern seas, the Mediterranean, and the Black Sea. I joined this initiative as a representative of Black Sea science.

[4] http://www.euro-argo.eu, last accessed on 03.02.2023.

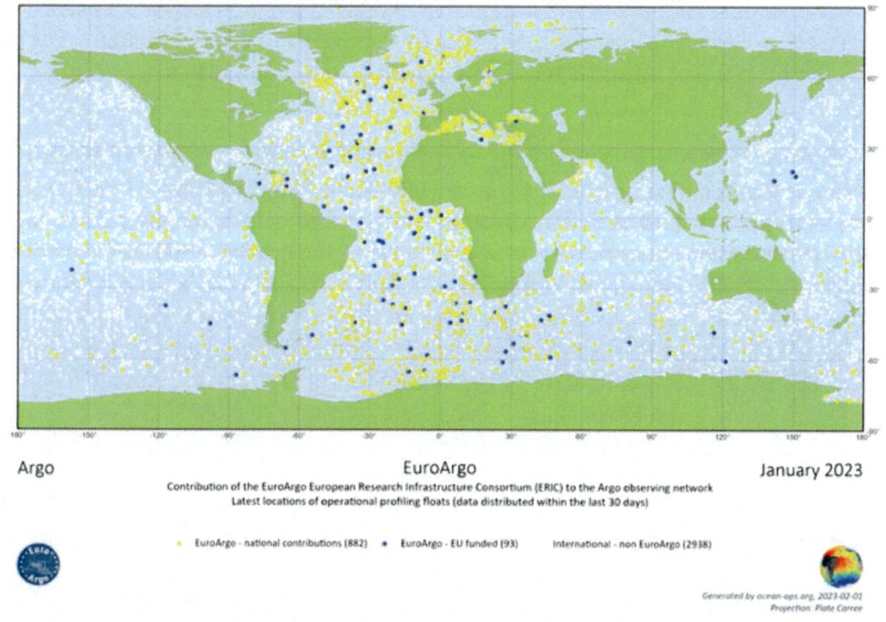

Fig. 3 Today, the ocean is densely seeded with Argo floats. The map shows their location in January 2023. Yellow, blue, and white dots and the respective numbers (in brackets) correspond to different float missions and the number of floats for each of them (see legend). Credit to Euro-Argo ERIC, European contribution to the Argo program, https://www.euro-argo.eu/, last accessed on 03.02.2023

Euro-Argo on the Road to ERIC

There were not only scientific interests behind the Euro-Argo initiative, but also the needs of the GMES/Copernicus[5] initiative and more specifically of operational oceanography. Euro-Argo started in January 2008 as a European project involving 25 organizations from 12 countries. Agreements were planned to be concluded at ministerial level, ensuring the long-term sustainability of the measurement process. Euro-Argo turned out to be the first European legal entity set up to develop the global ocean observing system. It is now a long-term European organization based in France. However, to get there, there was hard work to be done by dozens of scientists. This move to consolidating national activities was also financially supported by the EC in several international projects. The first project, between 1 January 2008 and 30 June 2011, under the name "Global Ocean Observing Infrastructure (Euro-Argo[6])", was the preparatory phase needed to ensure that by 2010, Europe would be able to operate a fleet of 800 floats simultaneously. An agreement had to be reached between Member States and other funding agencies for the long-term (more than 10 years) operation of Euro-Argo. In order to reach such an agreement, it was necessary to work both in the areas of technology and data management and delivery systems, as well as to organize the coordination of national contributions. The usefulness of the infrastructure for Europe had to be demonstrated.

University of Sofia was the only partner from Bulgaria. Its main task was to promote the new instrumentation among the Black Sea community. I proposed a cooperation to the Director of the Institute of Oceanology in Varna, Atanas Palazov. His institute had a research vessel, which was very important for the deployment of the floats. The deployment should be done in the open sea, because these devices are designed to work in deep water. We organized a workshop in Varna on 7 December 2009, which was attended, in addition to the 17 participants from the Black Sea countries, by P.-Y. Le Traon and Pierre-Marie Poulain, the latter from the National Institute of Oceanography and Experimental Geophysics (Istituto Nazionale di Oceanografia e di Geofisica Sperimentale, OGS), Trieste, Italy (Figs. 4 and 5). Pierre-Marie had to train the participants on the technical issues.

IFREMER provided one float, and the day after the workshop we set off with the vessel "Akademik" to deploy it. Violeta Slabakova from IO-BAS was also with us. She would play an important role later by deploying the floats. Early, the next morning we arrived at the chosen location. The float was of the PROVOR type.[7] On 8 December 2009, it made its way to the Black Sea, which lasted until 22 February 2013.

The project "Strengthening the International Dimension of the Euro-Argo Research Infrastructure (SIDERI)",[8] which started on 1 December 2011 and ended on 30 November 2013, was a natural extension of Euro-Argo and had as the main task to form the future ERIC Euro-Argo. The aim of SIDERI was to strengthen the links and integration of the Euro-Argo ERIC into the international ocean observing strategy. It was planned to initiate work on the development of oxygen and other BGC sensors that could be mounted on floats. Work was also needed on legal and political issues, because the trajectory of the floats could not be managed, and they might enter the waters of countries that did not want these measurements. Emphasis was also placed on organizing scientific and thematic (regional) workshops open to international partners. Soon after the completion of SIDERI, in May 2014, Euro-Argo acquired the status of ERIC.

In the initial years, I was a member of the Euro-Argo Governing Board, although Bulgaria was not yet a member. One major problem in joining was the high membership fee. My lobbying to the Bulgarian Ministry of Education, Youth and Science remained fruitless. Atanas Palazov was also seriously committed to this, but for a

[5] In December 2012, GMES was renamed Copernicus.

[6] https://cordis.europa.eu/project/id/211597/de, last accessed on 03.02.2023.

[7] https://nke-instrumentation.com/produit/provor-I/, last accessed on 03.02.2023.

[8] https://cordis.europa.eu/project/id/284391/de, last accessed on 03.02.2023.

Fig. 4 Author (on the right) in conversation with Pierre-Yves LeTraon after the workshop in Varna. Courtesy of L. Pashova

long time, there was no positive result. It was time to develop visible national activities. The answer came from BulArgo, a project funded by the Bulgarian National Science Fund.

BulArgo and MASRI

BulArgo aimed to create a national research infrastructure within the Euro-Argo activities. The partners of this project were the Bulgarian Institute of Oceanology, the National Institute of Meteorology and Hydrology and the University of Sofia "St. Kliment Ohridski". The main objective of BulArgo was to commission five floats during the three years of the project. Work was also to be done on validation and data assimilation into numerical models and their use for climate research. Because of my remoteness from Sofia and because E. Peneva was actively involved in Euro-Argo activities, she took over the main work at BulArgo.

Fig. 5 Pierre-Marie Poulain explains how the Argo float works. At the bottom in a white jumpsuit is A. Palazov. Courtesy of P. M. Poulain

In March 2011, three floats purchased by the BulArgo project (#6900803, #6900804 and #6900805) were deployed from aboard the research vessel "Akademik". In 2012, four Italian floats (#6901959, #6901960, #6901961, and #6901962) were deployed from the same ship. They were programmed to perform sampling between the surface and 1500 m every 5 days.

Incremental efforts and the involvement of more and more stakeholders led to the establishment of the "Infrastructure for Sustainable Development of Marine Research, Including the Participation of Bulgaria in the European Infrastructure Euro-Argo (MASRI)".[9] This is a project of the National Roadmap for Science Infrastructure (2017–2023), in which eight national institutions participate.

BulArgo and MASRI demonstrated the determination of Bulgaria to be involved in the operationalization of Argo practices in the Black Sea, and in 2018, Bulgaria became a regular member of the Euro-Argo ERIC with the obligation to ensure the deployment of at least three Argo floats per year. Meanwhile, I was slowly releasing myself from my organizational and administrative commitments, as I was more attracted by the analyzes and interpretation of novelties encoded in the data collected by the floats.

[9] http://masri.io-bas.bg/bulargo/, last accessed on 03.02.2023.

2 BGC-Argo and HYPOX

The ocean absorbs up to 30% of the carbon dioxide emitions, which is roughly equivalent to the amount taken up by the land biosphere. However, the price for ocean uptake of carbon is ocean oxygen depletion and ocean acidification. It has reached impressive dimensions and is considered nowadays one of the basic environmental stressors. The paucity of BGC measurements in the ocean is today the main reason for our limited understanding of the processes governing the ocean carbon cycle. Until recently, reliance has been mostly on shipboard observations and a small number of continuous records in limited areas. While satellites provide a wealth of information on some biological substances at the ocean surface, for example chlorophyll,[10] which can be seen as a proxy for phytoplankton, what we know about the ocean interior is very little. Poor spatio-temporal data coverage is a major problem that can be overcome by using floats equipped with appropriate sensors. The Biogeochemical-Argo (BGC-Argo[11]) floats proved to be a natural extension of Argo in the field of ocean biogeochemistry.

Which BGC parameters of the marine environment do we want to monitor? A simple answer would be: all of them. However, that is impossible now, and perhaps far into the future. The choice of the BGC-Argo program is the result of a trade-off between (1) what we are most interested in studying, (2) the optimal operation of the Argo fleet and their management, and (3) the ability of the available sensors to provide accurate, long-term, and reliable measurements. The set of such parameters around which scientists have consolidated includes the following key variables: oxygen concentration, nitrate concentration, pH,[12] chlorophyll concentration, suspended particles, and penetrating light. This set of parameters partially covers the BGC part of the so-called Essential Ocean Variables (EOV,[13] see also Sloyan et al. 2019; Capet et al. 2020; Sotillo and Wilkin 2022). The EOVs, in combination, allow the BGC cycles of carbon, oxygen, nitrogen, and biomass to be described.

Nitrate is an essential nutrient for phytoplankton, regulating its growth. Under conditions of global climate warming, the input of nitrate from deep waters may decrease, followed by a decrease in primary production in the ocean. In the Black Sea and many other places in the ocean, episodic "injections" of nitrate into surface waters are particularly important for phytoplankton development. The relationship between nitrate and oxygen is also important. A decreased amount of oxygen in the ocean can lead to an increased nitrate loss and altered life cycles.

---

[10] Chlorophyll (from the Old Greek χλωρός chlōrós - light green, fresh, and φύλλον phýllon - leaf) is formed during photosynthesis. Plants get their green color from chlorophyll. In the ocean, it gives a good idea of the concentration of phytoplankton. *Chlorophyll a* (Chl-a) which is a specific form of chlorophyll used in oxygenic photosynthesis is found primarily in the so-called cyanobacteria.

[11] https://biogeochemical-argo.org/, last accessed on 03.02.2023.

[12] In chemistry, pH is a number used to determine the acidity of an aqueous solution.

[13] One set of such parameters looks like this: oxygen, nutrients, inorganic carbon, transient tracers, particulate matter, nitrous oxide, stable carbon isotopes, dissolved organic carbon, see also https://www.goosocean.org/index.php?option=com_content&view=article&layout=edit&id=283&Itemid=441, last accessed on 03.02.2023.

Knowing the spatial distribution of chlorophyll and its changes over time is of paramount importance because of the key role of phytoplankton in the global cycling and transformation of biological material. The use of miniaturized sensors mounted on BGC-Argo floats now makes it possible to systematically collect vertical profiles of chlorophyll over large parts of the ocean.

Particles suspended in seawater are another important component of ocean biogeochemistry. These include microscopic algae and predators as well as bacteria that decompose organic material. The so-called *biological pump*, a key process for sequestering atmospheric carbon dioxide and sedimenting it to the bottom through the export of sinking particles, is the primary regulator of the planet's climate. Therefore, a good knowledge of the concentration of suspended particles is also required.

BGC Measurements

For the determination of the dissolved oxygen concentration, so-called oxygen optods are used. They use an oxygen-sensitive substance (luminophore) which, when excited with light of a certain wavelength (e.g., blue-green light), emits light of a different wavelength (e.g., red light). The phase shift of light associated with luminescence is reduced in the presence of oxygen, i.e., it is a measure of its concentration.

The principle of measuring chlorophyll is similar. Some of the photons[14] absorbed by a chlorophyll molecule in the blue part of the spectrum are emitted (fluoresced) as less energetic photons in the red part of the spectrum. The relationship between fluorescence and chlorophyll concentration depends on the physiological characteristics of the fluorescing substance, i.e., fluorescence is not a perfect proxy for chlorophyll, but the errors are not large, allowing a huge amount of good quality data to be collected.

Nitrate is determined by light analysis in the deep ultraviolet spectral range. This uses the fact that the nitrate ion absorbs ultraviolet (UV) light with a characteristic spectrum that has a maximum at 210 nm.

The concentration of suspended particles is calculated from the backscattered light emitted from a source mounted on the profiling float. Sensors measure the particle backscattering coefficient, bbp. This coefficient provides a proxy for the concentration of suspended particles. The backscattering, or light attenuation mode, allows us to estimate not only the concentration but also the characteristic particle size.

Due to the similar working principles of the sensors described above, we refer to these as bio-optical measurements.

In 2008, Antje Boetius, then from the Max Planck Institute for Marine Microbiology in Bremen, contacted me. "I am organizing a consortium for a project which will deal with the problems of low-oxic or anoxic lakes and oceans. I know about

[14] Light quanta or light particles.

your work in the Black Sea, would you like to join"? Antje Boetius graduated in biology at the University of Hamburg and continued with her studies in biological oceanography at the Scripps Institution of Oceanography. At the Max Planck Institute for Marine Microbiology in Bremen, she studied underwater gas sources and the microbiology of the methane cycle in the sea. Since November 2017, she has been director of the Alfred Wegener Institute in Bremerhaven.

Just a few names of scientists from the consortium alongside interesting scientific problems were enough for me to agree to participate in the project "In situ monitoring of oxygen depletion in hypoxic[15] ecosystems of coastal and open seas and land-locked water bodies", with the short name HYPOX. It started on 1 April 2009 and ended on 31 March 2012. The project was our scientific response to the observation that hypoxic conditions in aquatic ecosystems are increasing in number, duration, and extent due to global warming and eutrophication. In eutrophicated waters, the excess of produced algal biomass is generally not passed along the food chain. Instead, it sinks to the seabed where it is used by oxygen-consuming microorganisms. If oxygen at the bottom decreases significantly, biological systems and chemical conditions begin to change. Ecosystems undergo continuous degradation and eventually become permanently anoxic environments where microorganisms replace all higher trophic life forms. This collapse of biological communities leads to a dramatic decline in the quality of ecosystem functioning and a reduction in biodiversity, affecting fisheries, aquaculture, and tourism. Since the 1960s, the number of *dead zones* in the ocean have doubled every ten years.

Even without anthropogenic-induced eutrophication, climate change contributes to oxygen depletion: Warming seawater leads to oxygen degassing as its dissolution in water decreases with increasing temperature. Higher temperatures also imply greater oxygen consumption by biological systems. Changes in wind and precipitation can lead to changes in ocean stratification, which can reduce the vertical transport of oxygen to deeper waters and the seabed, and hence oxygen depletion.

Understanding Oxygen Depletion: HYPOX

Understanding the global changes associated with oxygen depletion requires a global system that continuously monitors oxygen at high resolution, including providing an assessment of the role of the seafloor in controlling the sensitivity of aquatic systems to and recovery from hypoxia. HYPOX has proposed such a system for European seas and lakes. To monitor oxygen depletion and related processes, regions were selected that differ in their oxygen status or sensitivity to changing hydro meteorological, oceanic, and anthropogenic forcing: open ocean, regions with high sensitivity to global warming (e.g., the Arctic), semi-enclosed basins with permanent anoxia (the Black Sea and Baltic Sea), and smaller but special regions such as fjords, lagoons, and lakes.

[15] Low oxygen conditions.

Continuous and adequate measurements of oxygen and related parameters are only an important first step to investigate the state of the ocean system. The knowledge gained from observations needs to be complemented by efforts to reproduce oxygen dynamics in models. Thus, numerical modeling is a key tool for translating observations into generalized knowledge that can be applied to other ecosystems. Most importantly, it sets the stage for predictive work that intelligently "extrapolates" current observations into the future. Such generalizations and predictive capabilities are essential for exploring the effects of future climate change and developing scenarios for ocean eutrophication. Modeling can also support decisions on adequate countermeasures to be taken in the event that ecosystem conditions deteriorate. The strategy of HYPOX was to complement oxygen monitoring with numerical modeling. This was my task in the work package "Modeling and prediction of short- and long-term factors affecting oxygen depletion in different systems", which I led together with the renowned Dutch biogeochemist Jack Middelburg.

My tasks, although important for the project, were in the realm of the familiar. My scientific interest was reaching farther than this. Even before HYPOX began, it was known that Argo floats equipped with oxygen sensors were providing valuable ocean data. With A. Boetius, we purchased two floats that would measure oxygen. At that time, they costed about 20 thousand dollars each. If the deployment was not successful, or the sensors malfunctioned, which had happened in the early years, it would have been money wasted. We took the risk and the risk paid off. The two floats were launched into the northern Black Sea on 7 May 2010 during an expedition by the German research vessel "Maria Merian". We received for the first time physical and oxygen data from continuous measurements in the Black Sea for more than a year and a half. The data turned out to be of very high quality.

My Chinese Ph.D. student Yunchang He was in charge of the data analysis. We discovered vigorous mesoscale events, "pushing" locally anoxic waters up to about 70 m below the sea surface. Conversely, ventilation of anoxic waters in the coastal zone was reaching depths of about 150–200 m. Apparently, mesoscale dynamics played the dominant role in the temporal and spatial evolution of the layer separating the oxygenated waters from the anoxic waters. Statistical analysis of the oxygen data allowed comparison of the amplitudes of variability with different periods. This enabled to verify the validity of the previous analyzes of climate trends over the last 50 years, which were derived from episodic shipboard measurements (i.e., with poor temporal and spatial sampling, compare Fig. 6 with Fig. 9 in chapter "Further into the Ocean"). The study of Stanev et al. (2013) demonstrated the need for continuous basin-wide sampling and gave impetus for later work aiming at understanding the dynamics of the zone separating oxic and anoxic waters. Evgeny Yakushev from the Norwegian Institute for Water Research (NIVA) and J. Staneva joined the modeling work. Evgeny had developed a Black Sea-specific model capable of reproducing the complex interactions of substances in aerobic and anaerobic environments. The 1D RedOx Layer Model (ROLM, Yakushev et al. 2007) included 24 parameters defining the chemical state of the sea, which described well the transformation of substances in the zone separating oxic from anoxic waters. ROLM was coupled with GETM (Stanev et al. 2014b). With this study, a step was taken toward linking the numerical models and observations from oxygen floats. It appeared that intense BGC reactions occured, which significantly distinguish the oxygen dynamics compared to those governing conservative tracers, such as temperature and salinity.

The study of Stanev et al. (2014 b) was not only interesting because of its BGC approach, but also because the analysis of data produced by the numerical model showed that the dynamics in the Black Sea undergo distinct seasonal changes. It was already known that the intensity of the horizontal circulation almost doubles in winter as a result of more intense winds (Stanev et al. 2000; Stanev 2005). The spatial characteristics of circulation from season to season were also very different. In winter, currents are structured in a single cyclonic circulation gyre covering the whole basin, whereas during the warmer part of the year, the eddies are more pronounced, and the general circulation is less structured (Fig. 18 in chapter " From regional to Coastal Oceanography"). Hence, the transition between summer and winter circulation is characterized by the changing rates of the gyre versus eddy motions (Stanev et al. 2014 b). This in turn is related to the seasonal variability of stratification in the upper layers of the sea (Staneva et al. 2001).

The ventilation of anoxic layers is regulated by two types of circulation that are very different in the cold and warm half of the year (Stanev et al. 2014b). In winter, the intense mixing processes are shaped by the dynamics of the cyclonic circulation gyre, covering the whole sea (Fig. 6 in chapter "Further into the Ocean" and Fig. 18 a in chapter "From regional to Coastal Oceanography"). This leads to greater dooming of the isopycnic surfaces, and they reach depths closer to sea surface, where the flux from the atmosphere tends to increase the oxygen content (top panel in Fig. 7). In the warm part of the year (bottom panel of Fig. 7), the dooming is not so pronounced, and the penetration of oxygen along isopycnic surfaces occurs in the area of coastal anticyclones (downward motion). Therefore, in summer, when the circulation weakens, water exchange between open and coastal regions is governed by ocean eddies,

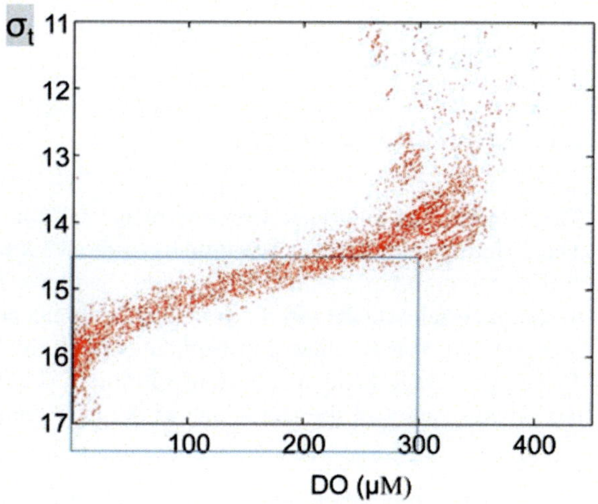

Fig. 6 Disolved oxygen (DO) versus density diagram from float WMO#7900466. From Stanev et al. (2013). Compare the abundance of data from multiple Argo-soundings with the limited number of data from ship measurements (blue symbols in Fig. 9 in the chapter "Further into the Ocean"). The rectangle shows the ranges of the graph in the cited figure. The strong gradient of oxygen depicts the transition from oxic to anoxic conditions

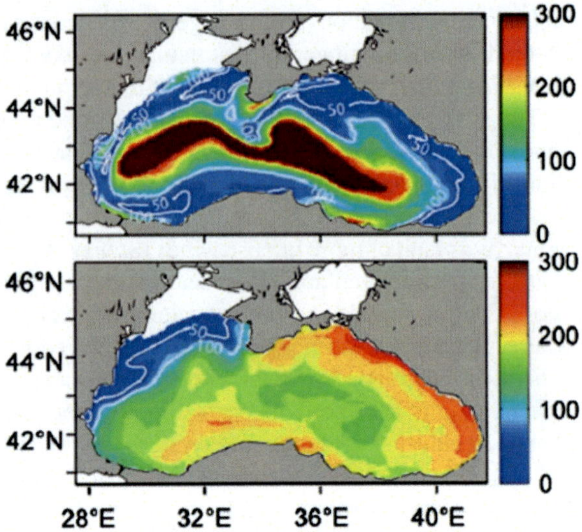

Fig. 7 Simulated concentrations of oxygen in μMol along density surface $\sigma_t = 14.5$ The panels on the top and bottom show oxygen concentrations on 15 February 2010 and 15 July 2010, respectively. Oxygen isolines of 50 and 100 μMol (white lines) are plotted to better illustrate the horizontal patterns. From Stanev et al. (2014b)

which largely impacts the BGC state of the sea. The results from numerical simulations showed that there is an alternating winter–summer pattern of along-isopycnal oxygen flux. Furthermore, the oxygen penetration pathways are characterized by multilayered exchange cells. This research provided evidence that, although atmospheric cooling in winter acts as a major source of cold and oxygen-rich waters, each of these two parameters, temperature, and dissolved oxygen, follows a different evolution. Hence, oxygen variability in the Black Sea is not just a simple response to physical processes and their changes in the surface layers but also undergoes its own evolution under the influence of complex BGC processes.

3 A New Look at the Black Sea

3.1 Physical Properties

The "seeding" of the Black Sea with floats has been increasing since 2013 (Fig. 8). Their lifetime varied from a few months to about six years. By 2020, in just a 15-year period,[16] more than 6000 vertical profiles had been recorded, which is comparable to the material collected in the previous 100-year period of ship measurements. It is impressive that automated instruments did all this! I recall the words of P.-Y. Le Traon about Argo being a second revolution in oceanography. Argo revolutionized the oceanography of the Black Sea as well by giving everyone full access to the

[16] The first four floats in the Black Sea were deployed in the framework of a US-Turkish observation activity in 2005 (Korotaev et al. 2006).

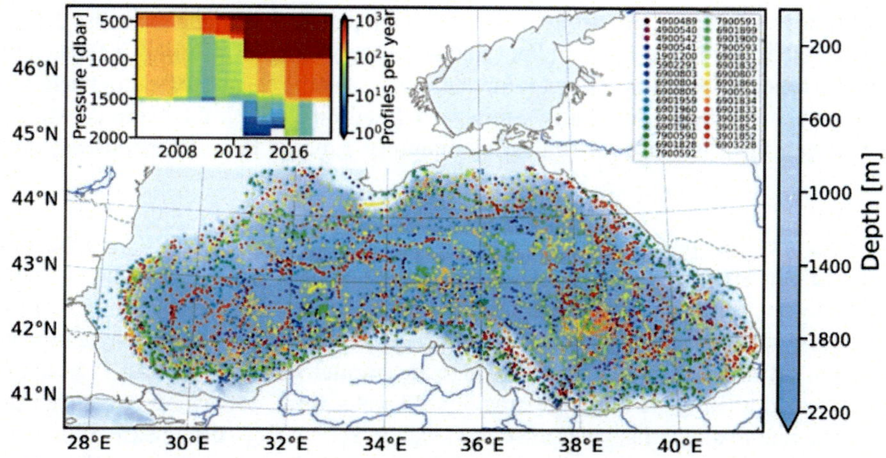

Fig. 8 Argo sampling positions from March 2005 to March 2020. Different colors correspond to individual floats (see the identification numbers in the legend inset in the top-right corner). The inset in the top left corner shows the number of profiles per year as a function of depth. The light blue background colors in the map show the depths. Replotted from Stanev et al. (2021). Courtesy of B. Chtirkova

data. Now, anyone could work on the data, and only the quality of the information extracted or synthesized from it in publications could give scientific priority to one or the other. Today, operators of Argo floats, through satellite communication, can change their mode of operation in real time, for example, to change the discretion of sampling. Sometimes, it is fun for data users to monitor the sounding of each individual vertical profile. You sit in front of a computer, but you see what is going on in the interior of the ocean. Typically, the floats are programmed to spend most of their life at great depths. If they reach shallow water, they stay there until the current brings them back to the deep. I watched anxiously as our first float #1901200 "jumped" the steep continental slope and was long stuck in the shallows opposite the Caucasus coast. Eventually, the current took float #1901200 back into the deep sea, and it continued its work there long after this "adventure".

Argo has helped us revisit the oceanography of the Black Sea. Twenty years ago, the following concepts prevailed: The water temperature in the CIL is lower than 8 °C; the temporal variability of temperature extends only to the lower boundary of the CIL; the waters of the Bosporus do not sink deeper than 600 m; the water in the deep layers of the Black Sea is almost uniform in temperature and salinity; geothermal fluxes[17] are not important for the Black Sea dynamics. The vast amount of data collected in only 15 years prompted us to test the above concepts.

In 2017, E. Peneva approached me: "I have a student, her name is Boriana Chtirkova and she is very ambitious and wants to work with us. I am thinking of engaging her to analyze Argo data. What would you advise me"? "Let's take a look at

[17] The heat from the interior of the Earth.

what has happened in the last fifteen years", I suggested. Analyzing the entire dataset was a laborious job, and I could not find someone to do it. B. Chtirkova showed up just in time, and she was computer literate. I again got the chance to be more involved in the work in Sofia. In addition, this work fitted nicely with the National Marine Research Infrastructure for Black Sea monitoring, based on autonomous profiling floats.

Earlier analyzes of relatively short Argo records gave first indications that a warming of the upper layer of the sea was underway (Stanev et al. 2013). Processing the 15-year datasets revealed a CIL warming trend of ~ 0.05 °C/yr, which is more than twice as high as the trend in previous decades (Fig. 9). The temperature of the CIL was approaching that of deeper waters (~ 9 °C), indicating that this layer is on its way to extinction. This evolution is due to the warmer winters of the last 15 years. In this period, there were only three major cold-water formation events, which appear insufficient to allow this layer to cool enough to keep it from disappearing. In addition to the observations from the floats, I also received direct impressions of one such event. On 17 and 18 January 2012, the first meeting of the PERSEUS project took place in Istanbul. The hosts organized an evening sailing along the Bosporus for us. We stopped for a while on the Asian coast, which was covered with snow. Before, I thought Istanbul was a warm city in winter. Actually, the situation was extreme, and as we later found out from the analysis of the Argo profiles, the intermediate layer had "filled up" with cold water (Fig. 10). The next such cooling event was not until 2017 and was not as strong. The disappearance of the cold layer seemed imminent (Stanev et al. 2019b).

Climate change in the Black Sea represents a precursor to what is expected to happen in the World Ocean. Since 2010, rhythmic fluctuations in salinity emerged. The relative influence of the two competing parameters, temperature and salinity, on the overall physical state of the sea was changing. Over the last 15 years, the role of temperature in the vertical stratification of surface layers has decreased threefold compared to the previous periods (Stanev and Chtirkova 2021). Such changes are

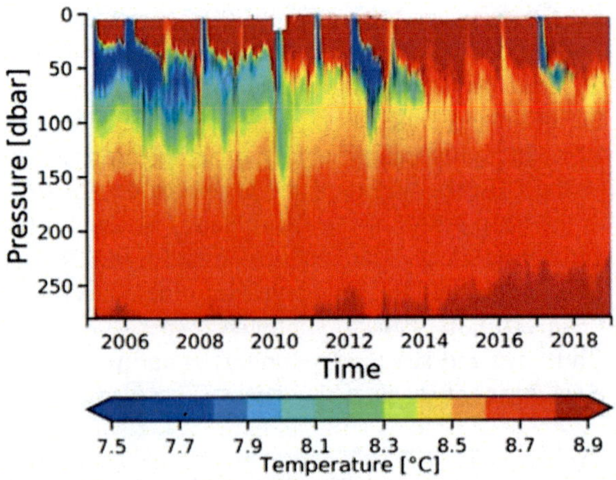

Fig. 9 Time versus depth diagram of temperature. The vertical markers of the x-axis correspond to 1 January of each year. White areas indicate no (or poor) data. All Argo data before 2020 are used. All profiles are plotted in chronological order independent of the position of floats. From Stanev et al. (2019b)

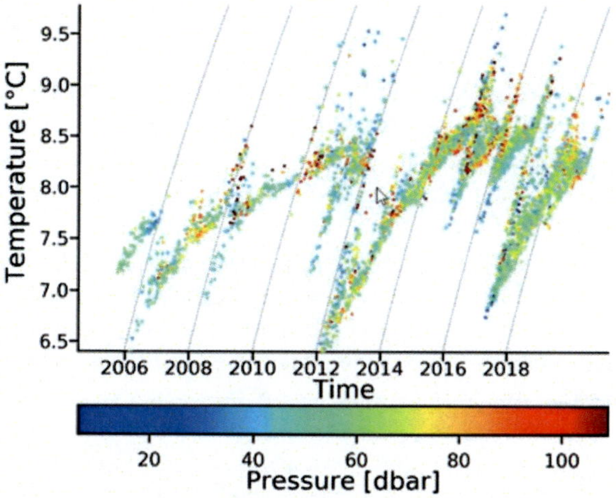

Fig. 10 Temporal evolution of T-S pairs (temperature and salinity) only for those data for which the seawater density is 1014.5 kg m^{-3}. Constant density means that the relative contributions of temperature and salinity to density must compensate each other (warm and salty or cold and less salty). Each individual observation is represented by a dot. The salinity of each successive observation is shifted to the right, this offset being proportional to the elapsed time between observations. In this way, we "unfold" the changes in the T-S pairs over time. The x-axis should therefore be understood as a modified time coordinate. The dashed lines are the lines of constant density. They are repeated offset every second year to illustrate the phase of events with respect to the times shown on the x-axis. The other parameter of the water state, pressure, is shown in colors. From Stanev et al. (2019b)

capable of affecting Black Sea circulation and eddies, as well as ventilation of deep water and mixing pathways. One striking discovery was the appearance of several different mixing pathways of the surface and deep waters that change after each cold-water formation event.

The surprises continued as we looked into the deeper half of the water column (Fig. 11). It is common knowledge that air-sea exchange, river runoff, and astronomical forces provide the main driver for the World Ocean. The role of geothermal fluxes, which are much smaller than heat fluxes at ocean surface, remains largely unknown and is thought to be negligible. However, this is not true for the Black Sea, which is one of the best places in the world to observe the effects of geothermal heating. A so-called bottom convective layer (BCL), approximately 400 m thick, is created in the depth interval between 1700 m and the bottom (Fig. 11c). This layer, which was discovered in 1991 (Murray et al. 1991), is the largest known convective layer in the World Ocean, which exists throughout the basin. It is amazing that no one had questioned why the temperature and salinity profiles were so different in the layers between 700 and 1700 m; compare the almost linear temperature profiles in Fig. 11b with the curved profiles of salinity in Fig. 11e. The explanation of Stanev et al. (2021) took into account the different distribution of heat and salt sources. It was found that this layer acts as a transition zone between the bottom convective layer and the layers above 700 m, where seawater temperature and salinity show significant changes in space and time.

A pair of deep intermediate layers, warm intermediate layer (WIL) and deep cold intermediate layer (DCIL), was observed at about 500 and 700 m depth, respectively (Fig. 11b). The temperature and salinity of these layers have shown strong trends over the last 15 years due to the warmer climate and increasing deep water intrusions from the Marmara Sea. It became clear, that it is not only the surface layers that are subject to strong temporal variations, as was assumed until recently. Even more striking was the first record of traces of the penetration of Marmara Sea waters down to the seafloor (Stanev et al. 2021), which substantially alters the notion that the deep waters are relict, created in the époque when the Black Sea and the Mediterranean Sea re-connected.

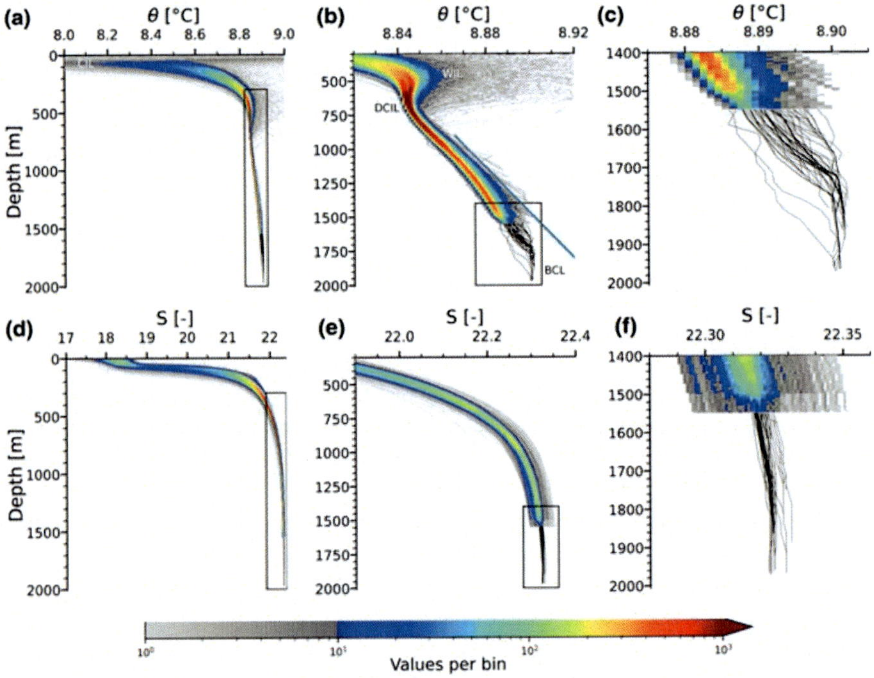

Fig. 11 Potential temperature (panels on top) and salinity (bottom panels). The panels on the left are for the entire water column; the ones in the middle are enlarged presentations below 400 m; the panels on the right are zoomed for the bottom convective layer (below 1400 m). The ranges of these zoomed plots are shown in the respective panels on the left (boxes). Data are presented as probability distributions (the number of data points in the individual coordinate bins). Because fewer observations are available below 1500 m, the individual profiles are shown with black lines. Temperatures lower than 8 °C and higher than 9 °C in the surface layer are excluded in (**a**) to provide a high-resolution plot of the deep layers. CIL, WIL, DCIL and BCL indicate cold intermediate layer, warm intermediate layer, deep cold intermediate layer, and bottom convective layer, respectively. Warm intrusions below 900 m are seen to the right of the blue line in (**b**). From Stanev et al. (2021)

3.2 Biogeochemical Properties

Biogeochemical Processes in the Suboxic Zone

Manganese reactions play a central role in the conversion process of various chemical and biological substances. These reactions can generate particles that sink. In redox reactions,[18] a substance known as a reductant releases electrons and is oxidized. The substance known as the oxidant accepts electrons and is reduced. Manganese acts as a redox shuttle that consumes both sulfide and oxygen (Fig. 12). An upward flux of dissolved Mn(II, III) passes through the suboxic zone into the oxic layer, where it is oxidized to particulate Mn(III, IV), which removes residual oxygen not consumed by organic matter. The particulate manganese oxides, Mn (III, IV), then sink back through the suboxic zone to the anoxic layer where they serve as an oxidant for the sulfide. The oxidized sulfide is converted to elemental sulfur (S°), which is reduced back to hydrogen sulfide by specific anaerobic bacteria. Mn(II, III) is regenerated, which closes the redox shuttle.

Working with Argo floats brought me together with Herve Claustre from the Laboratory of Oceanography in Villefranche-sur-Mer, France. He is one of the leading European scientists involved in the implementation of BGC-Argo in the ocean monitoring (Johnson and Claustre 2016). The study of the hydrochemistry of the Black Sea has become a scientific challenge not only for the two of us, but also for the co-Chairman of the BGC-Argo mission team, Ken Johnson.[19] He was very helpful in interpreting the hydrogen sulfide data (Fig. 13). The authorship of Stanev et al. (2018b) would not be complete without J. Murray, who can be considered the father of modern Black Sea biogeochemistry and Pierre-Marie Poulain, one of the most

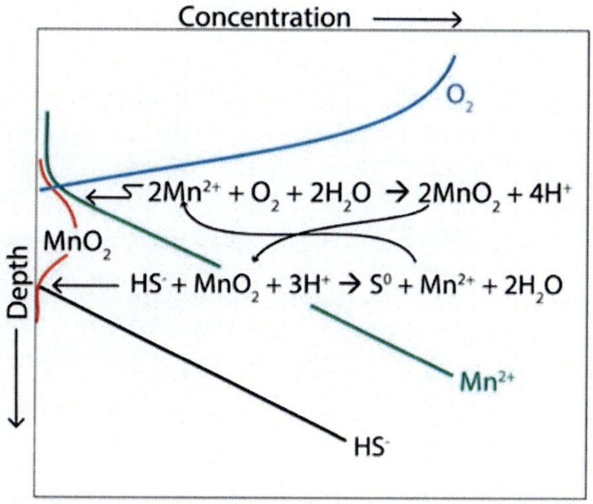

Fig. 12 Chemical reactions in the suboxic zone. For further details, see the original publication of Stanev et al. (2018b), from where the figure originates

[18] Redox (reduction–oxidation) is a type of chemical reaction in which the oxidation states of a substrate are changed.

[19] https://biogeochemical-argo.org/mission-team.phpt, last accessed on 03.02.2023.

active actors in Argo deployment in the Black Sea. It became evident that also some old paradigms of the Black Sea BGC needed substantial revision.

The data from profiling floats showed very clearly that in most of its volume below depths of 100–150 m, the Black Sea is filled with sulfidic deep water. The black and white isolines in Fig. 13c, d (zero concentrations of oxygen and sulfide) make possible to identify the boundaries of the suboxic zone, a layer where the concentrations of oxygen and free sulfide are below the reliable limits for their detection. This is a very thin layer, in which complex transformation reactions take place, characterized by the conversion of substances from soluble to insoluble form (Fig. 12). The deepest temperature intrusions are in the vicinity of the Bosporus Strait (location "B") and when the float enters the eastern basin (location "EB"). In the same areas, one observes a substantial change of vertical profiles of oxygen and

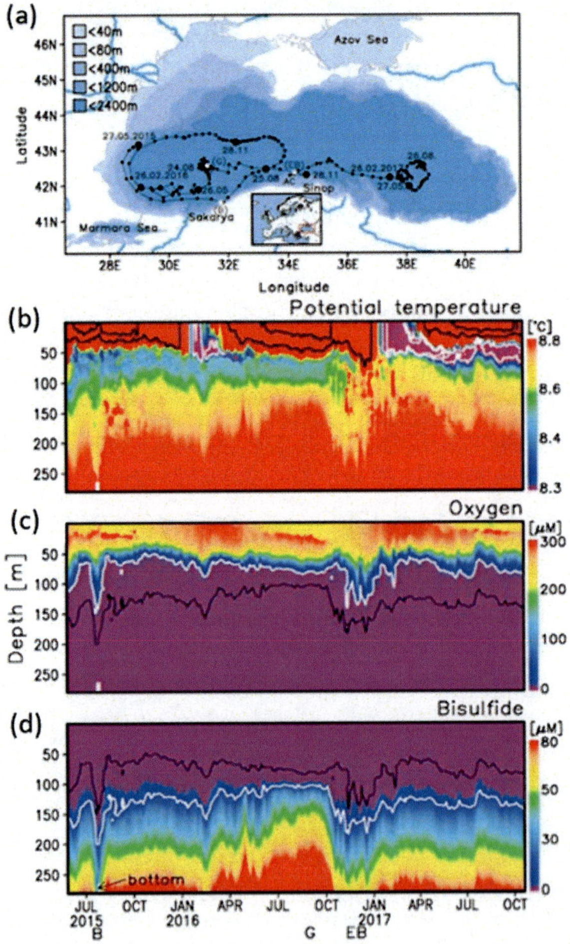

Fig. 13 **a** Trajectory of float WMO#6901866. The circles identify the surfacing position. The white circles (B, G, and EB) show specific positions mentioned below. The timing of observations is given approximately once per 3 months, respectively; the circle-symbol there is bigger. The inset shows the position of studied Black Sea region. **b–d** Vertical and temporal change of temperature, oxygen, and sulfide. The following isolines are plotted: in (**b**), 10 and 20 °C (black) in order to better see the stratification in the seasonal thermocline and 7.5 and 8 °C (white) in order to better see the core of CIL. In **c**, **d** black and white isolines are 0 μM for oxygen and sulfide, respectively. From Stanev et al. (2018b)

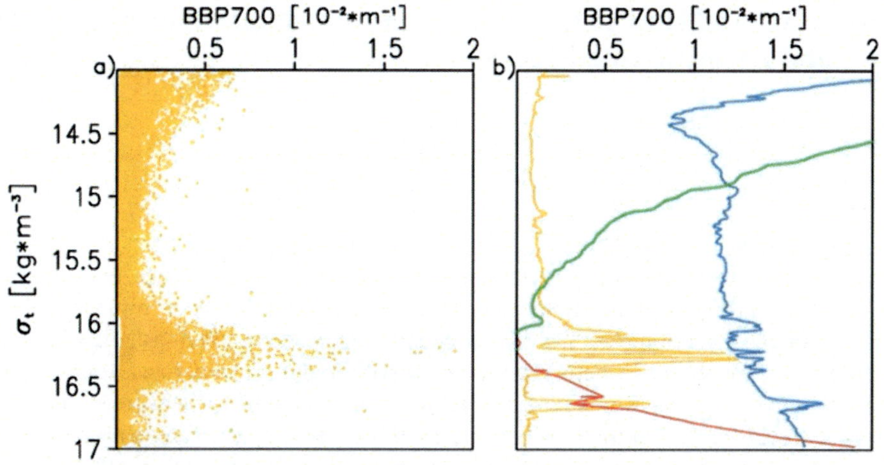

Fig. 14 Vertical profiles in density coordinates shown as **a** scatterplots of bbp at 700 nm (bbp 700), all data, **b** individual profile of temperature (blue), oxygen (green), sulfide (red), and bbp 700 (dark yellow) close to the Bosporus Strait. From Stanev et al. (2017b)

sulfide. Symbol "G" corresponds to a position in the center of Western gyre. There the thickness of suboxic layer reaches its minimum of ~ 20–30 m.

Alalyzes of BGC-Argo data showed a remarkably robust structure of BGC parameters, in contrast to temperature, for example. While temperature anomalies associated with Bosporus water intrusion are traced far from the Bosporus Strait, similar anomalies of BGC parameters disappear at much smaller distances (Stanev et al. 2017b). In explaining the dynamics of the processes in the suboxic zone, the view of the priority role of BGC processes starts to dominate (Fig. 14).

Peculiar intrusions in the deep layers of the Black Sea were reported by Stanev et al. (2017b), when analysis was carried out on the backscattering of light (bbp) by suspended particles. It turned out that in the area around the Bosporus Strait, the particle concentration was about five times higher than elsewhere, which may indicate an abundance of bacteria and possible biological involvement in the precipitation of Mn-containing particles. Notable is that these biological layers appear where thermal intrusions occur; however, the BGC intrusions are much more pronounced (Stanev et al. 2017b). The conclusion is that mixing associated with intrusions has a very strong response in the BGC system.

Biooptics has also made it possible to approximately distinguish particles by their size. It turned out that the relative proportion of larger particles in the anoxic layer increases with depth, which is interpreted as a change of particle composition and/or characteristics. This would suggest that it is the changes in the world of bacteria that BGC-Argo detects. The lower particle concentration during the extremely warm 2016 revealed the response of the BGC system to temperature changes. Work on this topic is currently ongoing.

4 What and How to Observe, or on Numerical Models and Artificial Intelligence

4.1 *Intuition and Artificial Intelligence*

Albert Einstein said: "Imagination is more important than knowledge" or "the only thing really valuable is intuition". You may ask how scientists know where and what to measure in the ocean. If they are talking about the surface of the ocean, one possible answer would be that they have observed the ocean from the shore or from ships, they have noticed something that has caught their attention and decided to clarify through targeted measurements whether this is a regularity, and if so, what rules it obeys. The harder question to answer is how they know where and what to measure in the deep ocean. It is even harder to know how adequately individual measurements represent the ocean system. It may be that some data are more informative than others. Here, the scientific intuition plays a major role. The Cambridge Dictionary defines intuition as: "Ability to understand or know something without needing to think about it or use reason to discover it, or a feeling that shows this ability". The question is whether there is intuition without knowledge. An affirmative answer would concede that intuition is godsend. However, there is another viewpoint. Is intuition not built on the previous experience, or knowledge that has been passed on to us through heredity, or has it been acquired without us noticing? Some scientist develop important theories on the basis of a small amount and perhaps not very qualitative data (e.g., Millikan[20]), whereas others look at perfect data and do not see the important pattern in it. Why? Because the second one lacks the "tool for understanding" and that tool, besides being perhaps innate, has grown and developed over the years.

I will leave the subject of intuition for a moment and talk about the "tool". The flood of information, not only scientific, presents one of the greatest challenges today. As the amount of information increases, the capacity of our tool, the brain, to fully comprehend information decreases. Recently, I have been working on issues of artificial intelligence in my scientific field. Here, again, we see the striking similarity of information processing by the brain and artificial neural networks.

The brain is made of neurons. The neuron (nerve cell) is the basic structural unit of the nervous system. The concept of neurons was developed by the Spanish anatomist Santiago Ramon y Cajal, who won the Nobel Prize in Physiology or Medicine in 1906. A neuron is excitable and processes and transmits information through electrical and chemical signals. Neurons communicate with each other through specialized nodes. The theory of S. Ramon y Cajal is not only the cornerstone of modern neurobiology but also the basis for the development of computer neural networks (Fig. 15). In a neural network, there are layers of input and output neurons. In the former layer, information is input to the network. In the real brain, the information can come from receptors, in other words, the input layer is something like a collection of senses.

[20] John Waller questions the goodness of the experimental data that Millikan used in determining the charge of the electron (Waller 2004).

The signals from the input neurons pass through intermediate (hidden) neural layers and finally reach the output layer. This means that neurons can receive and transmit signals. The interesting question is what happens to the signals that are received and transmitted? Received signals are summed up, and this summation goes through an activation function. Information about the level of activation (excitation) is passed on to other neurons. The connections between neurons have certain weights that determine the importance of the signals transmitted along the circuit. The weights are determined when training the neural network iteratively so that the error decreases the next time the network is activated. Thus, neurons in the hidden layer can respond to different features of the input and ignore others simply by adjusting the weights. Ultimately, training the neural network comes down to remembering events and phenomena from the dataset used in training and the patterns between them. Thus, the trained network can be activated with independent data and reproduce the state of a system corresponding to this independent dataset. The new (independent) dataset may be much smaller than the dataset used in the training process, but it may very well activate the neural network so that it optimally describes the new system state.

Now let's go back to intuition. In light of what has been described above, it can be thought of as the ability of the brain, based on a minimal amount of input, to produce an outcome for which a less well trained brain would need much more information. To me, the intuition of classical oceanographers is based on a combination of previously known information and the analytical abilities of the brain, fueled by a large dose of curiosity.

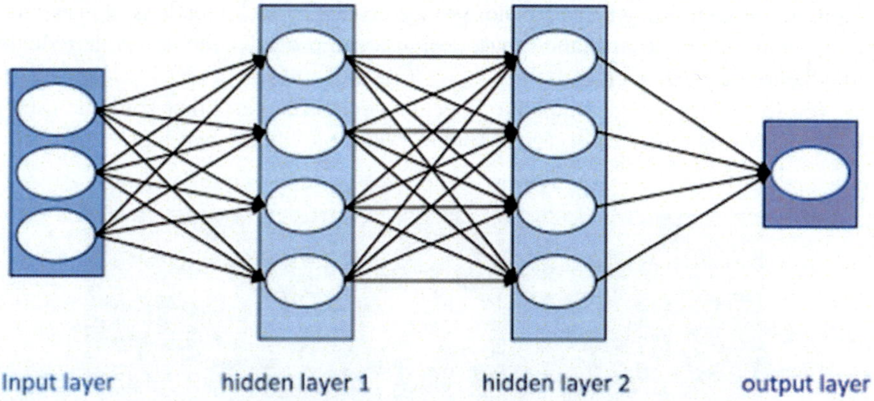

Fig. 15 Idea of how neural networks work. Courtesy of S. Hartmann

4.2 The Ecological Fairways: BalticWay and the Inverse Problems

Tarmo Soomere (Fig. 16) is a mathematician and professor of coastal engineering at the Tallinn Institute of Technology. He is very calm, considerate, sociable, and friendly person. He has been president of the Estonian Academy of Sciences since 2014. "I am putting together a consortium for a project that I would like to submit to the Baltic Organisations Network for Funding Science (BONUS)", he wrote to me in 2008. The project will be called "BalticWay: The potential of currents for environmental management of the Baltic Sea maritime industry", would you like to join"? I replied: "You know I have never worked on the Baltic Sea", and he answered: "You know it is very similar to the Black Sea. You'll be fine". For lack of sufficient experience, it was too early for me to get involved in the numerical modeling of this new sea, but I knew, being at GKSS, it would happen sooner or later. Therefore, I decided not to miss the opportunity to start cooperating with colleagues who were familiar with numerical modeling of the Baltic Sea.

The increasing intensity of maritime traffic and the associated adverse impacts on the marine ecosystem requires new solutions to manage the potential harm of these impacts. The situation is particularly complex in areas such as the Baltic Sea, where intense shipping traffic passes close to vulnerable to pollution areas. Actually, all ships traveling between the North Sea and the Baltic Sea pass through the Danish Straits. In the Baltic Sea (including Kattegat), around 76 ports handle over 1 million tons of cargo per year. There, the consequences of marine accidents, can be extremely high, because shipping lanes are close to the coasts. In such locations, it is essential to create a comprehensive and robust design for improved safety aimed at reducing the likelihood of an accident.

The BalticWay project aimed to develop a scientific basis that could be used to reduce environmental risks, for example, pollution caused by ships, which often

Fig. 16 In September 2011, Tarmo Soomere (on the left), who was visiting my department, joined us on a field trip in the vicinity of Hamburg (Kikeberg). Courtesy of Vihang Bhatt

affects vulnerable areas away from the accident site. Our approach was based on the intelligent use of known statistical characteristics of the currents that determine the propagation of pollutants with a focus on the straits, which are the highest risk areas. Optimizing passage through such vulnerable areas is an important challenge from an environmental point of view.

The work needed a model for the Baltic Sea, but I did not have one. Colleagues from the DMI, Jens Murawski and Jun She, the latter and I worked together at ECOOP as work package leaders, provided me with their model, which was known as the HBM[21] (Funkquist 2001). My new Chinese Ph.D. student Lu Xi joined the project.

The HBM system used three models with different resolutions that worked in parallel and exchanged information with each other. The model with the coarsest resolution covered large parts of the Atlantic Ocean. It transmitted one-way information to a finer resolution model operating in the North Sea and Baltic Sea area. The latter model exchanged two-way information[22] with a finer resolution model operating in the area of the straits connecting the two seas.[23] The resolution of the finest model was half a nautical mile (about 900 m). This was the best of its time in terms of representing the geometry of the straits connecting the Baltic and North Seas (Fig. 17).

Xi et al. (2012) identified optimal fairways by calculating the spatial probability distributions of a particle dropped by a ship (an oil slick) reaching the coast. They were recording current velocities from the HBM at small time intervals. In doing so, a set of proxy data (I remind that there is no data from direct measurements over large regions) was created. Then, the inverse problem[24] was solved. The direct problem is when one drops a float in the sea and keeps track of when or where it will reach the shore. In the inverse problem, one looks for where to drop the particle, i.e., where the fairway is, from which the dropped particle will travel the longest to shore, leaving time for the emergency services to do their job. By mapping the endangered and less endangered areas, it appeared that the safety of the straits is highly dependent on the regime of water flowing in or out of that sea. With judicious selection of the navigational regime, the danger to the coasts in these sensitive areas could be reduced by as much as 10–30% (Xi et al. 2012).

There were also spin-off results from the work in BalticWay. A study was launched on marine physics in narrow and shallow sea straits using an adequate ocean model with a horizontal resolution unprecedented for that time (Stanev et al. 2015b).

---

[21] The predecessor of the HBM was the BSHcmod code, which was developed in the early 1990'ies at BSH in Hamburg, Germany. It has undergone extensive revision in the DMI implementation. This model is known as HIROMB-BOOS Model (HBM)—High-resolution Model for the Baltic Sea—Baltic Operational Oceanography System.

[22] "One way" is when the big model (parent model) sends information to the small model (child model). Two-way concept necessitated that also child model returns information to parent model.

[23] The operation of three models presents an illustration of the downscaling approach (see Sect. 4.2 in chapter "In Search of a Destination" for more detail).

[24] The inverse problem in science is the process of calculating from a set of observations the causal factors that gave rise to them.

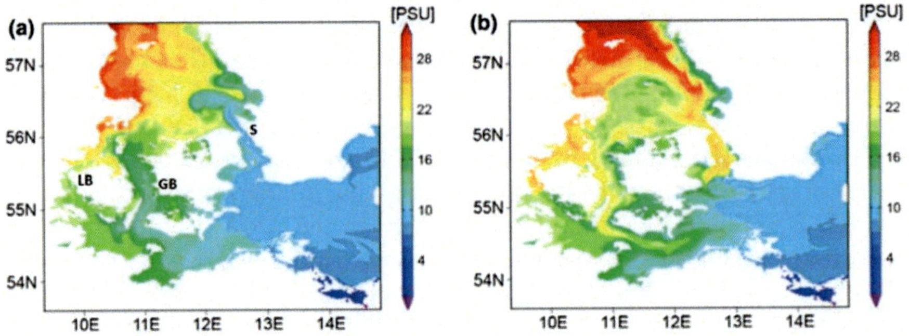

Fig. 17 Results from the fine-resolution (900 m) model of Baltic Sea straits, from left to right, the Little Belt, the Great Belt, and the Sound. The abbreviations LB, GB, and S in (**a**) show the position of the respective straits. Salinity distribution at the sea surface is shown under outflow conditions (28 December 1992, **a**) and inflow conditions (12 January 1993, **b**). The two panels illustrate the strong salinity contrast between Baltic Sea (low salinity) and North Sea (high salinity) waters. During outflow conditions, **a** low salinity Baltic Sea water flows to the Kattegat. When the wind changes direction (**b**), the North Sea water penetrates the Baltic Sea along the three straits. From Stanev et al. (2015b), with permission from Elsevier

4.3 Reconstruction of Ocean Fields

The example of the inverse problem above brings us closer to the question of how to optimally use the measurement data to get the most complete picture of the system we are studying. The perfect situation would be if we had a complete dataset of what we observe and then conduct an analysis that allows us to determine what combination of data best describes the entire system. Such measurement data does not exist and will not exist soon, so, again, we have to work with data produced by numerical models, which we will use as "real data". In Schulz-Stellenfleth and Stanev (2010), the following questions were addressed: (1) How well can we reconstruct the state of the ocean from a limited number of measurements? (2) How informative are measurements about future states of the ocean? and (3) Which components of the measurement network are important? These questions are relevant to the issue of data assimilation. Central to this mathematical discipline is the question of how to extract the maximum amount of information from observational data. Here, we face a problem related to control theory. Control theory was first developed in engineering and deals with the control of dynamic systems. The goal is to observe the evolution of the system in an optimal way so that when it "goes astray, it is put back on the right track". Optimal control systems monitor errors in behavior, and when they exceed a certain value, a correction is introduced, i.e., signals and errors are always handled, and the latter are kept within certain limits.

The theoretical foundations of what we wanted to do led us to the application of the Kalman filter,[25] which is a mathematical method for describing the states of an observed system based on insufficiently accurate and complete observations. The linear filter proposed by Kalman became established in science and applications after its successful use by NASA for flight control of the Apollo space program.

The German Bay model (Staneva et al. 2009) was employed as a data generator. Using the Kalman filter, an algorithm to weigh the used observations was developed, such that these observations described the system as accurately as possible. The method allowed to formulate a priori criteria by which to evaluate the quality of the reconstruction of the true ocean system. A scheme describing how this method can be used in prediction mode was also suggested. Of particular importance for subsequent applications was the new approach for assessing the relative importance of different sensors operating in the observing network.

What We Learned from Our Works on Reconstructing Ocean Fields

In the demonstrations, we did on reconstructing ocean fields, we used synthetic tide gauge, satellite altimeters, and ground-based high-frequency radar systems (HF radars). Some unexpected results emerged such as that the orientation of satellite tracks relative to the shoreline is important for the information value of altimetry data. Furthermore, the inclusion in the observing network of continuous water level measurements, albeit from single points in the open ocean, for example, from the permanent research platform FINO-1,[26] improves the quality of estimates even more than the use of altimeters on board the Envisat and Jason-1/2 satellites.

We were already able to communicate to our colleagues conducting measurements the value of each observing platform. Some of them even asked, "It costs a lot of money and effort to maintain the X-system. Say, can't we stop operating it without adversely affecting the overall information about the area we are observing". Obviously, observationalists and modelers were well on their way to shortening the distance between their fields of scientific activities.

[25] Rudolf Kálmán (1930–2016) was born in Budapest but emigrated to the USA with his parents in 1943. He was a professor at Stanford University.

[26] FINO-1—Forschungsplattform in Nord- und Ostsee № 1 (Research platform in the North and Baltic Seas № 1). It is located in close proximity to areas where wind farms are being built or are already in operation (see Fig. 26 in chapter "From Regional to Coastal Oceanography"). The construction of this platform was intended to enable measurements and research in order to reduce costs, increase electricity yields, and ensure more reliable operation of wind turbines.

4.4 Experiments with Observing Systems

4.4.1 WIMO

The impact of individual data or observational networks on the overall description of the state of the observed area of the ocean is traditionally assessed using OSE. Targeted numerical experiments can determine whether existing ocean observations are adequate, or whether observing systems need to be improved (Oke et al. 2009). In other words, OSE evaluates the collective or individual impact of (existing) observations on ocean analyzes and predictions. In OSSE, we use synthetic observational data as a proxy for real observations. These proxy data are generated by simulations using numerical models, assuming that they reproduce the true ocean state. Since a complete set of real observations is not available, OSSE provides a useful tool to assess the quality of the (synthetic) observing system. OSSE can show where certain measurements are redundant, or where the observing system needs to be enhanced with new measurement platforms. The project "Scientific Monitoring Concepts for the German Bight" (*Wissenschaftliche Monitoringkonzepte für die Deutsche Bucht*, WIMO) which started in 2010 was in line with the above concepts.

WIMO and the Common Maritime Policy of the EU

Monitoring climate and ecosystem changes in the North Sea resulting from natural and anthropogenic impacts requires the use of reliable observing and data interpretation systems. The scientific focus of WIMO was the changes in seawater temperature and biodiversity, the appearance of harmful plankton blooms,[27] the decline in fish stocks, the change in seabed sediments, etc. Underpinning the idea of sustainable use of coastal waters is the need to monitor EOVs and their possible interdependence, as well as the monitoring of their evolution over time and space. Decision-makers, regional organizations, and individuals need science-based advice and presentation of scientific results for the North Sea in a format, which is suitable for understanding and convincing the public of the need for the measures being taken. Accessibility to the scientific information had to be ensured through Web-based information systems. All this required considerable effort, research, and development. One of the starting points in WIMO was to ensure that the work was in line with the new requirements arising from the development of the Common Maritime Policy of the EU (see Sect. 9.2 in chapter "From regional to Coastal Oceanography").

WIMO used advanced observational, modeling, and information processing methods. It involved eleven research institutes and universities, including GKSS

[27] The development of a level of phytoplankton biomass that is uncharacteristically high for a given water body. This can cause harmful conditions that negatively impact people, aquatic ecosystems, and economies.

and ICBM. The structure of the project envisaged the following four research areas: ecology, remote sensing of sediment dynamics, marine technology and scientific basis for decision-making, policy advice, and coordination. The main part of the work, which I coordinated, was on hydrodynamics and sediment dynamics and used the tools of numerical modeling, as well as datasets provided by the COSYNA project. The time scales of processes to study were set in a wide range, reaching up to decades. For the most part, our work in WIMO was a transfer and adaptation of methodologies we developed earlier in COSYNA and ECOOP. The basic question to address was "with the help of what available dataset can we best describe the sediment dynamics in the southern North Sea"?

Willi Petersen, who was responsible in GKSS for the operation of FerryBox, ensured the use of high quality data. This enabled addressing the question to what extent the assessment of the hydrodynamic and thermal state of the German Bight was improved by using temperature (from satellites and FerryBox) and salinity (from FerryBox) measurements in parallel. Grayek et al. (2011) showed that assimilation of sea surface temperature (SST) data from FerryBox leads to a qualitative improvement of estimates of the state of the marine environment. However, this could not be achieved for salinity, which proved to be much more difficult to predict. This is partly explained by the fact that the quality of model predictions of salinity is highly dependent on river flow and precipitation; the quality of the corresponding data is not good enough. Interestingly, assimilating the temperature data improved the salinity estimates.

It was not long before we were able to quantify the usefulness of data from COSYNA to assess the overall hydrological state of the German Bight (Stanev et al. 2011). The data package we used included in situ observations from fixed stations, HF radars measuring surface currents, the FerryBox system, and data from satellites. The examples provided in this study can be considered as initial steps in the creation of new information products for the coastal ocean, enhanced by integrated observations and numerical models of COSYNA. One example is shown in Fig. 18. Thus, the KSD team launched a pre-operational system that significantly improved the quality of forecasts both in space and in time. This was achieved because of the synergism of ocean observations of different types. H. Poincaré said, "It is better to predict without certainty than not to predict at all". Our predictions also gave an objective estimate of certainty. For me, the COSYNA pre-operational system was the major contribution of the GKSS not only to WIMO but also to coastal operational oceanography.

4.4.2 E-AIMS

The WIMO project had not yet finished in 2012 and I was "whirled into another vortex" in the field of observing system experiments. Work in the frame of Euro-Argo continued with the project "Euro-Argo Improvements for the GMES Marine Service (E-AIMS)". The main aim was to prepare the next phase of Argo with extensions to biogeochemistry, polar oceans, coastal and near-coastal seas, and the deep ocean. Significant improvements in Argo float technology were required, as

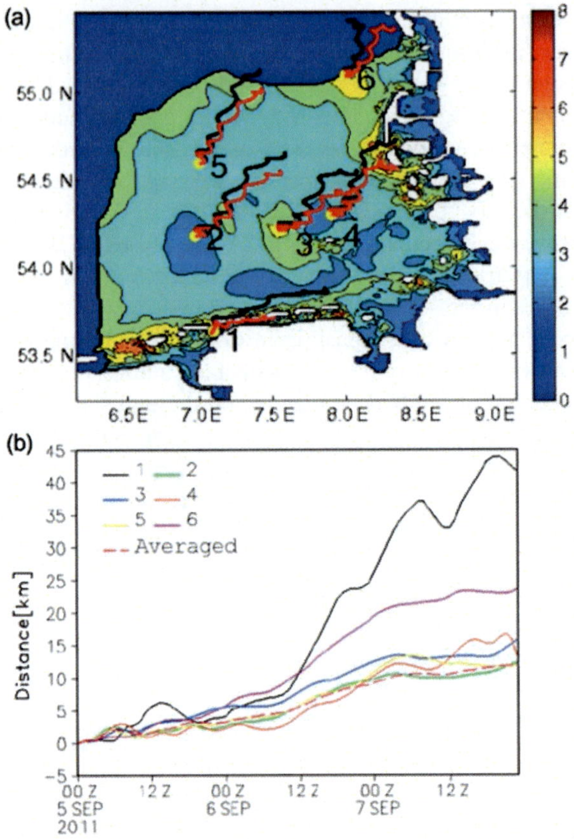

Fig. 18 This figure gives an illustration of using pre-operational forecasting in the field of search and rescue operations. COSYNA model is used in two versions: as a free run and with assimilation of HF radar data. Several trajectories (black and red lines) of lost objects (in this case Lagrangian particles) are compared in the panel on the top. With time progressing, the trajectories separate one from another. Black lines represent the results from the model free run, red trajectories visualize the results in the data assimilation run. Color coding gives the mean distance in km between position of Lagrangian particles in the analysis and free run for September 2011 after 24 h of integration. (bottom) Distance of drifting particles (1–6, see in (**a**) their initial positions) in the data assimilation run and the free run. With time progressing, distance increases, which makes the success of the operation uncertain. Therefore, it is preferable to use models which assimilate data from observations. From Stanev et al. (2016)

well as extensive testing of data before it could be used for operational monitoring. Systematic evaluations and experiments on the performance of the floats and the usefulness of the information, which was made available, had to also be carried out in order to provide reliable recommendations for the use of Argo data for GMES purposes.

The E-AIMS project started on 1 January 2013 and ran until 31 December 2015. The interest to include the Black Sea in E-AIMS was explained by its unique

BGC features. This sea represented an optimal laboratory for testing new sensors measuring BGC parameters. The Bulgarian Institute of Oceanology committed to deploy two BGC floats funded through E-AIMS. My main involvement in the project was to contribute to providing a synthesized summary of how Euro-Argo is contributing to the improvement of the marine services of GMES. We had to define optimal horizontal and vertical sampling characteristics of the Argo floats in the Black Sea, as well as scientifically justify what should be a sufficient coverage of the sea with Argo-observations. Our main data sources included satellite altimetry data, satellite SST data, and of course, Argo temperature and salinity profiles (Grayek et al. 2015). Investigating the synergy between different observation platforms was our main interest on the road to optimizing the design of the forecast systems. For the numerical modeling, we used NEMO and the statistical techniques we had developed with Johannes Schulz-Stellenfleth, a mathematician, who was employed at KSD shortly after I moved to GKSS. In short, we used the OSE and OSSE approaches, which allowed us to evaluate the performance of the observational networks in terms of errors in the reconstruction of the Black Sea hydrological state. Additionally, the contribution of different information sources was quantified. Argo data provided an indispensable contribution. One useful practical outcome from this research was the conclusion that about ten floats operating simultaneously represent an optimal measurement system for operational purposes for the Black Sea.

4.5 Again on Artificial Intelligence

To conclude this section, I will return to where I started, sharing my experience using artificial intelligence, in particular on how seemingly, incomplete information about the ocean can be successfully used to fully decifer (reconstruct) its state. The specific task that led me to turn to artificial intelligence for help was the reconstruction of the spatial (basin-wide) and temporal changes of the sea level in the entire North Sea from limited information measured by tide gauge stations operating along its shores. Some might say that we can use existing observations from satellite altimeters for this purpose. I have to recall that these data are not very accurate in coastal areas. Furthermore, their temporal resolution is not good enough to capture the short-period variability of the sea. The task I set myself was to employ machine learning to reconstruct the basin-wide sea level using data from only 19 coastal stations. Whatever reconstruction one does, it requires knowing the "true" sea level in space and time, for which proxy data from an operational numerical model were used. A particular kind of neural network, which learns how to generate datasets with the same statistical characteristics as the training dataset, made it possible to reconstruct future states unknown to the neural network from the training period with unexpectedly high accuracy (Zhang et al. 2020) (Figs. 19 and 20).

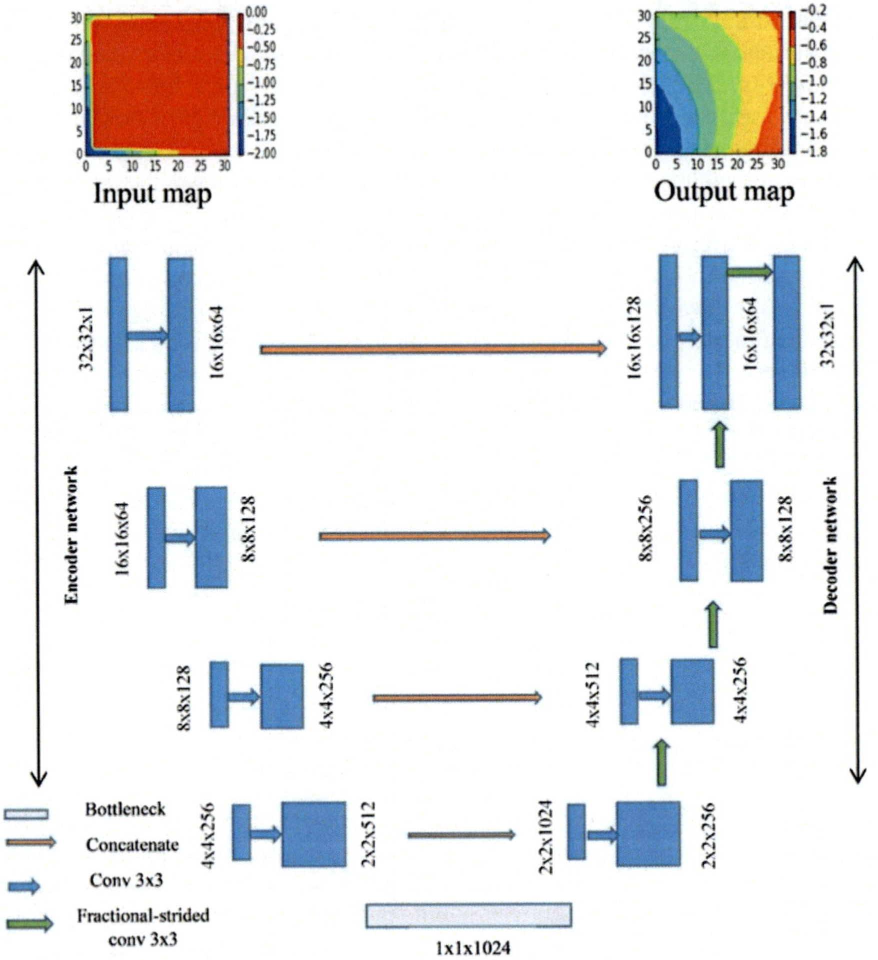

Fig. 19 Schematic presentation of the generator part of the deep neural network model for sea surface height map reconstruction. From Zhang et al. (2020)

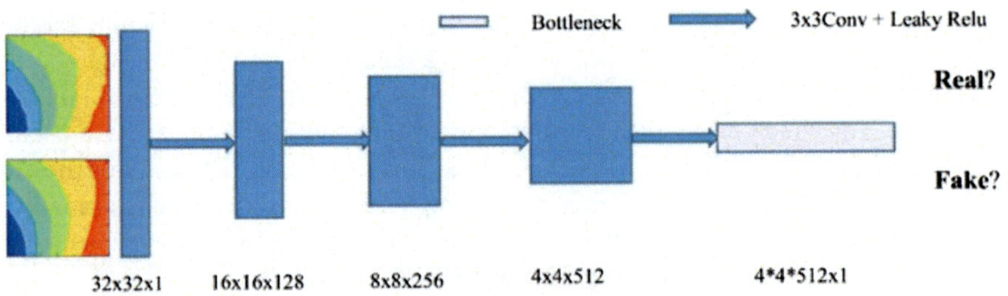

Fig. 20 Discriminator part of the reconstruction model. From Zhang et al. (2020)

Machine Learning and Deep Learning

Neural networks are the basis of artificial intelligence. Deep learning is a special method of information processing and a subfield of machine learning. The adjective "deep" refers to the use of multiple neural layers in a network. The increasingly abstract transmission of information down the layers enables the effective use of deep learning to analyze large datasets. A key feature of deep learning models is that they can learn on their own. This happens when systems repeatedly associate what they have learned with new content and thus learn further. This is also the principle difference with machine learning, in which humans intervene in the data analysis and decision-making process.

One method of self-learning can be achieved by teaching two models. The first model is a generative model (Fig. 19). It captures the underlying characteristics of the data. The second model is discriminative, and its role is to estimate the probability that a given data sample comes from the training data rather than the generative model. Simply put, data is generated and one model asks the other: "Can you guess whether this data is artificial or natural"? This process continues until the artificial data, which is continuously refined as the "game" progresses, and the natural data become indistinguishable. In other words, this continues until you get a perfect recovery of the training data from the generative model (Fig. 20).

After submitting the manuscript of Zhang et al. (2020) for peer review, one of the anonymous reviewers, who was familiar with our earlier works, wrote approximately the following, "two of the co-authors have successfully used in the past the Kalman filter to solve similar problems. Why not compare the two methods and demonstrate the advantage of using neural networks"? I must admit that I had not thought of this idea, perhaps, because I did not expect another method to give such good results. It turned out that the Kalman filter was as good *for the problem we were solving* as neural networks.[28] In fact, the Kalman filter is similar to machine learning; only it does not use neural networks, but simpler mathematical approaches. I arrived to the same conclusion that I have had to make many times in my scientific path: check all aspects of what you are doing.

[28] This is not to belittle the qualities of neural networks. It was just that the particular task did not allow for a full expression of their abilities.

5 Further Inland

5.1 Water Cycle, Estuaries, and Deltas

The hydrosphere of planet Earth consists of several reservoirs (or subsystems) that are connected by the water exchanged between them. In descending order of the amount of water contained in each reservoir, these are the World Ocean, ice masses and snow, surface waters, the atmosphere, and the biosphere. The exchange of water between the different reservoirs essentially represents the water cycle of the planet (Fig. 21). It is sustained by the energy coming from the Sun. At the heart of the water cycle is evaporation, the process by which water transforms from a liquid to a gaseous state, i.e., it moves from the ocean to the atmosphere. The amount of water entering each year the atmosphere from the ocean in the form of vapor is estimated at 361×10^{12} m^3. Along with this huge amount of water, heat is released into the atmosphere as the vapor changes by condensation to a liquid state. This contributes to the intensification of some atmospheric motions, for example, hurricanes.

If the total amount of water in the atmosphere (mostly vapor), approximately 13,000 km^3, were distributed uniformly over the surface of the planet in the form of liquid water, a layer only 2.5 cm thick would result. Assuming that the average annual rainfall for the whole Earth equals approximately a 1 m water column, it follows that the amount of water available in the atmosphere must be removed from the atmosphere forty times a year, i.e., the typical residence time of water in the atmosphere is only about 9 days.

Globally, evaporation equals roughly precipitation. On continents, precipitation usually exceeds evaporation and vice versa, over the oceans, evaporation exceeds

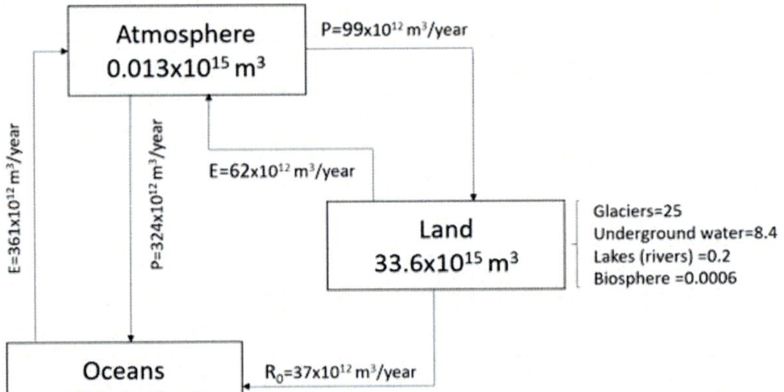

Fig. 21 Earth's water cycle represented by the amount of water held in the ocean, on solid land, and in the atmosphere, and by the amount of water exchanged between different reservoirs through evaporation (E), precipitation (P), and river runoff (R_0). Numbers are taken from Peixoto and Oort (1992)

precipitation. When precipitation falls on the land surface, water follows different pathways. Some of it evaporates, returning to the atmosphere; some seeps into the soil as soil moisture or groundwater; and most of it runs off through rivers into the oceans (see Sect. 8.2 in chapter "From regional to Coastal Oceanography"). This amount is about ten times less than the amount of water exchanged directly between the atmosphere and the ocean. Although, globally, river runoff appears to be small, it is crucial to the functioning of coastal marine systems as estuaries and river deltas concentrate huge inflows of freshwater in small areas. In the early years of numerical ocean modeling, little attention was paid to the quality of river flow data. With increasing demands on the quality of the models, and especially their adequate performance in coastal areas, the issue of river runoff is becoming increasingly relevant.

When rivers reach the ocean, the flow velocity slows down dramatically, which favors sedimentation. Thus, sediments are deposited in a small area near the river mouth. The continuous influx of sediment into the coastal area and its deposition there causes the river mouths to shift toward the ocean. The seabed there is lined by numerous channels (deltas) allowing river water to drain efficiently into the sea. The size of the delta depends on the specific balance between the flow of sediments and the rate at which they settle or are carried away from the coast.

The situation in estuaries is the opposite. There, the sediment input from the land is less than the sediment flux that can be transported to the sea. For the process of sediment transport to the ocean to prevail, energy is needed to hold suspension in the water column; otherwise, it will settle to the bottom (as it does in deltas). This energy can be at the expense of both wind waves and strong tides. While deltas advance seaward, in estuaries, due to tides, for example, the sea periodically enters estuaries deep into the land, sometimes a hundred kilometers or more, as in the case of the river Elbe. The strong currents erode the sediment from the bottom, which is carried into the ocean at low tide. Thus, ocean tides cut through the land in the estuarine area, creating favorable conditions for shipping. Cities such as Shanghai, New York, London, Hamburg, Rotterdam, Montevideo, and Buenos Aires have arisen on the shores of major estuaries.

5.2 *Waves in Shallow Water Environment*

5.2.1 Waves in the Open Ocean and in a Shallow Water Environment

Like all other waves, wind-generated waves are associated with deviations of the hydrodynamic system from an equilibrium state. In the absence of disturbing forces, the equilibrium state of the ocean is the horizontal sea surface. Wind can create disturbances, and waves are the mechanism that propagates these disturbances away from the area in which they are created, thereby facilitating the sea level to become horizontal again.

Wind is capable of generating disturbances over a wide spectral range, i.e., shorter and longer waves. When wind blows continuously, the energy of all waves increases because they absorb the wind energy, which we notice by their increasing height. In doing so, the waves become steeper, more unstable, and the shorter ones are the first ones to break. Their energy is transmitted to the longer waves, which also become steep and collapse over time. Thus, the short wind waves cannot withstand the strong wind, and because they break they cannot absorb much of its energy. This process of wave generation continues as long as the wind, which we assume has not changed its speed in the meantime, has created a situation where waves of a certain length carry a maximum fraction of the energy of the whole wave spectrum. That is, these waves are long enough, withstand the force of the wind and do not further collapse. What happens is that a certain wind speed corresponds to a wave of a certain energy, or more simply, wave height (Fig. 22).

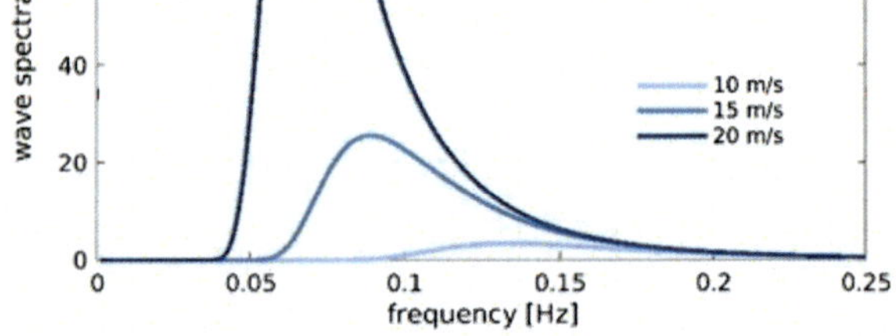

Fig. 22 Wave spectra of a fully developed sea for different wind speeds based on Pierson and Moskowitz (1964). Different curves correspond to different wind speeds. Figure courtesy of K. Wahle

Beaufort Scale and Wave Spectrum

The relationship between wind speed and wave height is the basis of the Beaufort scale, which was proposed in 1805 by the Irish hydrographer Francis Beaufort. It is still used today by meteorologists and sailors to characterize wind speed and wave height. The movement of air masses is categorized from 0 to 12th scale. Here is an example of what some of these scales look like: 0—no wind, wave height—0 m; 4—moderate wind (5.5–8 m s^{-1}), wave height 1–2 m; 7—quite strong wind (13.9–17.1 m s^{-1}), wave height 4–5.5 m; 10—storm (24.5–28.4 m s^{-1}), wave height 9–12.5 m; 12—hurricane (> 32.7 m s^{-1}), wave height over 14 m (intermediate scales are omitted here). The concepts of Beaufort, based on practical reasoning, found its theoretical support in the second half of twentieth century. Several international projects and observational campaigns helped develop sophisticated mathematical theories explaining waves dynamics. The Joint North Sea Wave Observation Project (JONSWAP) made clear that the wave spectrum is never fully developed. It continues to evolve through non-linear, wave-wave interactions (Hasselmann et al. 1973). The JONSWAP spectrum is actually a modified version of the known Pierson-Moskowitz spectrum (Fig. 22). This figure illustrates two fundamental properties of wind waves: (1) Lower energy corresponds to weaker wind and (2) a shift of the spectral maximum to lower frequencies with increasing wind speed. Fundamental here is the idea proposed by Kitaigorodskii (1962) based on a similarity hypothesis that, when plotted in a certain dimensionless way, the power spectrums for all fully developed seas should be of the same shape.

Parallel with development of theoretical concepts of wind waves, numerical modelers did in 1979, in the frame of Sea Wave Modeling Project (SWAMP), a wave model intercomparison study to clarify the performance of various wave models for real-time wave forecasting. One problem in existing theories was that real wind fields are typically neither stationary nor uniform. This justifies the attempt to carry out intercomparisons of models for selected real wind cases, for which adequate wind and wave field observations existed (see the review of Cavaleri et al. 2007).

It was soon realized that while the wind wave models applied to the open ocean (e.g., the WAM[29]) had reached a good level of maturity, coastal wave research was lagging behind. In 1993, scientists organized the Waves in Shallow Environments (WISE) group with the aim of developing wave models for coastal regions. The scope of interests in the WISE group extends from fundamental research associated with hydrodynamics to numerical modeling for wave forecasting and prediction, with

[29] Wave model (WAM) is the first numerical model that solves the energy balance equation, including non-linear wave interactions (WAMDI Group 1988).

applications ranging from climate change impact studies to coastal engineering (see the review of Cavaleri et al. 2018).

5.2.2 FIELD_AC

In the ECOOP, A. Sánchez-Arquila launched the idea of conducting focused research on flows and interactions at the land–ocean interface. The scientific interest was driven by the understanding that wave behavior in the coastal domain is of great importance for both physical and BGC processes. In KSD, Heinz Günther was one of the world leaders in developing numerical models for sea waves, but mostly on a global scale. The functioning of the WAM, which he developed as part of an international team, needed to be improved, particularly in terms of the quality of the coastal forecasts. The project "Fluxes, Interactions, and Environment at the Land–Ocean Boundary. Downscaling, Assimilation, and Coupling, (FIELD_AC)" has given impetus to this work. Motivation came from the realization that oceanographic predictions in the coastal zone, based on the use of numerical models, rarely take full account of the enormous importance of processes at the interface between land and sea, and in particular the impact of wave-current interactions.

Not only was the wave modeling in the coastal zone an issue. River fluxes were accounted for in many numerical models, but their three-dimensional nature was often overlooked at that time. Furthermore, many coastal processes, especially those in specific geographical areas (coasts with harbors or estuaries), were not well represented in regional and global models. For this reason, local projections were burdened with significant errors and were not sufficiently reliable. An example of this is the uncertainties in coastal flood forecasting.

The FIELD_AC project has taken a step toward modeling coastal processes. The work aimed at developing the scientific basis for improving the quality of operational forecasts in coastal areas. Of particular importance was the full use of measurement data. An objective assessment of model errors was needed, so that a reliable service could be offered to public and private entities. The test areas to be worked on covered the Catalan coast, the Gulf of Venice, Liverpool Bay, and the German Bight. The challenge was to overcome the missing links between shelf forecasts on the one hand and estuarine forecasts on the other. FIELD_AC started on 1 January 2010 and continued until the end of December 2012. A few months after the start of the project, on 7 and 8 June 2010, I hosted the first project workshop in Hamburg. It was there that I first met Fabrice Ardhuin, a young French scientist (then from from SHOM) who over the next ten years became a world leader in the study of ocean waves and their various effects on ocean circulation.

KSD Contribution to FIELD_AC

At the time when FIELD_AC started, there were separate models for waves and models for currents, but they did not interact, something that did not correspond to what was happening in nature. Heinz Günther made the WAM work well in the coastal zone. Joanna Staneva, who had been working mostly with ocean circulation models up to that point, began to take first steps toward understanding and evaluating the effects of wave-current interactions. In coastal areas, this proved to be essential in both improving the quality of predicted currents and improving the quality of predicted waves. This was confirmed by validating the models against an extensive dataset of dedicated sea buoys and satellite data. I was personally more involved with investigating the sensitivity of the models in terms of how they account for continental runoff. It proved particularly important to address the three-dimensional structure of water flows when specifying river runoff. We also applied a previously developed scheme for accounting for freshwater sources from groundwater and those discharged to the sea in a controlled manner from coastal freshwater lakes (Staneva et al. 2009).

On 6 October 2021, the day after the 2021 Nobel Prize in Physics was announced (see Sect. 4.1 in chapter "In Searsh of a Destination"), F. Ardhuin sent an email to the WISE group.[30] Here is what he wrote: …However, the same person [Klaus Hasselmann] has created our wave modeling community and many other things! … Basically Klaus was the main driving force to bring people together on in situ experiments (JONSWAP), models (SWAMP and later WAM, from which the WISE meetings developed), and remote sensing, including assimilation (today, we would have no Sentinel-1[31] "wave mode", were it not for Klaus et al.). As for Fabrice's frustration that the Nobel Committee did not also credit the contributions of K. Hasselmann to sea wave theory, I share it, but still the contributions appreciated by the Committee are, in my opinion, more fundamental.

5.3 The Practical Need for Estuarine Modeling

The potential users of scientific information on the state of estuaries are the companies managing navigation in the lower reaches of rivers. This management has a long history. The date 7 May 1189 marks the official foundation of the port of Hamburg. On this day, Emperor Friedrich Barbarossa issued a document granting Hamburg a

[30] Now F. Ardhuin is the leader of WISE group.

[31] Sentinels are Earth observation satellites under the European Space Agency's Copernicus programme. Sentinel-1 is a pair of radar satellites, Sentinel-2 is a pair of passive-optical satellites, and Sentinel-3 determines the topography of the sea surface, the temperature of the sea and the Earth, and the color of the ocean. See Fig. 37 in chapter "Further into the Ocean".

number of rights and privileges guaranteeing free navigation from the mouth of the Elbe to Hamburg. Since the foundation of the port, administrative structures have been put in place, tasked with ensuring optimal and efficient navigation along this part of the river. The main problem is to control and maintain within appropriate limits the depth of the navigable channel, which is particularly imperative in view of the fact that ever larger and deeper-draft vessels are being used to carry cargo. The bottom has to be dredged, but this does not always solve the problem, because much of the dredged sediment is soon replaced by new sediment.

The Hamburg Port Authority

The Hamburg Port Authority (HPA) is responsible for port operations, development, and maintenance of the port infrastructure. The port area occupies about one tenth of Hamburg's total area. The responsibility of HPA includes also shipping safety, port rail facilities, real estate management, water and land infrastructure as well as the economic conditions in the port. The HPA monitors sea level, outflow, and currents in the Elbe Estuary. This includes sediment management, i.e., the transport of dredged material in the harbor area. The HPA is also charged with carrying out engineering measures to control tidal waves. As I have mentioned elsewhere, there have been several catastrophic floods in the area around Hamburg. Medium and small-scale floods occur at least once every few years. That is why protection, and early warning systems are being developed for coming floods.

By the period I describe here, 2007–2014, much of the management of the harbor and its environment was based on engineering developments. The science that provided the methods for working in this field is called coastal engineering. The methods of oceanography were less well covered. Shortly before I moved to GKSS, H. von Storch had written to me that I was to take over the supervision of the topic "Tidal Monitoring on the Elbe using Numerical Modeling" (model-based monitoring for short). He had taken on the task of consolidating the scientific interests and available experience in several German institutes and organizations in order to lay a deeper scientific foundation for the problem of the optimal use of the Elbe estuary. The combined effort was to be directed toward short-term sediment management, which required future monitoring steps in the area.

Heinz Glindemann, the engineer, who had developed the concept of "Tidal Elbe (Tideelbe)", was the head of the River Department at the HPA. When I first met him, he began by saying: "The HPA invests around € 50 million a year to remove the sediment from the harbor basins and fairways that is deposited there by the Elbe current. We dump it offshore, south of the island of Helgoland, but it is not long before the dredged areas are filled with silt again. I am beginning to understand that we should not work against nature, because it maintains conditions that are optimal for it. However, big ships should be able to navigate on the Elbe too. What a conflict"!

Drawing on my modest experience in optimization problems, I explained to him that they needed to have a strategy that "soft-touches" nature rather than doing brute-force interference. Iteratively, Hans von Storch and I were trying to launch an Elbe monitoring project based on the use of data from numerical models. We had to clarify which estuarine states were subject to change with minor effort and exactly what kind of change. Hans von Storch was pushing the idea to start exploring the potential of a numerical model used as a "monitoring platform" without requiring a logistically complex measurement network. We wanted to replace the real measurement network with data derived from a numerical model, assuming that there was a model that was so good as to serve as an "ideal substitute for the real nature". As such, we would use the model of the Federal Institute for Hydraulic Engineering (Bundesanstalt für Wasserbau, BAW). My team had to develop a coarser and hence less accurate model and investigate its "receptivity" to assimilate the data from the "perfect model" in order to improve the estimates of the coarse model. This in fact constituted an OSSE exercise. The main question was what were the sensitive places in the estuary where, if we put in external data (from another model), we would substantially improve our model. In other words, the question was of investigating the success of data assimilation in estuarine models. Such experiments were also intended to explore the possibilities for the optimization of the measurement network. In the longer term, the results were to be directed toward solving practical problems, e.g., the optimization of dredging and, more generally, the development of a scientifically sound pre-operational system for the Elbe estuary.

Everything up to that point was fine, the bad thing was that I did not have enough experience in estuarine physics at that time, I did not have a proper model, and I did not have a coworker to do the work. For H. von Storch, things seemed easy: "With the Ph.D. position you get, I guess you'll be fine", he used to say. Not one to be easily startled in unfamiliar situations, I accepted the challenge. It turned out that the physics of estuaries is a rather peculiar science; it required a lot of reading and creating of own experiences. I made the decision about what model to use pragmatically. There was no time to learn a new model, so I decided to go with GETM. Initially, a lot of numerical modeling work had to be done, so the assimilation of observational data in tidal estuaries was postponed until later.

The Change of Tides in the German Bight and its Estuaries

Numerical experiments made it possible to clarify a number of issues in relation to tidal wave transformation in the German Bight and its estuaries (Stanev et al. 2014a). In the open sea, tidal waves are characterized by a period of 12 h and 25 min, which is known as an M2 tidal period. The modification of tidal wave appeared most significant in the eastern part of the German Bight, where Kelvin waves lose most of their energy after reflection or refraction. The loss of energy leads to a change in the spectral composition of the wave and the generation of tides with two times higher frequency (overtides, M4 tides, Fig. 23). Submarine channels prove to be particularly important for this tidal transformation. It turned out that the ratio between the amplitude of the tides during neap and

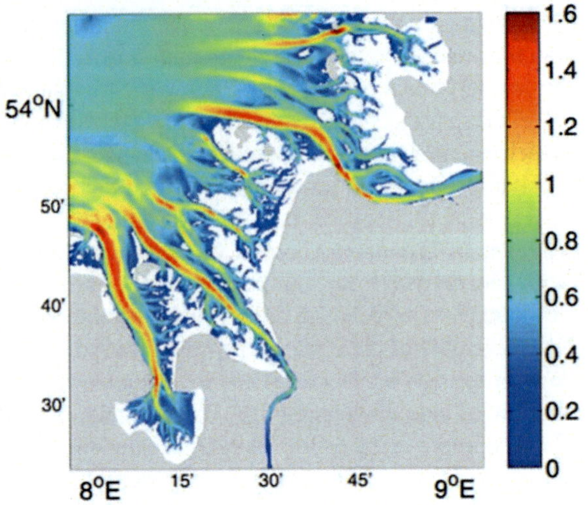

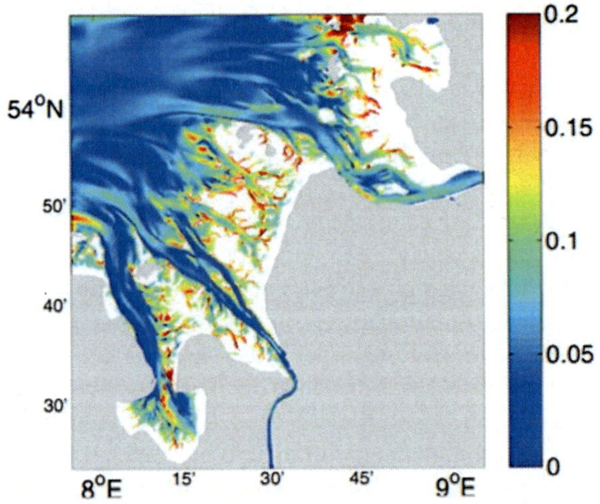

Fig. 23 Amplitudes in meters per second of the major axes currents for spring tide periods in April/ May 2011. Two tidal constituents are presented: M2 (top) and M4 (bottom). In the white areas (shallower than 0.8 m), which fall dry during relatively long time, tidal analysis has not been performed to avoid possible misinterpretations due to the inhomogeneity of the data series. From Stanev et al. (2014a)

spring tide decreases with decreasing distance to shore (1.3 in the open sea and 1.15 in coastal areas). This is explained by the enhanced "extinction" of waves in the shallow water. The spatially and temporally variable stratification of water in the coastal zone also affects the hydrodynamics in the estuarine area (Stanev et al. 2015a). The differences in stratification could be considered as a major control mechanism for sediment dynamics.

6 Numerical Models Using Unstructured Grids

6.1 How to Discretize the Space

The American architect Richard Buckminster Fuller (1895–1983) said, "You never change things by fighting the existing reality. To change something, build a new model that makes the existing model obsolete".[32] In the late 1920s, Fuller created new building concepts that he introduced to the public under the name Dymaxion (the Dymaxion globe). It became known for the geodesic dome of the American Biosphère pavilion for Expo '67 (Fig. 24). Fuller's geometric structures are based on the use of the simplest geometric basic solids. There are five so-called Platonic solids.[33] The simplest of these is the tetrahedron, which has four surfaces, and the most complex is the icosahedron (see the figure) with twenty surfaces. According to Plato, regular polyhedrons are best suited to explain the nature of matter. He made the following assignments: tetrahedron—fire, octahedron—air, cube—Earth, and icosahedron—water. Plato regarded the dodecahedron as representing the kosmos. This intellectual effort to identify the fundamental constituents of matter, although unsuccessful, cannot be ignored because the description of nature profited a lot from geometry. In the example given in the bottom inset of Fig. 24, we plot the continents and oceans on the surface of an icosahedron. When we unfold the three-dimensional figure and reduce it to one plane, we get the inset on the left composed of multiple triangular elements. This map does not suffer from the distortions, which are characteristic of other types of geodesic projections (it preserves shapes and sizes). In turn, the use of these triangular elements in architecture and construction guarantees a particular stability of structures.

Recently, it has been realized that grids based on triangular or hexagonal elements (Fig. 3b, c in chapter "In Search of a Destination") open up new horizons for numerical modeling of the ocean. This branch of numerical modeling is about 50–60 years old, but its development has remained in the shadow of so-called structured grid methods, in which the horizontal grid elements have simple geometric shapes, such as lines that follow parallels and meridians, or just rectangles (bottom-left panel in Fig. 3 in chapter "In Search of a Destination"). These structured grids suffer from some drawbacks. For example, when grid elements in structured grid models are small enough and can describe complex coastlines in detail, they cannot practically

[32] Perhaps the best example in the history of science is the replacement of the well-tuned geocentric celestial system of Ptolemy, which was actually giving good predictions, by the heliocentric system of Copernicus, which, at the beginning (before Kepler), was not so perfect in predictions, but avoided unnecessary fine tuning.

[33] The Platonic solid is a convex regular polyhedron. The faces of the Platonic solid are polygons of the same shape and size. The same number of faces meet at each vertex. For any convex polyhedron, $V - E + F = 2$, which says that the number of vertices (V) minus the number of edges (E) plus the number of faces (F) is always equal to two. This topological relationship (invariant) is discovered by L. Euler.

Fig. 24 In the background is a view of the Biosphère pavilion at Expo 67 in Montreal. The insets on the top show the Platonic solids with their respective elements, as drawn by Johannes Kepler in Harmonices Mundi from 1619 (*Images in the public domain,* via *Wikimedia Commons).* From left: tetrahedron, cube, octahedron, icosahedron, and dodecahedron. Below, the respective unfolded images of each Platonic solid are shown. The insets at the bottom give an idea of the Dymaxion map or Fuller map, which is a projection of a world map onto the surface of an icosahedron (right inset) using equilateral triangles. The map which can be unfolded and flattened to two dimensions is in the left inset (bottom)

be spread over large areas in the ocean due to the limited computational capabilities at present. Conversely, when the grid is coarse and computationally manageable in the entire ocean, the coastlines are broken, rough, and do not correspond to the real ones. These computational problems explain why models based on unstructured grids, where individual elements can have different shapes and sizes, find an important application. In these models, grid elements are smaller in areas requiring a dense grid (estuaries, straits, and coastlines), and their size increases toward the open ocean (Fig. 25). The wide acceptance of unstructured grids is largely due to their performance in modeling estuarine and coastal processes.

In 2011, Anna Zorndt, then a Ph.D. student from Hannover, called me to tell me that a young American scientist of Chinese origin was going to give a talk at the Technical University of Darmstadt on the use of unstructured grids in ocean modeling. The official host was prof. Ulrich Zanke, a specialist in hydraulic engineering, coastal engineering, sediment transport, and numerical modeling. The organizer was his collaborator Aron Roland, a young scientist of Croatian origin. The rapporteur was Yinglong Joseph Zhang. After completing his undergraduate degree at Peking University, he continued his studies in Australia, in the field of applied mathematics with a major in fluid mechanics. After several years in Singapore, he moved to Oregon, USA, where he and Antonio Baptista developed the SELFE (Semi-implicit

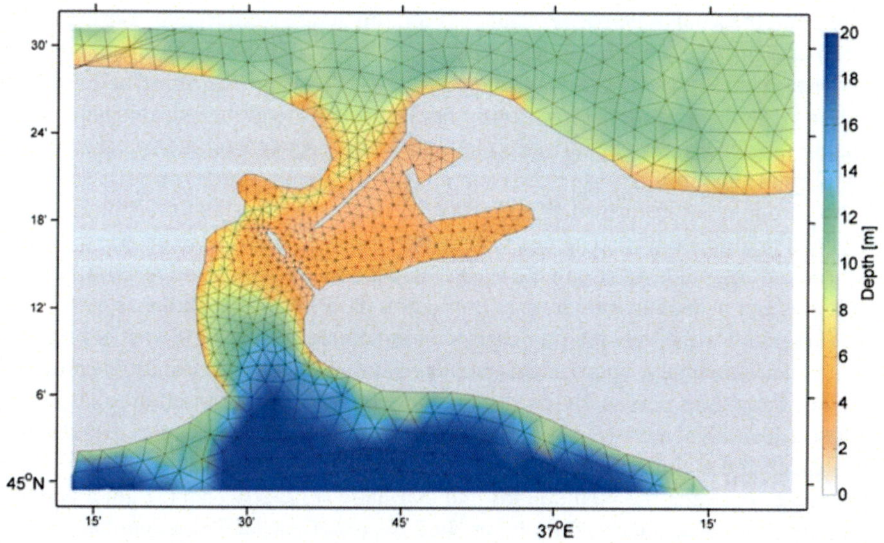

Fig. 25 Kerch Strait is represented by smaller triangular elements in the model for the entire Black Sea. From Stanev et al. (2017a)

Eulerian–Lagrangian Finite-Element model for cross-scale ocean circulation, Zhang and Baptista 2008).

The seminar room was small, there were too many listeners, and there were numerous questions after the interesting talk. Mine were a bit provocative and after his report, J. Zhang found time to talk to me. The conversation started to get interesting. After having lunch, we both went out for a walk, I explained my interests in coastal and estuarine modeling; he was looking for new users of his model. This is how a strong collaboration between us started. He visited my coworkers and me many times. In 2014, he received a fellowship from HWK[34] for several months to work with me on the topic "Large-scale numerical study of the North and Baltic Sea using unstructured grids". Again, I turned to something new.

6.2 *Ems-Dollart Estuary*

The Netherlands Organization for Scientific Research (NWO) and the German Federal Ministry of Education and Research (Bundesministerium für Bildung und Forschung, BMBF) launched in 2010 their first international call for proposals for collaborative research between scientists from both the Netherlands and Germany in the Wadden Sea Region (WSR). This region includes major tourist destinations, ports, and a population of around 3.5 million inhabitants living in low-lying coastal areas as

[34] Hanse-Wissenschaftskolleg (Institute for Advanced Study), Delmenhorst.

well as on the sand dune barrier islands belonging to the Netherlands, Germany, and Denmark. In 2009, UNESCO declared most of the WSR a World Heritage Site. Part of these coastal areas are below the mean sea level; therefore, the future of this unique region is highly dependent on various processes related to global warming and accelerating sea level rise.[35] Efforts are required to manage coastal areas and to ensure the normal functioning of the ecosystem. This implies in-depth research and international cooperation to further protection of intertidal areas and their biodiversity. The two organizations had agreed to fund two projects, one on "Georisks: responses to accelerating sea level rise in the WSR", and the second on "Biorisks: implications of biological invasions for nature conservation in the WSR coastal ecosystem".

Shortly after the call for proposals was announced, Huib de Swart, a professor at Utrecht University, approached me suggesting that we prepare a joint project. I did not know him personally until then, but I knew from the literature that he was a world-leading scientist on the research of tidal and seabed change. He was also the leader of a major Dutch study on the evolution of sediments in the river Ems, on the border between the Netherlands and Germany (Fig. 26). Later, I got a personal impression of how successfully he applied his excellent mathematical background in hydrodynamics.

We focused on the topic "Georisks" and called the project "Impact of climate change and human intervention on hydrodynamics and environmental conditions in the Ems-Dollart[36] estuary: an integrated data-modeling approach". The short name of the project was "Future Ems". The central theme was the rapid deterioration of environmental conditions and the amplification of tides due to human intervention.

In the lower reaches of the river Ems, massive dredging works are being carried out at a total cost of € 24 million per year between the towns of Emden and Papenburg and an additional € 7 million for dredging from Emden to the North Sea. This is necessary because large tourist liners are built in the shipyard Meyer Werft in Papenburg, which is about 40 km upriver. For the big ocean liners (top-right panel in Fig. 27), the river bed is shallow, and in order to reach the sea, the bottom has to be deepened. Deepening of estuary results in changes in the amplitude of the tides (Fig. 28). As they increase in height, the currents increase too, and the regime of erosion and sediment deposition changes. Sediment concentration at the surface exceeds grams per liter, and near the bottom, it can exceed 100 g per liter. These waters are called fluid mud. Zones of hypoxia are created, and the concentration of phytoplankton is reduced. The implications for society are profound, including the loss of the ecological value of the estuary.

---

[35] The global mean sea level increased by 1.7 mm per year during the period 1900–2009. The sea level rise for the period 1993–2009 was 3.2 mm per year. Scenario estimates give a range of 75–190 cm for the period 1990–2100.

[36] The Dollart is a large tidal basin at the mouth of the River Ems. For this reason, the estuary is often referred to as the Ems-Dolart Estuary. The Dolart lies to the left of the estuary and is part of the Netherlands.

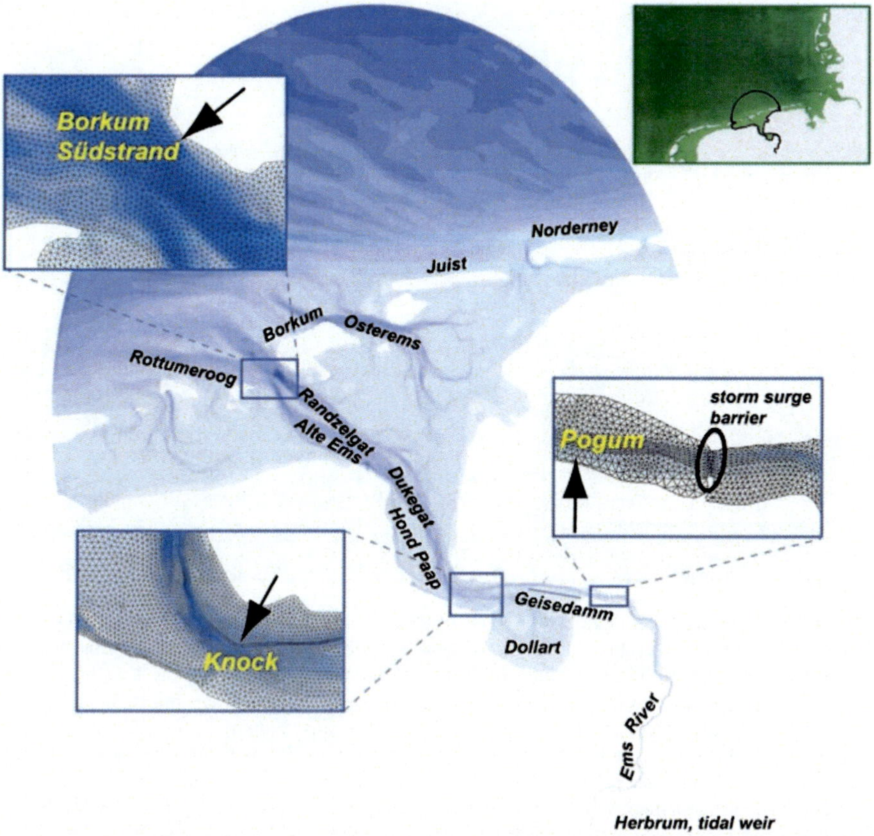

Fig. 26 Ems Estuary as part of the German Bight (top-right inset). The black line in the inset shows the model area, which is magnified in the central plot. This main plot displays details in the bathymetry along with the major geographical names referred to in the text. Three further magnified insets are also given with a model grid illustrating the resolution in different areas. The position of the storm surge barrier is identified by the ellipse. From Pein et al. (2014)

To reduce the costs required for optimal estuarine management, reliable models are needed, which can predict the response of the estuarine system to changing external conditions and human interventions. Addressing these challenges in the Ems-Dollart estuary required a transnational approach. The need to work in a multidisciplinary way necessitated to invite scientists with different expertise to join (oceanographers, mathematicians, biochemists, geologists). The first steps were targeted at developing a forecasting tool based on integration between models and data. The tool had to provide a solid scientific basis for the ongoing debate on the design of sound, low-cost, and environmentally friendly measures aiming to improve the ecological value of the river-estuarine system while maintaining its important role for the local economy.

Fig. 27 Ems barrage near Gandersum is one of the most modern barrages in Europe and has been in operation since September 2002. The 476-m-long structure (bottom panel) fulfills two main tasks. On the one hand, it significantly improves storm surge protection on the Ems by reversing storm surges that are two meters higher than the normal high tide, when the barrage is closed. The dam's stowage function ensures the flexibility of the Ems shipping route between Papenburg (see position of tidal weir in Fig. 26) and Dollart Bay. The maximum water level of 2.70 m above sea level allows the passage of ships with a draft of up to 8.50 m. Figure courtesy of Lower Saxony Water Management, Coastal Protection and Nature Conservation Agency (Niedersächsische Landesbetrieb für Wasserwirtschaft, Küsten- und Naturschutz, NLWKN) and of Hans-Jürgen Zietz

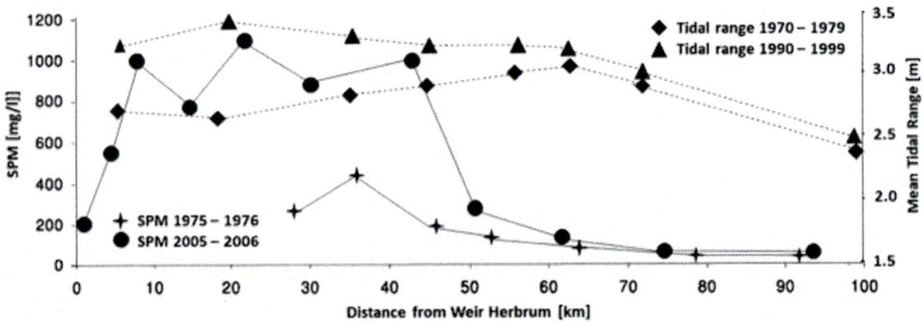

Fig. 28 Long-term changes in the tidal range (dotted lines) and sediment characteristics (full lines) in the Ems. Pogum (see also Fig. 26) is at ~ 50 km. The position of tidal weir in Herbrum is shown in Fig. 26. Replotted using data from from Schuttelaars et al. (2013)

In 2012, J. Zhang got a position at the Virginia Institute of Marine Science (VIMS) at Virginia State University and moved from the West Coast to the East Coast of the United States. Although the time was very busy for him during this transition period, he visited me in Geesthacht. In two weeks, with his great help, we configured several versions of SELFE for different areas of the North Sea. One of my Ph.D. students at that time, Johannes Pein, started to work in the frame of "Future Ems" with the set up for the Ems estuary (the model area and grid are shown in Fig. 26).

Some Results in "Future Ems"

Pein et al. (2014) showed that SELFE reproduces very well not only the general features of estuarine circulation, but also a number of details specific to the Ems-Dollart Estuary. The attenuation or amplification of the tidal wave amplitude along the length of the estuary predicted by the model agreed perfectly with observations. The asymmetry of the tides was reflected in the fact that in most of the estuary, the currents during high tide were stronger than during low tide (Fig. 29). This was of great significance for the regime of bottom sediment deposition and its subsequent erosion. The Dollart Bay was an exception to this rule. It appeared that the complex bathymetry of the area created a very complex interaction between different physical mechanisms in different parts of the estuary, i.e., simplistic models had little relevance to the real estuary. Moreover, the density distribution turned out to be of decisive importance for the currents.

In Pein et al. (2016), it was shown how a limited amount of salinity measurements in the Ems estuary could help reconstruct the full details of the salinity

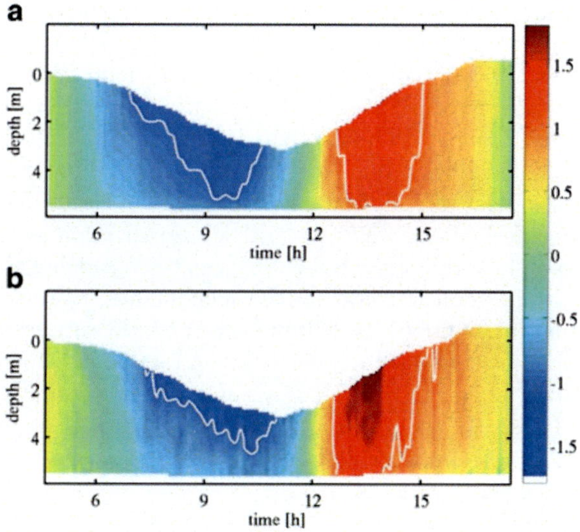

Fig. 29 Time versus depth diagrams of along-channel velocity [m s^{-1}] as simulated by the model (**a**) and observed by an ADCP (**b**). Data are taken in the channel in front of Pogum, see for the position the inset in Fig. 26. Positive velocities are directed toward the River Ems (flood). Observations were taken on the 12 June 2012 from 05:00 h to 17:30 h. The isolines (white) of $|u| =$ 1 m s^{-1} are also given. From Pein et al. (2014)

field. To my knowledge, this was the first study applying the tools of statistical ocean reconstruction to the estuarine physics. What we have planned earlier for the Elbe estuary was realized first for the Ems estuary.

Meanwhile, the model of J. Zhang was gaining a growing community of users. Since moving to VIMS, Joseph has significantly updated and improved the model, which is why he raised with the user community the question of changing its name. The proposed name was Semi-implicit Cross-scale Hydroscience Integrated System Model (SCHISM, Zhang et al. 2016a), as this new model is known now.

7 The Sea Straits

7.1 At the Border Between the Baltic Sea and the North Sea

The North Sea and Baltic Sea (Fig. 18 in chapter "Further into the Ocean") are coupled tidal and non-tidal basins of approximately equal size connected through a system of straits. This straits system, hereafter the North Sea–Baltic Sea Transition Zone (NBTZ), includes the Kattegat, the Danish straits, and the Western part of the Baltic Sea (Fig. 30). The surplus of freshwater in the Baltic Sea and the limited water exchange with the North Sea support a two-layer exchange flow in the NBTZ, which explains the low salinity (brackish water[37]) of the Baltic Sea. The deep part of the Baltic Sea known as the Baltic proper is separated from the straits area by a number of sills: the Darss Sill (depth of 18 m), the Drogden Sill (depth of 8 m), and the Stolpe Channel (depth of 63–64 m) further east. In these shallow areas, the inflowing North Sea water is subjected to substantial mixing with the highly stratified Baltic Sea water.

The North Sea–Baltic Sea system represents a challenge for the numerical modeling because of several reasons. The first one is that it consists of a tidal and non-tidal basin with different dominating physical balances in each individual basin. The tidal basin (the North Sea) is very shallow, except for the Norwegian trench. The relatively high salinity values in the North Sea are typical for the ocean; the vertical mixing which is mostly due to tides dominates the hydrophysical fields. The non-tidal basin (the Baltic Sea) can be considered as a huge estuary with extremely strong vertical stratification, which inhibits the vertical mixing. The diffusion coefficients approach to their molecular values, resulting in the extremely weak mixing between the surface and deep layers (see the explanation in Sect. 1.1 in chapter "Further into the Ocean"). This specific vertical stratification is maintained by two major factors: (1) the river runoff and precipitation–evaporation balance at sea surface and

[37] Brackish water is water that is saltier than freshwater, but not as salty as seawater.

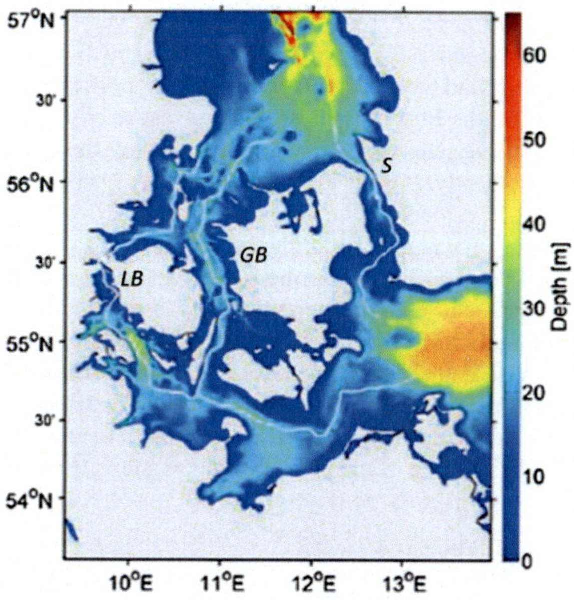

Fig. 30 NBTZ and talwegs along which model results were analyzed by Zhang et al. (2016b). Abbreviations LB, GB, and S show the position of the three straits Little Belt, Great Belt, and Sound. From Zhang et al. (2016b), with permission from Elsevier

(2) periodic intrusions of the saltier North Sea water triggered by extreme atmospheric conditions with an approximately decadal periodicity. All this suggests that there are large similarities between the Baltic Sea and the Black Sea.

The second challenge in the modeling of the Baltic Sea and the area of Skagerrak and the Norwegian Trench stems from the fact that the processes there are strongly dependent upon the exchanges in the NBTZ. Compared to other similar transition zones (e.g., Mediterranean–Atlantic Ocean), the NBTZ is more complex because of the presence of multiple straits. Under different weather or circulation conditions, their individual contribution to the exchange between the basins varies (Stanev et al. 2015b). Thus, the partitioning of flows and recirculation in the belts (which includes three major waterways: the Great Belt, the Little Belt, and the Sound) are central to the problem of the ventilation of deep Baltic basins by the North Sea water (see also, Meier 2005, 2007). An extremely high resolution is needed there to resolve the complex coastal line as well as the bottom topography in the three straits (note that the Little Belt is only 1 km wide). Without an adequate resolution, the hydrodynamics of the inflow–outflow system cannot be accurately simulated.

Taking SCHISM on board, Zhang et al. (2016b) developed a circulation model for the Baltic and North Sea. With this model, it was possible to achieve a resolution in the straits region of about 100 m (the minimum grid size of 60 m was specified in the narrow Little Belt). The model design was such that we prescribed finer resolution (blue to green colors in Fig. 31a) along the coasts, as well as in the area of the straits connecting the North Sea and the Baltic Sea. With such an unprecedented resolution for that time, all important details of the geometry of the straits were resolved, and as a result, the exchange of water between the two basins was very successfully assessed

(Stanev et al. 2018a). This exchange is very similar to the case of the Bosporus, i.e., the surface current is directed toward the North Sea and the bottom current (saltier and hence denser) flows in the opposite direction, toward the estuarine basin. Unlike the Bosporus, these currents, especially in the parts of the straits close to the North Sea, are very strongly influenced by the tides and therefore change direction periodically. It turned out that due to the shallow depths of the straits, the currents are also strongly influenced by the wind. Extreme wind conditions can significantly change the regime of filling the deep parts of the Baltic Sea with ocean water.

The SCHISM-based model of the Baltic and North Sea became the basis for several studies performed by my new postdoc, Sebastian Grashorn. DMI colleagues, with whom we have had an active collaboration since the ECOOP days, provided us with data from their measurements in the Danish Straits. Comparisons with these observations showed that the model reproduces the currents well. This gave confidence in the analysis and interpretation of the model data and contributed to a better understanding of the processes in the straits (Stanev et al. 2018a). It turned out that

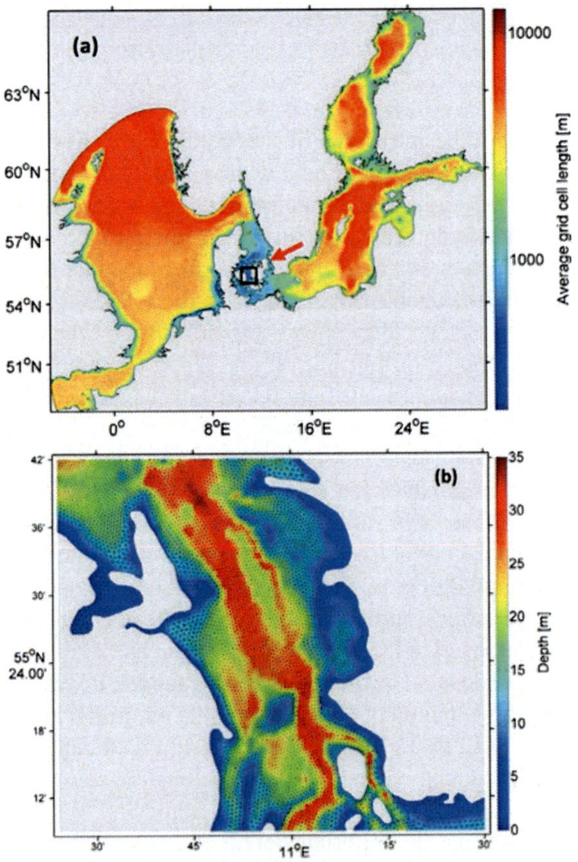

Fig. 31 Model grid size (**a**), and illustration of the fine grid resolution in the Great Belt (**b**), which is a zoom-in view of the rectangular box in (**a**). The small red arrow shows the position of a cross-section discussed further in Fig. 32. The bottom panel shows the bathymetry of the Great Belt and the model grid. From Zhang et al. (2016b), with permission from Elsevier

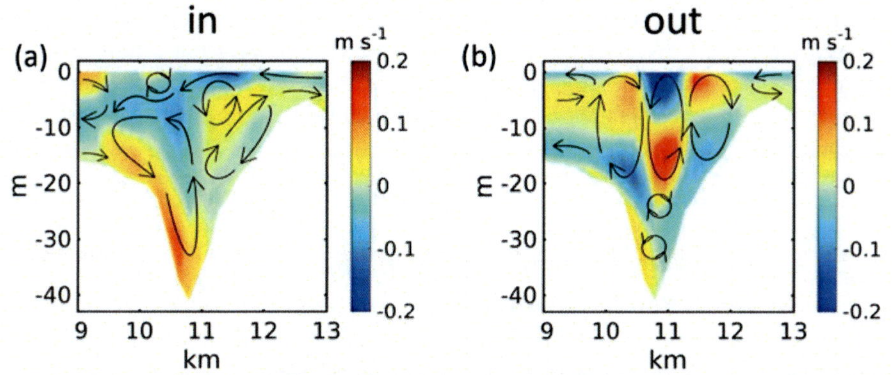

Fig. 32 Lateral velocity (velocity component perpendicular to the axis of the strait) along the section shown by the arrow in Fig. 31a during inflow (25–30 October 2014, **a**) and outflow (12–17 November 2014, **b**) conditions. The values are positive when velocity is directed toward the eastern shore and negative when directed toward the Western shore. From Haid et al. (2020), with permission from Elsevier

in the system consisting of three straits, the Sound, the Great Belt, and the Little Belt, the differences between the dynamics in the individual straits were significant. Comparing the results with those from a parallel model, in which we deliberately used a coarser resolution of 500 m, revealed that the resolution changed the results significantly. This presented a major challenge for scientists, who had been using in the past models with a very coarse resolution.

Numerical modeling helped ask questions that have not been explored in the straits before. It was known from the literature that in many estuaries, secondary circulation cells (Chant 2010) are created in a plane perpendicular to the direction of the main (along-channel) currents. Haid et al. (2020) investigated whether similar secondary circulation cells exist in the straits. One could expect that they must exist because the Baltic Sea can be seen as a huge estuary, but until then this question had not been investigated. Observational data provided by the Danish scientists revealed movements perpendicular to the main current in the Danish straits. These observations fitted well with the model predictions of secondary circulation. In addition, the model provided much more data than observations, which helped build a more complete and reliable picture of the dynamics. It was particularly important that the ultra-high resolution of the model allowed for the full development of these small cells of the secondary circulation, something that does not exist in coarse resolution models. The use of coarser computational grids resulted also in the creation of stronger stratification and misrepresentation of flow velocities. All these drawbacks in coarse resolution models are due to the missing secondary circulation cells, which provide an important mechanism for mixing and water mass exchange.

The dynamics in the straits and estuaries are similar, but there are also significant differences due to the fact that tidal waves are the main driver of circulation in the estuaries, whereas in the straits connecting the Baltic and North Seas, wind appears to be the main mechanical driver. The second difference is that there is a sea on one side of the estuary and a river on the other. In the case of the straits, there is sea on both sides. For these and other reasons elucidated by Haid et al. (2020), there are significant differences in the persistence of vertical stratification of seawater density in estuaries and straits.

7.2 *At the Border Between the Black Sea and the Mediterranean Sea*

One interesting difference between the straits of the Baltic Sea and the Black Sea is that while the Baltic and the North Sea are connected by three straits, the same number of straits (the Kerch, the Bosporus and the Dardanelles) connect four seas, the Azov Sea, the Black Sea, the Marmara Sea, and the Aegean Sea. This constitutes a huge sea cascade (Fig. 33), in which the salinity of the water is constantly increasing as we approach the Mediterranean: The salinity in the Sea of Azov is about 10–12; in the Black Sea, it varies from 17 to 18 at the sea surface (excluding areas of river plumes) to ~ 22.3 at the bottom; in the Sea of Marmara, the salinity increases from ~ 22 to more than 38 within a surface layer of only ~ 50 m thick; in the Aegean, the salinity is ~ 39, and its vertical gradient is much smaller than in the other (estuarine) seas.

The success with the implementation of the Baltic-North Sea model motivated a similar numerical modeling study of the cascade from the Sea of Azov to the Mediterranean Sea (Stanev et al. 2017a).[38] The model developed in this study provided comparable or better results than others, until then existing models, which do not assimilate observational data.[39] Currents and vertical stratification of temperature and salinity, as well as their seasonal variations, agreed very well with existing observations. Eddies in coastal regions, particularly important for the mixing of waters, were also represented realistically. The superiority of the new simulations compared to earlier numerical studies was confirmed by the model‘s ability to describe the dynamics of straits and denser waters coming from the Sea of Marmara and spreading along the bottom of the Black Sea (Fig. 35). This S-shape pathway of propagation of mixed Marmara Sea water is in agreement with observational data previously reported by Latif et al. (1991), compare also Fig. 35 with Fig. 41 in chapter "Further into the Ocean". The maximum depth of the main intrusions of the Marmara Sea waters, up to about 700 m, corresponded exactly to the available at that time observations.

[38] See Sect. 6 and Fig. 18 in chapter "From Regional to Coastal Oceanography" describing an earlier numerical model aimed at studying the cascade from the Azov Sea to the Marmara Sea.

[39] It is ment here that the realism of models in which data are assimilated is largely due to good data quality. The model described here does not assimilate data, i.e., its realism is due to an adequate model system.

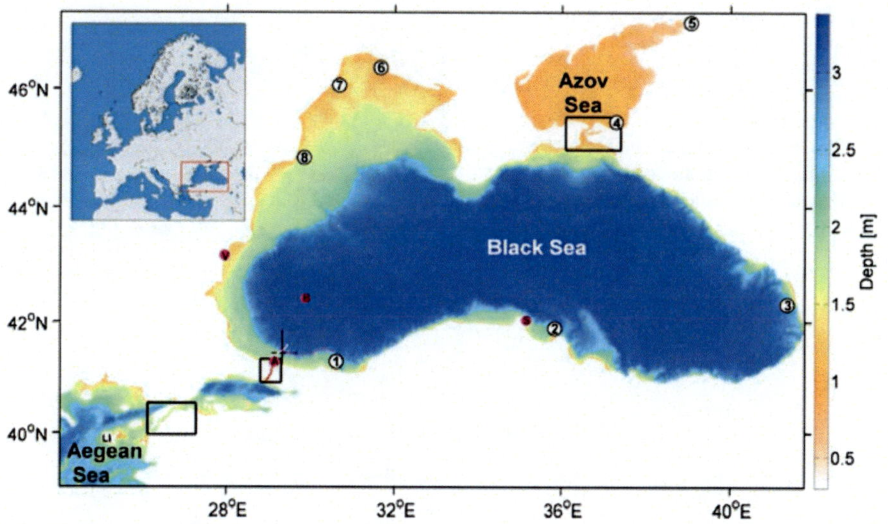

Fig. 33 Cascade extending from the Azov Sea to Aegean Sea. The map shows the model area and bathymetry in log scale (2 means 10^2 m). Black boxes are the areas around the Kerch (Fig. 25), the Bosporus (Fig. 34), and the Dardanelles Straits, where a very fine resolution (down to 70 m in the Bosporus Strait) has been used. Geographic names, locations, and transect lines where analyzes of model simulations are discussed by Stanev et al. (2017a) are also given. The following notations have been used: LI Limnos Island, V Varna station, S Synop. The mouths of the following rivers, from which run off is used in the model, are shown with numbers: 1 Sakarya, 2 Kizilirmak, 3 Rioni, 4 Kuban, 5 Don, 6 Dnieper, 7 Dniester, 8 Danube. The position of the Black Sea is shown as an inset in the upper-left corner. From Stanev et al. (2017a)

Fig. 34 Bathymetry in the area of the Bosporus Strait where a very fine resolution has been used in the model. Geographic names, locations, and transect lines where analyzes of model simulations are discussed by Stanev et al. (2017a) are also given.The inset with the model grid gives an example of enhanced resolution in the box just to the North of the BSC section. From Stanev et al. (2017a)

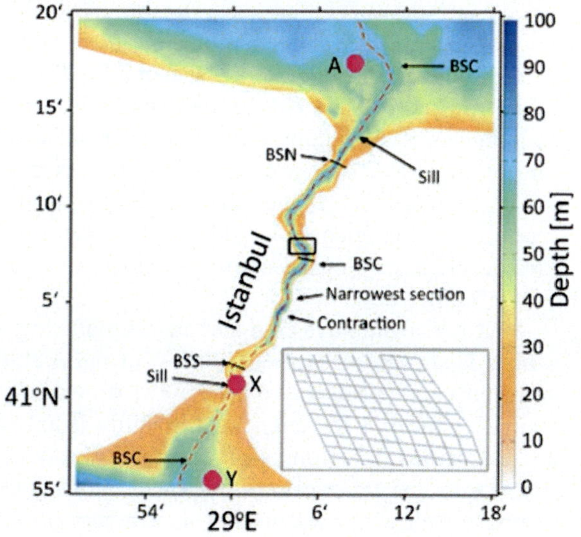

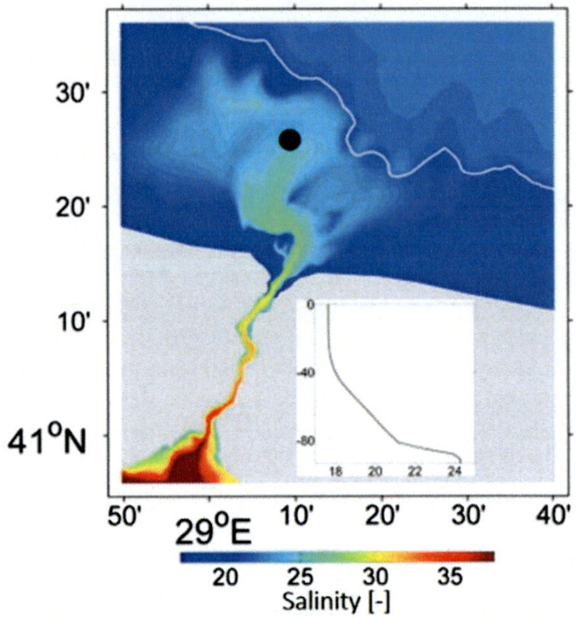

Fig. 35 Bottom salinity during the inflow event on 27 November 2008. White isoline is the 250-m isobath. The inset shows a vertical profile of salinity in the location marked on the horizontal plot. From Stanev et al. (2017a)

The period of model integration was chosen as to coincide with the observational periods during which data from long-term observations are available (Jarosz et al. 2011). Simulated and observed along-channel currents at location X (Fig. 34) are shown in Fig. 36. Timings of the major simulated events corresponding to the large inflows into the Black Sea (e.g., October 07, 2008, from November 20, 2008, to November 27, 2008, and some others) agree between the simulations and observations. Velocity magnitudes are also similar. During these events, the upper layer current was completely blocked. At this analysis position (the Marmara Sea exit), no instances of the bottom current being blocked were observed during the whole simulation period. Blocking of the bottom current occurred rather at the northern exit of the strait and resulted in a one-layer transport. The coherence between the appearance of anomalies of wind and strait transport suggests that the former plays a major role in shaping the short-term variability of strait transport (compare the two curves in Fig. 36c).

A further quantitative presentation of simulations and their agreement with observations is shown in Fig. 36d, e. Most of the profiles and, in particular, the time-averaged ones reveal a two-layer flow, which is typical for estuarine-dominated regimes. However, the temporal variability is extremely large, changing the state of the estuarine flow from one to two layers and vice versa. The mean profiles from simulations and observations, presented by the green curves in Fig. 36d, e, agree relatively well with each other. This is a very positive result given the fact of the strong dependence of the solution on the local depths which are not error-free in

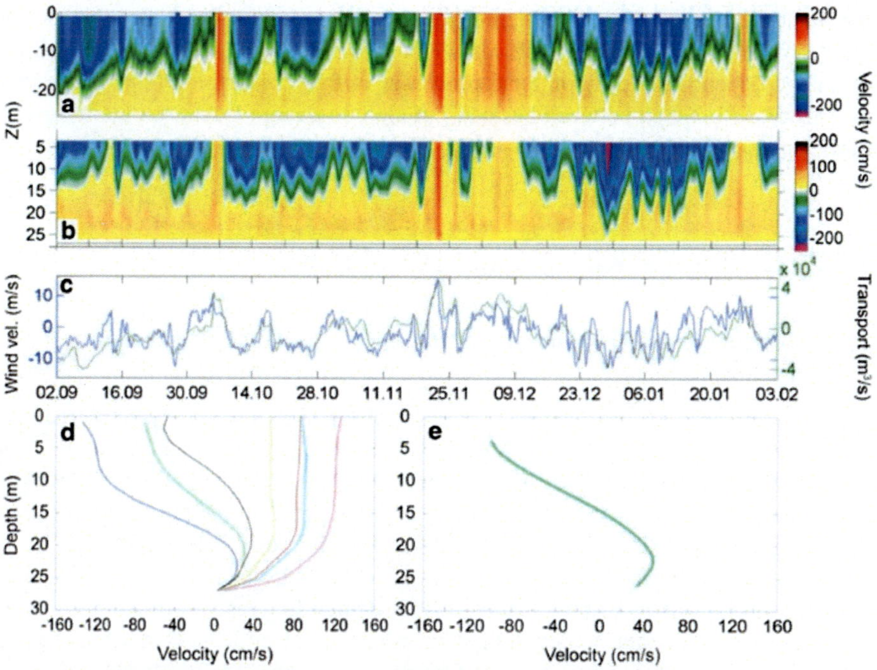

Fig. 36 Along-strait components of currents (positive toward the Black Sea) in the southern Bosporus at the position X shown in Fig. 34 during September 2008–March 2009. **a** Numerical simulations, **b** Replotted from Jarosz et al. (2011). **c** The along-channel wind velocity (positive is along the channel directed to the Black Sea, that is, roughly to the north) and the net transport (positive to the Black Sea) across the section BSS (see Fig. 34 for its position). Velocity profiles at the same location as in **a** and **b** are shown from the simulations (**d**) and observations (**e**, replotted from Jarosz et al. 2011). The individual lines in d correspond to: red: 06 October 2008 00:00, blue: 26 October 2008 12:00, black: 02 November 2008 00:00, magenta: 22 November 2008 18:00, cyan: 06 December 2008 00:00, yellow: 21 January 2009 00:00. The green lines in d and e show time-averaged profiles for the period in **a–b**. From Stanev et al. (2017a)

the underlying digital elevation map (DEM), and the fact that model forcing is not perfect.

It is well known that the sea level in each basin up the cascade is higher than the level of the adjacent basin further down. However, this is true for the basin mean sea level. Prior to the study of Stanev et al. (2017a), no attention was paid to the fact that the sea level in the central part of the Black Sea was lower than the average sea level in the Marmara basin. This is due to the cyclonic circulation in the Black Sea, which raises the sea level in the coastal areas and lowers it in the interior of the basin. The sea level near the Bosporus Strait determines largely the magnitude of water transport in the Strait. Since the local sea level in coastal areas is dependent on the general circulation in the Black Sea, the latter controls the exchange of waters with the Sea of Marmara (Stanev et al. 2017a). This turns out to be very important, since the seasonal increase in the intensity of the Black Sea circulation in winter raises the sea level in coastal areas, which enhances the outflow of Black Sea waters into the Sea of Marmara. These results allowed to formulate the concept that interbasin exchange depends not only on local processes in the Strait region (e.g., local wind or density differenced between basins), but also, to a large extent, on the dynamics of the basins as a whole.[40]

8 Predicting the Known and the Unknown

8.1 Earth System Knowledge Platform

The work in many of the Centers of the Helmholtz Association is complementary and often requires a collective approach to certain issues. In 2011, it was realized that coordination between Research Centers was needed in terms of how knowledge about the different domains of the Earth is disseminated to the public and to users of scientific information. The main objective of the initiative "Earth System Knowledge Platform (ESKP)", which was launched in 2012, was to clarify the interlinkages between research at the different Helmholtz Research Centers working in the field of Earth and environmental sciences. Focus was put on the collection of scientific data and, on this basis, on the development of appropriate products informing society, politics, and economy about global environmental change, its effects, and the associated risks. Ways had to be found to disseminate this information and to propose strategies for adapting to the changes, as well as to advise different economic sectors and social groups. The ESKP relied on the involvement of external partners, mainly universities. Topics of significant socio-political importance were identified as priorities, such as

(i) Earth system dynamics and risks;
(ii) Climate change;
(iii) Water availability and management;
(iv) Ecosystem dynamics and biodiversity; and
(v) Sustainable use of resources.

This initiative continued until the end of 2020. Its first phase aimed at developing a concept, methodology, and strategic orientation, which were to ultimately lead to an operational interface between the different Helmholtz Research Centers, social actors, and decision-makers. All this had to be directed toward developing a mechanism for the provision and dissemination of scientifically sound and reliable information to mitigate the risk of extreme natural events and avoid the effects of

[40] While working on this book, I realized that I could have developed this theory as early as 2005. Figure 18 in chapter "From Regional to Coastal Oceanography" shows that (1) the sea level in the central Black Sea is lower than the sea level in the Sea of Marmara and (2) that its slope undergoes large seasonal fluctuations (compare this figure with Fig. 3 of Stanev et al. 2017a). This result could motivate me at that time to do a more thorough analysis of the dependence of the currents in the strait on the intensity of the general circulation in the Black Sea. Obviously, I failed to notice it.

adverse environmental changes, or to propose strategies to adapt to them. Two of the pillars on which the knowledge platform was build were (1) observations (integrating observing systems needed to manage the effects of environmental change) and (2) predictions (improving benefits through forecasts). Because of the involvement of KSD in coastal ocean forecasting, H. von Storch commissioned me to organize the participation of HZG in ESKP. This initiative was expected to work closely with the European GMES program. Cooperation with other international programs such as the Global Earth Observing System of Systems (GEOSS) was also to be established. The strategy of ESKP implied the involvement of institutions outside the Helmholtz Association to work together, and ICBM was included in the research project "Identification, quantification, and prediction of coastal risks".

Hans von Storch asked scientific and practical questions in an intelligent way. "Can we study the response of the environment to impacts that have not been observed before"? Volcanoes, tsunamis, or melting glaciers can have catastrophic effects, but they have had them before, so we can more or less expect their consequences. Can human activity create catastrophic events in nature, the likes of which have not happened before, and can we assess their risks and consequences? This was unconventional because it was not just about forecasting in the familiar sense (forecasting similar situations as we have seen repeatedly in the past and present) using numerical models and observational data. It was also about discovering the unknown using the available scientific tools. This is where science has a say, and in particular, numerical modeling of the environment, which enables us to experiment safely with nature. Weather forecasting is a good example of predicting the known. As far as predicting the unknown is concerned, here we can refer to anything that could potentially happen in nature but that we have not yet discovered, e.g., hitherto unknown changes in the environment due to human activity.

8.2 Straits and Possible Future Ecological Catastrophes

You will ask what straits have to do with ESKP. Is not ESKP supposed to be dealing with catastrophes in nature? I have already examined the catastrophic re-connection of the Black Sea and the Sea of Marmara (Sect. 5 in chapter "From Regional to Coastal Oceanography"). I have also outlined some problems in relation to shipping through the Bosporus (Sect. 16 in chapter "Further into the Ocean"). I will add here something that is happening or may happen in the near future. Stanev et al. (2017b) showed that the concentration of microscopic particles is highest in the Bosporus area (Fig. 14). There are at least two reasons for that: (1) the megapolis of Istanbul being the cause of pollution, and (2) the particles are created due to active biological processes related to the mixing of the waters of the Sea of Marmara and the Black Sea. In both cases, water originating from the Bosporus Strait descends as denser water along the seabed (gravity current) and penetrates into the interior of the basin in thin intrusive layers. The depth of intrusions and their magnitude can change under the influence of natural factors (Stanev et al. 2019b), e.g., global warming (Fig. 9). This could lead to unexpected changes in the state of the BGC in the Black Sea.

More drastic anthropogenic influences may soon be added to climatic factors. The Istanbul Canal, planned by the Turkish Government, aims to connect the Black Sea with the Sea of Marmara (Kundak and Baypınar 2011). It will run parallel to the Bosporus, in the Western part of Istanbul, and is expected to relieve heavy traffic. The idea is not new. It was considered as early as the sixteenth century by Sultan Suleiman the Magnificent.

There are reasons for building the Istanbul Canal. While in 1936, an average of 13 ships per day passed through the Bosporus that number has increased to an average of 255 ships per day in 2012. In that time, the population of Istanbul has increased from 750,000 to 13.9 million. Avoiding traffic accidents is a big concern of the Turkish government, since there have been many. On 14 December 1960, a Yugoslav and a Greek tanker collided, resulting in 20 deaths and a large oil spill. On March 1, 1966, two Soviet oil tankers collided. Again, large quantities of oil leaked, causing a fire on the ferry connecting the European and Asian parts of Istanbul. On 1 July 1970, the Italian oil tanker Ancona collided with a building on the coast during which five people were killed. There are other similar cases.

On 27 April 2011, Prime Minister Recep Tayyip Erdogan announced his intention to build a new link between the Black Sea and the Sea of Marmara. The canal will be about 40 km long, 150 m wide, and 25 m deep. The cost of the project varies, according to various sources, from € 8 to 14 billion. There is also a political issue. Some believe that the Turkish authorities could use loopholes in the Montreux Treaty[41] to force shipping to use the new canal or to change the status quo created by this treaty with regard to the transit of military equipment and warships.

The problems associated with environmental change are not less important, and I, along with several Turkish colleagues, have taken an interest in them (Saçu et al. 2020). In practice, adding the canal will change the total cross-section area, which will change the balance of water flowing in and out. This will affect the stratification of the waters of the Black Sea and its ecosystems (Şen et al. 2022). The consequences are difficult to predict, but the model systems we have created with J. Zhang, S. Grasshorn, and Turkish colleagues can help determine whether the Black Sea could be facing an ecological disaster once the new navigation channel is built. I believe that the relevance of modeling works to the ESKP theme of catastrophic phenomena in nature is beyond doubt.

[41] The Treaty of Montreux (officially in French Convention sur le régime des détroits) of July 20, 1936, restored to Turkey full sovereignty over the Dardanelles, the Sea of Marmara, and the Bosporus. It regulates the free movement of ships through these waters. The convention allows Turkey to close the strait to all warships in time of war.

8.3 Offshore Wind Turbines Swirl Water

Wind energy is about to become an unavoidable component of the energy and climate policies. Offshore wind turbines are significantly more efficient than those over land because the wind blows more strongly over the smooth sea surface than over land. Not to mention, that in densely populated areas, wind turbines are not welcome to the population. The North Sea, which has a high wind potential, is shallow, and this is conducive to the expansion of wind energy use. By the end of 2020, 7760 MW of wind turbines had been installed in German waters, and 20,000 MW of wind power capacity should be built by 2030.

Wind farms are clusters of tens to several hundred wind turbines. The turbines take part of the wind energy. Wind speed deficits of up to 20% have been measured in strong winds, which is accompanied by mixing of the air on the leeward side. Behind the turbines, trails (plumes) of air with different characteristics from those of the surrounding air are formed, which are easily observed by aircrafts or satellites. Their lengths can exceed 30 km. At longer distances, the perturbed flow gradually dissolves and adapts to the surrounding conditions. In general, wind farms can leave footprints in the atmosphere over distances of more than 70 km. Although these effects have long been known, the local and global climate change associated with wind farms is still underestimated.

Wind turbines are located in constantly moving seawater. Ther foundations act as obstacles to the water flow. By flood, water flows between the poles of individual turbines in one direction; by ebb, the direction reverses. In the North Sea, the speed of the tidal currents reaches 1 m s^{-1}. Obviously, the effect of this interference with nature will not be negligible. It is not only the currents that are expected to be affected by wind turbines, but also the turbulence of seawater. Everyone can see this for himself or herself. Take a stick, stand on a small bridge over a relatively homogeneously flowing stream, and dip the stick. If your stick is thin, you will see only a slight trace. If it is thicker, eddies will begin to spread downstream, i.e., what you will see is similar to what I will talk about below.

The question is how the turbine foundations affect the marine environment, a question just for ESKP and not least of great public interest. Eventually, the increasing use of wind power will show whether there are any sensitive changes in the environment. However, there is a danger that the coastal environment will undergo irreversible changes. Therefore, before future measurements show any adverse effects, it would be more practical to investigate and predict them theoretically. The large-scale expansion and operation of offshore wind farms in coastal waters have confronted us with new scientific challenges.

How did it all start? At a conference, I listened to a talk by the Belgian scientist Kevin Ruddick, who was exploring the ocean using space imagery. From these images, the concentration of sediments in seawater can be determined. Landsat-8[42] images, which K. Ruddick was using, showed both trails left by passing ships and by

[42] Landsat satellites are NASA's Earth observation satellites. They observe the Earth's continental surface and coastal regions. Eight satellites have been launched since 1972.

individual wind turbines. The sediment trails started from the location of each individual turbine, the sediment concentrations were about five times higher than in the surrounding undisturbed space, and the trails were well oriented in the direction of the tidal currents (Vanhellemont and Ruddick 2014). The foundations of the turbines were apparently the reason for the creation of sediment trails. At the same time, my coworkers and I were taking the first steps in modeling marine sediment dynamics using a sediment block in SCHISM. Was it possible with the model we were using to reproduce what satellite was seeing?

Kármán Streets

In hydrodynamics, the problem of flow around cylindrical obstacle has been known for a long time. Repeated measurements have been made in the laboratory, and the theory is well understood. The occurrence of a series of vortices behind the obstacle depends on the flow velocity, the dimensions of obstacle, and the viscosity of fluid. Depending on these parameters, different configurations of vortex "streets" can be observed behind a cylinder immersed in a moving fluid. At very low velocities or very high viscosities, no disturbances are observed. As the flow velocity gradually increases, a laminar[43] trace is created. When the velocity increases above a certain critical value, vortices arise, which are named after the great hydrodynamicist Theodor von Kármán.[44]

The laboratory channels where measurements have been made have dimensions many times smaller (meters) than wind parks, and it is not at all certain that at larger scales, the vortices will manifest themselves in the same way as in the laboratory channels. Expectations that this would happen were supported by satellite observations of disturbances in the currents around some islands where eddy "streets" were visible. Furthermore, it was not sure whether SCHISM would be able to realistically reproduce processes on scales many times smaller than those previously modeled. What we did first was to reproduce the results of some laboratory experiments known from the literature. The model did a great job. We had a modeling tool to study the effect of wind turbine foundation on water flow! Due to the fact that the diameter of these foundations is only a few meters, the model grid around the foundations had to have cells of no bigger than approximately 1 m. Along with this, the ocean flows had to be simulated over a larger area, which we achieved by coarsening computational mesh with a distance from the turbine.

[43] Laminar flow (lat. lamina "plate") is a movement of fluids in which no visible turbulence occurs, that is, the fluid flows in layers that do not mix with each other.

[44] Theodore von Kármán (1881–1963) was born in Budapest. Until the 1930s, he did much of his theoretical research in Germany, including the discovery of Kármán vortex streets. After Jews in Germany began to be persecuted, he moved to the United States, where he established himself as a pioneer of modern aeronautics. In Pasadena, California, Von Kármán founded the Jet Propulsion Laboratory. After 1956, and for the rest of his life, he was the head of the Institute of Fluid Mechanics in Belgium that bears his name and was involved in the development of rocket propulsion technology.

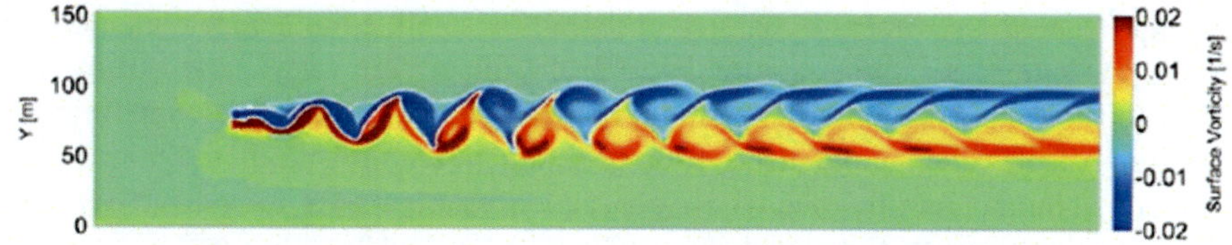

Fig. 37 Vorticity of flow around a single cylindrical obstacle. Above the axis of the vortex street, the water rotates in a clockwise direction and below it in the opposite direction

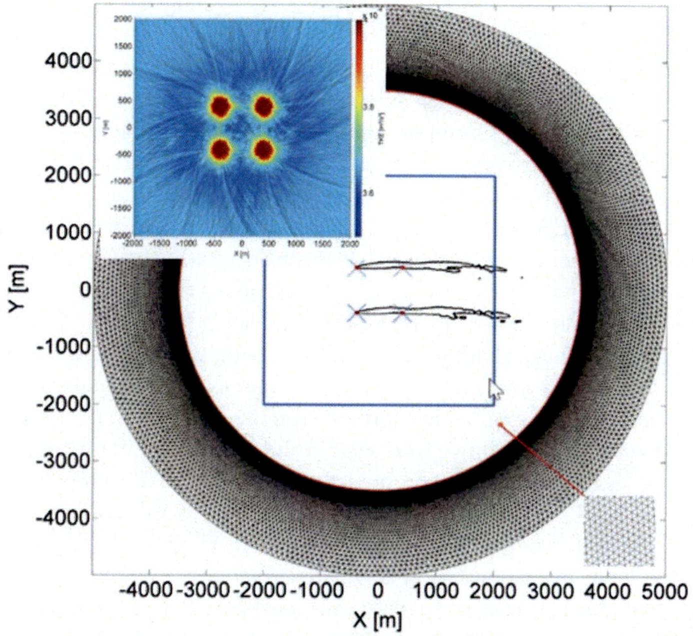

Fig. 38 Unstructured grid used in the area of an experimental wind park consisting of four turbines placed under the influence of tidal currents. Inward from the red circle model resolution is uniform (5 m) which is exemplified by the small grid-inset shown in the bottom-right corner. Resolution around the foundations increases further to approximately 1 m. The blue crosses show the positions of the piles, and the blue rectangle is the area shown in the in the upper-left corner. In this inset, the turbulent kinetic energy (TKE) averaged over one tidal period is shown. High turbulent kinetic energy values make sediment erosion possible. To visualize the extent of the wake behind the piles, isolines of the surface sediment concentration of 12.5 g m^{-3} are plotted using the results of numerical simulations. From Grashorn and Stanev (2016)

Numerical experiments have shown that obstacles of the shape and size of wind turbine foundations give rise to eddies of the Kármán type (Fig. 37), which significantly alter the turbulence of the ocean (Grashorn and Stanev 2016). The latter leads to erosion of sediments from the ocean floor (Fig. 38), thus spreading them into the water column and making them visible to satellites. The model proved to be able to reproduce the sediment dynamics observed by satellites in a very realistic way. What the satellites cannot capture, because the images are very sparse in time, but the models can, is the temporal evolution of the sediment behind the wind turbines. These obstacles change the distribution of sediment on the bottom and in the water column, a human interference whose consequences are not yet fully clear. Together with the new research, the road was paved to investigating how the operation of wind farms affects the environment, an issue, which is a priority for HZG/Hereon today.

8.4 Sea Storms and the Changing Seabed

The history of the eastern coastal areas of the North Sea, from the Netherlands to Denmark, can be described as an ongoing struggle between the people who wanted to settle there and the adverse nature that often flooded the area during sea storms. During the ice ages, when the ocean level was more than a hundred meters lower than it is today (see upper-left panel of Fig. 16 in chapter "From Regional to Coastal Oceanography"), the British Isles were connected to the European mainland, and the coast was far from its present position. The continuous rise in ocean level due to the melting of glaciers is causing the shoreline to advance landwards. The coastal zone in the North Sea is very shallow and the tides high, therefore the water inundates vast areas. Flooding is amplified by adverse coincidences of events, when tides are high, and sea storms strong, which further raise the sea level in the coastal zone.

The Roman historian Pliny, who visited the Germanic North Sea coast in the middle of the first century AD, referred to a landscape that was flooded twice a day, being part sea and part land. At that time, people built hillocks near their homes that remained unflooded during storms, and there they found refuge. In the last millennium, there has been a slowdown in sea level rise, and people have begun building small dikes at these sites. These were the first acts of human intervention in the morphodynamics of coastal areas. Today, the entire coast is lined by dikes about 10 m high, i.e. the coast is no longer natural but man-made. Human intervention in coastal processes over the last few hundred years can be seen as an act of coastal engineering. The objectives of coastal protection present the driving force for the further development of knowledge in the field of coastal hydro- and morphodynamics. In the beginning, trial-and-error methods were used, but today, the most advanced technologies and theories are applied.

The Intergovernmental Panel on Climate Change (IPCC[45]) provides strong evidence that the ongoing global climate change will lead to future risks in many coastal areas. There is no doubt that the North Sea is one such area at risk of flooding due to extreme water level during severe storms. Storms usually last no more than 12–24 h but can raise the sea level significantly (up to 3.5 m in the North Sea). Dikes

[45] The IPCC was created in 1988 with the support of the *United Nations Environment Programme* (UNEP) and the WMO to assess climate conditions of planet.

do not provide complete protection because large rivers, along the course of which the sea enters deep into the land, cannot be closed during a storm.[46]

The impact of ocean level rise on potential risks, which is a major concern in the German coastal zone, can be assessed using numerical models. What was usually done was very simple. In the numerical models, scientists increased by a constant number, say two meters, the bottom depths at all grid points in the study area. This was supposed to mimic future ocean level rise by two meters. However, a number of issues arise. Typically, in the previous scenarios with present and future bottom depths, the same boundary conditions were used at the open boundary of the model. This is incorrect because the modern boundary conditions are not applicable to future periods. One solution would be to use the model boundary data from global future scenarios.[47] The problem is that the global models are too corse, and it is hard to expect that, if we took data from them to construct boundary conditions, the results would have a reasonable representativeness for the southern part of the North Sea. The main conceptual problem was that we cannot expect that, as sea level changes, the change in depths will be the same everywhere, i.e., we do not know what the seabed will look like in future.

In 2012–2013, initial reports emerged of an annual seafloor mapping off the German coast over a period of ~ 30 years. My new Ph.D. student, Benjamin Jacob, within the framework of his thesis, which was supported by ESKP, performed the first computations aimed at determining the influence of the different bottom topographies on the overall hydrodynamic state of the southern North Sea. The model that he used was the first unstructured grid model that J. Zhang and I developed. It covered the southern part of the North Sea, was computationally efficient and suitable for use by students.

The study of Jacob et al. (2016) examined the response of tides in the southern North Sea to morphodynamic changes. What proved interesting was that the model response to the use of different bathymetric data was not only local, but traces were found even as far as the British Isles. The explanation is that tidal waves in the North Sea propagate local disturbances throughout the basin, which creates an effective physical teleconnection mechanism. The distortion of the tide, as a result of relatively small changes in bathymetry, is shown to be crucial for sediment transport, i.e., the feedback between hydrodynamics and sediment dynamics is at work.

The hydrodynamics-morphodynamics system is not symmetric. While hydrodynamic processes respond instantaneously to morphological changes, morphological changes require a longer time for sediment redistribution. Because sediment settling and compaction at the bottom are a slow process, a delay in the morphological response to hydrodynamic forcing is observed. That is, in the hydrodynamics-morphodynamics systems, we have slow (morphodynamics) and fast (hydrodynamics) processes, which are interconnected by the processes of erosion and sediment

[46] The Ems barrage has been mentioned already, see Fig. 27, but Ems is a relatively small river; therefore, it can be closed during severe storms or when there is a need to increase depth of the estuary.

[47] There are no open boundaries in global models, so there are no problems with open boundary conditions either.

settling and transport. In other words, the sediment dynamics is the connecting link between hydrodynamics and morphodynamics.

Bathymetric evolution is a very complex issue. The great challenge is to use a reliable morphodynamic model that adequately reproduces the evolution (over all time scales) of the morphological state. Such a numerical model does not yet exist. Only some idealized model studies have been published so far with limited application to large-scale natural systems. Hence, Jacob and Stanev (2021) performed a diagnostic study, as a first step toward understanding the control of bathymetry. 31 bathymetric maps were available for the period 1982–2012 with an extremely high resolution of 50 m (Fig. 39). A numerical model was used to compare the hydrodynamics in the coastal zone resulting from the use of these bathymetric data. As a first step, the computational block for sediment was turned off. The analysis of the model simulations showed that the changes in tidal currents caused by changes in bathymetry are such that they can reverse the direction of erosion and sedimentation processes in some areas (if the sediment transport module is enabled in the model). In a model accounting for morphodynamics, this would result in sediment recharge to the tidal basins, which can be considered a natural response to the rapid deepening indicated by the bathymetric data for the period before 1995.

Results from the numerical experiments with different bathymetries provided clear indications that one of the conditions for adequate numerical simulations in the coastal ocean is the use of adequate (time-varying, as it is in the real world) bathymetric information. This would require that modern bathymetric charts and the data used to calculate them be revised and that, where possible, temporal variability be accounted for. More importantly, campaigns need to be organized to regularly monitor future changes in coastal bathymetry and to produce datasets for individual years. This is very important for future morphodynamic modeling, where appropriate

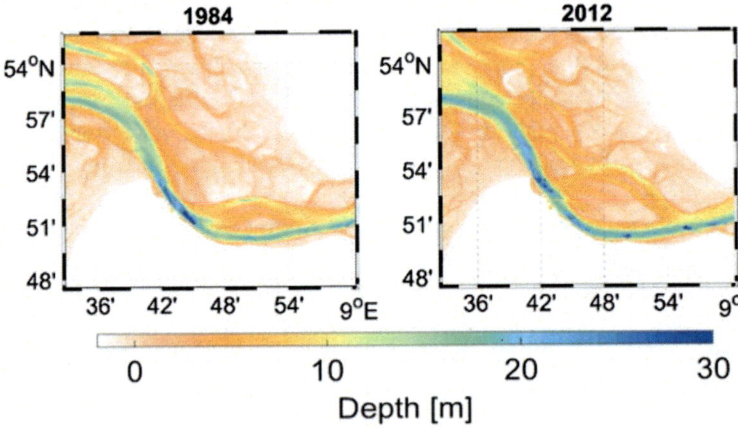

Fig. 39 This is how different the seabed can look in the estuarine area of river Elbe. For further details, see Jacob and Stanev (2021) from where this illustration originates

data are needed for both model initialization and validation. The lack of such data remains an obstacle to the development of coastal morphodynamic models.

9 Marine Litter

9.1 *How it All Started*

During the Christmas-New Year period of 2016–2017, I was visiting Vietnam and Cambodia with my family. We stayed for a few days in a hotel on the coast of southern Vietnam. One day, I was walking along a nice beach, and when I got to the mouth of a small stream, the view became very unpleasant. The creek bed was full of trash, and the waves on the shore uniformly lifted and lowered a thick cover of plastic debris. The picture was repeated in even more horrifying proportions a few days later when we were navigating down the river Mekong. There, the scale of the pollution was considerably greater. One would think that this is only the case in densely populated areas and mainly along rivers and their estuaries. My observations showed the opposite. On the previous tourist trip, exactly two years earlier, my family and I had an unforgettable time on Cayo Largo del Sur, a small holiday island about 50 km off the southern coast of Cuba. It was probably the most beautiful tropical place I have ever visited. The island is famous for its long white sandy beaches and huge lagoons. Unfortunately, the sea had brought a lot of litter here too, away from the populated areas.

Various scientific and media reports and announcements herald threatening predictions about the level of pollution in the World Ocean. Estimates show that around 10% of the plastic produced globally reaches the oceans, which in 2018 alone amounted to approximately 36 million tons. Plastic pollution has been declared by the United Nations as one of the greatest environmental challenges of our time. Although the situation in Southeast Asia is the most critical, there is a concern that many Western countries also contribute significantly to marine pollution.

In recent years, the number of scientific publications on marine plastic litter has increased exponentially. With the growing number of publications, more and more negative effects on the ecology and economy have become known, some of them significant and irreversible. These range across a wide spectrum and include fatal consequences for marine life due to the ingestion of plastic materials, destruction of the tourism industry, and toxins entering the sea. It has recently been shown that nanoplastics[48] can pass through selective membranes such as the blood–brain barrier, which separates blood circulation and brain extracellular fluid in the central nervous system, or it can pass through the placenta. The influence on metabolism has already been demonstrated.

[48] Depending on the size, the waste is divided into macro-, micro-, and nanoplastics with sizes cm, μm, and nm, respectively.

In early 2015, the Lower Saxony Ministry of Science and Culture issued a call for research projects in areas related to the development and use of marine resources and the study of the resulting ecological consequences. Among some other priority aspects, ocean pollution, including plastic waste, was mentioned in the call. Jörg-Olaf Wolff gave the idea to prepare a joint research project on "Macroplastics pollution in the southern North Sea—sources, pathways, and abatement strategies". An interdisciplinary consortium of physical oceanographers and ecologists was formed to synergistically use observational and numerical modeling methods and to seek input from so-called Citizen-Science.[49] It was known that members of the general public often work in collaboration with or under the guidance of professional scientists or scientific institutions, and we wanted to test how useful this collaboration could be for our work. Our objectives were to clearly identify the main sources of plastic pollution in the North Sea (e.g., from ships, rivers, tourism) and the main pathways for the spread of contaminants. The project, which was to run between 1 April 2016 and 31 March 2020, included several subprojects; the one led by me aimed to use the tools of numerical modeling.

9.2 *Marine Litters in Shelf and Semi-enclosed Seas*

In the computer simulations performed by my Ph.D. student, Marcel Ricker, we used data of currents produced by a numerical model, "seeding" the model domain with virtual particles and exploring their propagation and accumulation pathways. There had been many similar works up to then, but they applied mainly to the global ocean. The realistic numerical reconstruction of the famous subtropical "garbage gyres[50]" in the Pacific had proved the usefulness of numerical modeling in dealing with marine debris. Simulations of their dynamics in shelf seas, and the North Sea in particular, were unknown until then, even though the North Sea is one of the most studied seas in the world.

Numerical modeling gives a wide scope for experimentation. In some simulations, the properties of the virtual garbage were simplified, so that it only moved under the action of the ocean current. In other simulations, the influence of wind and waves was added. It turned out that, unlike the case of the Great Pacific Garbage Patch, macroplastics in the North Sea "prefer" to accumulate in the area of ocean fronts (Ricker and Stanev 2020), narrow regions separating water masses with different properties (Fig. 40).

---

[49] Citizen-Science refers to methods and fields of science in which research is conducted with the help of or entirely by interested lay people. They formulate research questions, conduct research, report observations, make measurements, and publish or evaluate data.

[50] The Great Pacific Garbage Patch is located in the North Pacific Ocean. The reason garbage accumulates there is due to ocean currents. The North Pacific subtropical circulation gyre, which rotates clockwise, makes it difficult for the garbage to escape. This is the largest and most famous garbage patch in the World Ocean; there are others in the South Pacific, North and South Atlantic, and Indian Ocean.

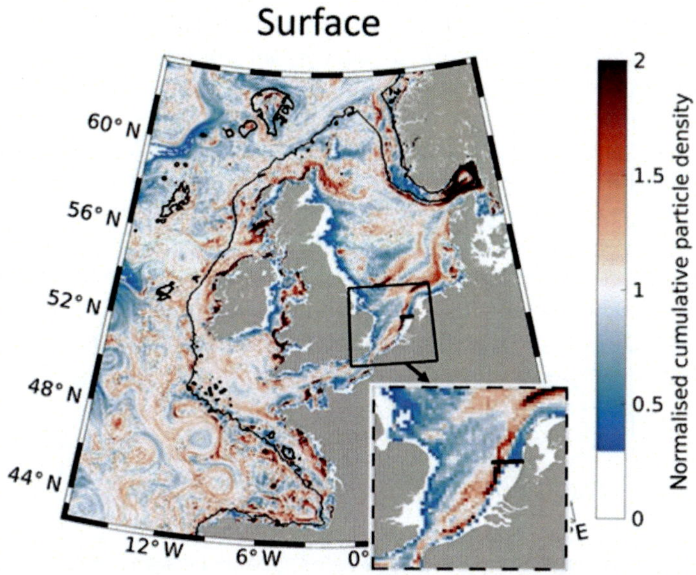

Fig. 40 Tendencies of accumulation shown as the January mean normalized cumulative particle density (NCPD). This characteristics measures the compaction (> 1) and divergence (< 1) of particles in model grid boxes. From Ricker and Stanev (2020)

Step to the Black Sea

In 2018, the Permanent International Secretariat of the Black Sea Economic Cooperation (BSEC) and the International Center for Black Sea Studies (ICBSS[51]) have taken a step toward promoting cooperation and coordinated action among Black Sea countries to address environmental challenges. With the support of the Ministry of Environment and Water of the Republic of Bulgaria, they organized on 4 April 2019 in Sofia the International High-level Conference on "Interbasin Cooperation on Marine Litter: a focus on the Danube River and the Black Sea". It was realized that the Danube River and the Black Sea basin are not only valuable natural resources of great ecological and economic importance, but they are unfortunately also largely endangered. Data showed that the Danube brings 530–1500 tons of plastic into the Black Sea annually. Unsustainable waste management, a lack of infrastructure and of coordinated national and regional policies, in addition to insufficient awareness, exacerbate the problem, thereby undermining any attempts at sustainable growth. Marine litter management plans already existed for the Mediterranean and the Baltic Sea. A step in this direction had to be taken in the Black Sea as well.

Marcel Ricker and I repeated for the Black Sea what had already been done for the North Sea (Stanev and Ricker 2019). The interesting discovery was that, similarly to the North Sea, in the Black Sea, there was no pronounced accumulation of floating debris resembling the Great Pacific Garbage Patch. This was explained by the great variability of the intensity of circulation in the Black Sea, which makes it impossible to consolidate the litter in one particular area. It turned out that the characteristic time for litter to “wash” from the sea to the shore is only about 20 days, i.e., an object floating in the sea will visit the coastal area every 20 days, increasing the probability that it will strand at some point. This indicated that collection of floating litter in coastal areas would be an effective tool to reduce pollution of the Black Sea. The calculations also showed that the areas most at risk extended from the Kerch Strait to the west coast.

9.3 *Unexpectedly Big Help from Citizen Science*

Following this detour into the Black Sea area, I come back to the North Sea project. Already when the project "Macroplastics Pollution in the Southern North Sea" was being designed, the idea arose to produce a huge number of wooden tiles (drifters, Fig. 41) to be released at certain locations in the sea to track where the sea was bringing them ashore. These drifters, cut from solid spruce wood, were 10 × 12 cm big. During cruise HE503 of the research vessel “Heincke”, a total of 1600 wooden drifters were released at sea. The first release of a group of 800 drifters was on 24 February 2018 (south circle in Fig. 41), and the second group was released on 27 February 2018 (north circle in the same figure). It was expected that people would find them somewhere on the beaches. Therefore, they were pyrographed with a text asking the person who found the drifter to report via the project website the location and time when it was found. The message, which was in German and English, also included an identification number. Each report of a found drifter could also contain additional comments that the finder could leave if they wished.

Now, this is the place to explain the usefulness of Citizen-Science. By 5 June 2018, in less than four months after the release of the drifters, 782 detection reports (out of 1600 drifters released) had been logged. All these reports came from the British northeastern coast (right part of Fig. 41). The surprise was great because normally the currents in the North Sea are such that drifters were expected to strand on the Danish and Norwegian coasts. Colleagues from one of the experimental groups had also released a large drifter equipped with a GPS device that informed us of its exact trajectory (Stanev et al. 2019a). It had also headed west (the line in Fig. 41).

---

[51] A BSEC-affiliated body based in Athens that serves as its think tank.

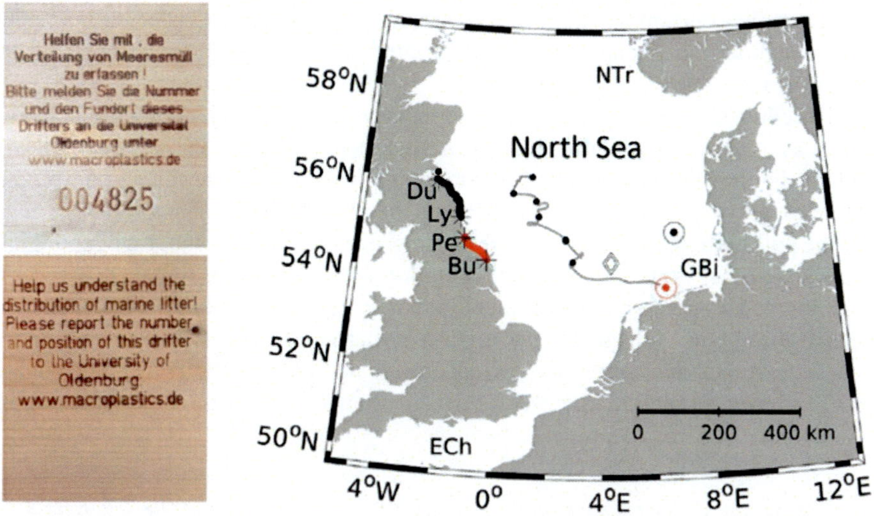

Fig. 41 On the left, the front and reverse of the wooden drifters, on the right, the locations of their release into the sea and their discovery on shore. The line shows the trajectory of a GPS drifter. The following abbreviations are used for some geographical terms: Dunbar (Du), Lynemouth (Ly), Peterlee (Pe), Burniston (Bu), English Channel (ECh), German Bight (GBi), Norwegian Trench (NTr). From Stanev et al. (2019a), with permission from Elsevier

The Contribution of Citizen-Science

Of all the wooden drifters released, 426 were registered from the first batch, with the first report of ten drifters from this group submitted on 9 March 2018, just 13 days after release. The majority of the finds, i.e., 288 reports (67.61%) were made in the second week after the first discovery, i.e., from 16 to 23 March 2018. The shortest distance traveled by the detected drifters was 451.51 km, and the longest was 559.45 km. We knew these details and many more thanks to the voluntary involvement of members of the public in our project.

It was M. Ricker's turn to check whether the model predictions confirmed the observations. It turned out that the model replicated this atypical propagation perfectly, too. Its explanation is given by a rare phenomenon in the North Sea associated with strong and persistent easterly winds that carry cold air from Russia to central Europe. They call it the "beast of the east". Coincidentally, it happened at the time when we had sent the Ph.D. students on the research vessel "Heincke" to release the wooden drifters. The data from the measurements allowed M. Ricker to calibrate and validate his model, which was a guarantee of a significant improvement in the quality of future simulations. The more in-depth research showed that the frequency of the anomalous atmospheric phenomenon was about once per decade. It turned out that the deeper ocean layers were also strongly influenced by the easterly wind, with the North Sea current having reversed direction throughout the water column, down to the seafloor. This happens in science, sometimes you plan an experiment and you discover something you did not expect.

Recently, there has been very strong media coverage of projects aimed at clearing the seas of litter. Our research has shown that there are places in the North Sea with relatively higher concentrations of floating litter (Fig. 40) where collection would be more effective. Personally, I am convinced that it is more effective to fight the cause of the pollution, that is, to fight to keep the waste out of the ocean. I will end with a conclusion that we come to in Gutow et al. (2018). The sea is a strong spreader of floating waste. Garbage brought into German waters very probably comes from the Dutch coasts, and garbage dropped on the German coasts ends up in Danish waters or on the Scandinavian coasts. We are responsible not only for our own coasts, but also for foreign coasts and for their sensitive ecosystems. This is another global problem.

10 The Sea and the Light

10.1 The Euphotic Zone

Several important scientific developments in recent decades have expanded our knowledge of ocean biogeochemistry and its evolution. The first is due to satellites providing data on some BGC properties of seawater at the ocean surface. The second is the development of sensors for in situ measurements and, in particular, the new observation platforms such as BGC-Argo. The third is advances in numerical modeling of BGC. There are many problems with each of these scientific developments (Stanev et al. 2022). With satellite data, we only "see" the ocean surface, but the layers below can be much more representative of BGC systems than the thin surface layer (Chai et al. 2020). BGC-Argo "sees" properties over large depth ranges, but the sampling position cannot be determined in advance, and sensor coverage of the World Ocean is still insufficient in most parts of it. Furthermore, the number of sensors is limited. Numerical models provide an enormous amount of data, but their quality is not yet optimal (Gehlen et al. 2015; Fennel et al. 2019; Wang et al. 2020). One major problem is that the description of the influence of BGC processes on light absorption requires substantial improvement and further development of numerical models.

The euphotic zone (or photic zone; meaning well lit in Greek) is the uppermost layer of the ocean that is exposed to intense sunlight. Two competing processes occur in this layer: oxygen production by phytoplankton photosynthesis, which absorbs carbon dioxide, and carbon dioxide production by biological processes associated with oxygen consumption. At a given depth below the ocean surface, the rate of carbon dioxide absorption is equal to the rate of its generation. The lower limit of the euphotic zone is located at the same depth, and there the net assimilation of carbon dioxide is zero. Processes in the euphotic zone are highly dependent on the amount of light penetrating to a certain depth because light regulates the rate of photosynthesis. Approximately, the euphotic zone extends from the surface to a depth where the light intensity is about 1% of that at the surface. In the transparent waters of the Sargasso Sea, the thickness of the euphotic zone reaches 200 m, whereas in the turbid waters of the North Sea, it is no more than 20–30 m, and in the Black Sea, it varies from 30 to 70 m.

The interest in the euphotic zone of the ocean is great because about 90% of marine inhabitants are concentrated there. Light stimulates the primary production of biological substances; its spectral composition influences physical, biological, and chemical processes at all scales. In particular, the intensity and spectral composition of light in water depend on the concentration of sediments, plankton, or colored dissolved organic matter (CDOM[52]), as well as other oceanographic parameters.

BGC-Argo floats, measuring a number of essential properties of marine environment, as well as light in the ocean provide an excellent illustration of the BGC dynamics. We take as an example data measured in the Black Sea by the BGC float #6901866 (Fig. 42, see also Sect. 3.2). In many marine basins, the nitrate maximum occurs below the euphotic layer and supplies this layer with nutrients, thereby stimulating plankton growth (Fig. 42b). Hence, there are two limiting elements in the ocean controlling plankton concentration. Optimum conditions for its growth are created in the interior of the euphotic zone, and there concentrations are at their maximum. Above the level of maximum concentrations, biomass accumulation is limited by nutrient reserves (Fig. 42b); below the layer of maximum concentrations, it is light that limits growth rates (Fig. 42g). The above short presentation of the basic BGC processes in the euphotic zone are well represented in Fig. 42 depicting the variability of profiles of temperature, salinity, *photosynthetically active* radiation (PAR),[53] and the concentrations of nitrates, oxygen, bbp, and Chl-a.

In addition to the plankton content, light has a direct thermal influence in the upper ocean. Depending on the concentration and vertical distribution of plankton (more generally, depending on turbidity), the same amount of light incident on the upper ocean surface penetrates to different depths, thus heats a thinner or thicker layer. Thus, the BGC processes affect the heat distribution, i.e., the physical state of the ocean.

10.2 Ocean Color and Coastal Ocean Darkening

In 2016, together with prof. Oliver Zielinski from ICBM, I started working on a project on Coastal Ocean Darkening (COD). Oliver Zielinski graduated in physics at the University of Oldenburg, and after defending his thesis at the same university,

---

[52] CDOM is the optically measurable component of dissolved organic matter in water. It is also known as yellow substance, and gelbstoff.

[53] PAR is light of wavelengths 400–700 nm, the portion of the light spectrum utilized by plants for photosynthesis.

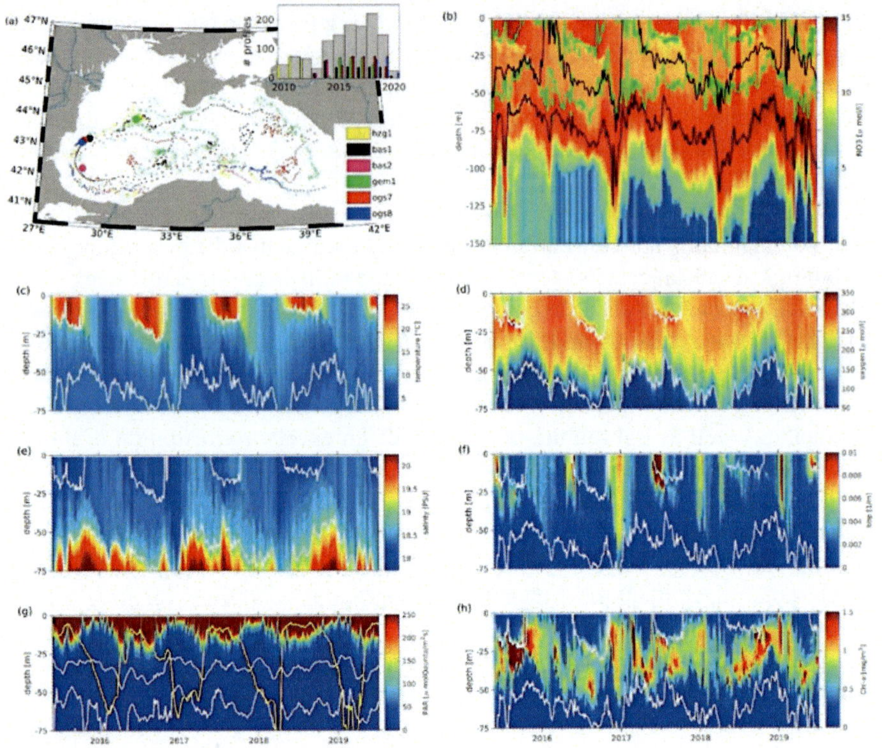

Fig. 42 Euphotic zone in the Black Sea as seen by BioArgo float #6901866. **a** Sampling positions (colored dots) of the BGC-Argo floats operating from June 2009 to March 2020. The deployment positions are shown with larger symbols. The histogram in the inset shows the number of profiles per year sampled by each float as well as the total number of profiles per year (gray). The colors assigned to individual floats are shown in the inset, bottom-right. Time versus depth diagrams (**b–h**) show observed by float #6901866 marine characteristics. **b** represents the (detrended) nitrate concentration. The 0.5 mg m^{-3} Chl-a isolines are superimposed with green lines to show the approximate core of the subsurface chlorophyll maximum layer, which is between these two isolines. Black lines represent the positions of $\sigma_t = 13.75$ and $\sigma_t = 15.0$. **c**, **e**, **g** time versus depth diagrams of temperature (**c**), salinity (**e**), and PAR (**g**) in the euphotic zone. White isolines in the temperature and salinity plots denote $\sigma_t = 12$ and 14.5, respectively. In the PAR panel, white isolines denote 1% and 0.1% of its surface value, respectively; the yellow line denotes the depth of the mixed layer. **d**, **f**, **h** time versus depth diagrams of dissolved oxygen (**d**), bbp at 700 nm (**f**) and Chl-a (**h**) in the euphotic zone. White isolines denote $\sigma_t = 12$ and 14.5. The marks on the time axes of time versus depth diagrams correspond to 1 January (longer marks) and every third month afterward (shorter marks). From Stanev et al. (2022)

turned to the field of marine technology. In 2007, he became the founding director of the Institut für Marine Ressourcen (IMARE) in Bremerhaven. In the period, I will report on below, 2017–2019, O. Zielinski was the director of ICBM; in 2023, he became the director of the IOW. He was involved in numerous research expeditions, and his area of research was the detection of processes in the marine environment using innovative sensors, e.g., for measuring nitrate, dissolved organics, and hydrocarbons.

The Idea Behind COD

The project Coastal Ocean Darkening aimed to investigate the process of potential reduction of available light in the coastal ocean over the last 100 years and into the next 100 years. The idea was to combine historical observations and hyperspectral observations in the modern ocean with future scenarios played out based on numerical modeling. The working hypothesis was that the future coastal ocean is darkening, i.e., light is expected to be reduced in intensity and to spectrally shift in depth. This in turn could have serious implications for biological processes and biodiversity, as well as for the the coastal ecosystem as a whole. Some observations of ocean transparency estimate a global decline in phytoplankton of 1% per year. However, there are other studies, for example in the North Atlantic, that show the opposite. Coastal areas were known to be highly sensitive to changes in abiotic conditions, but the paucity of information regarding light spectrum observations, as well as possible future development of estuarine and coastal systems, presented a major challenge.

The world-renowned marine optics expert Marcel Vernand from the Royal Netherlands Institute for Marine Research in Texel, a large island north of Amsterdam, joined the project. M. Vernand had a unique database of historical watercolor data. Many will ask what the difference is between the color of the water and the color of the sea. The color of the sea refers to the visible color of its surface. As a reminder, many seas are named after colors: White, Black, Red, and Yellow. The actual color of the water is a consequence of the selective absorption and scattering of light, i.e., it depends on the optical properties of the water and the thickness of the visible layer of water, but it does not depend on external factors. The color of the sea, on the other hand, is strongly dependent not only on the optical properties of the water but also on external factors (e.g., illumination, turbulence, etc.). An observer standing on the shore, or on board a ship, will see not the color of the water but the color of the sea. The latter is determined by the ratio of the spectral compositions of two main light fluxes, the light flux falling from the Sun, and the sky reflected from the surface of the sea, versus the light flux coming from depth. That the color of the sea also depends on the angle of observation may easily be seen. When we look vertically downwards toward the surface, the reflected flux will be small, and the color of the sea will be saturated. As we move our gaze toward the horizon, the color of the sea will become less saturated, it will fade, approaching the color of the sky due to the increase of the reflected flux. In a swell, the surface of the sea is slopped. This favors the perception of the luminous flux from the interior, and the saturation of the color increases and the horizon is seen much more clearly.

In clear ocean water, at a depth of 25 m, the red spectrum of sunlight completely disappears, then as the depth increases, the yellow spectrum drops away, the color of the water is seen in green, and by a depth of 100 m, only the blue color remains.

The oceans appearing dark blue in color indicates the absence of admixtures in the water. As one approaches the shore, or when the depth of the ocean decreases, there is a gradual transition to green, and in the immediate vicinity of the shore the color changes to yellow. This is because the admixtures in the water significantly change the color of the sea, i.e., the spectral composition of the scattered light coming out of the water changes.

The Forel-Ule scale is used to determine the color of water in oceanography. François-Alphonse Forel was a Swiss physician who pioneered the study of lakes and is therefore considered the founder of limnology.[54] He was the first to propose the method, and three years later the German limnologist Wilhem Ule extended it to the consideration of greenish-brown and brown colors. The method is simple but reliable. By mixing different chemicals, a standard color scale is prepared in a series of numerically labeled vials (1–21), which is compared with the color of the water body. The result is a color index that helps to classify waters with different biological activities.

10.3 The Future Evolution of Light Climate in the North Sea

The main objective of the studies in the COD project was to assess the sensitivity of the light climate in the North Sea to various natural and anthropologically induced changes. More sophisticated light absorption schemes, compared to the ones previously used, had to be implemented in a numerical model, accounting for light spectral composition. Finally, yet importantly, we wanted to account for the influence of mineral sediments, which was not accounted for in many models until then (Fig. 43).

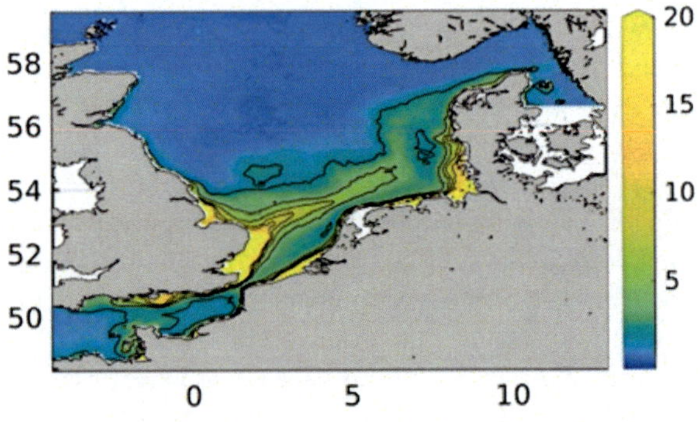

Fig. 43 20-year average (non-living) suspended particulate matter in g/m^{-3} in the North Sea. Contour intervals are at 2, 4, 6, 8, and 10 g/m^{-3}. Note that the color map is saturated above 20 g/m^{-3}. From Thewes et al. (2021)

[54] Limnology deals with the study of freshwater bodies and their biological, chemical, and physical characteristics.

I gave my Ph.D. student Daniel Teves the freedom to decide which hydrodynamic model to use, and he chose the Regional Ocean Modeling System (ROMS, Haidvogel et al. 2000; 2008), one of the most widely used models for regional and coastal simulations. It belongs to the class of models using so-called terrain-following vertical coordinates. As a BGC engine, we decided to use the Carbon, Silicon, and Nitrogen Ecosystem model, CoSiNE (Chai et al. 2002; Xiu and Chai 2011). This choice was in line with my ongoing works with F. Chai, the author of this model.

Together with D. Thewes, and with O. Zielinski, we have shown that optical properties of water in the North Sea, which vary widely in space and time, have a significant influence on the performance of ecosystem models (Thewes et al. 2020). It appeared that relatively simple changes to the model description of light absorption lead to significant improvements in model predictions. The modifications that D. Thewes introduced proved successful in the shallow regions, which was of major interest to the project. The reduction in available light in those regions led to a strong reduction in phytoplankton growth in the areas of high turbidity (Thewes et al. 2021). The spatial variability of light attenuation proved to be crucial for modeling such a heterogeneous area as the North Sea. Concerning the future evolution of the light climate under different scenarios of the Earth's climate evolution, the conclusions were ambiguous.

References

Capet A, Fernández V, She J, Dabrowski T, Umgiesser G, Staneva J, Mészáros L, Campuzano F, Ursella L, Nolan G, El Serafy G (2020) Operational modeling capacity in European Seas—an EuroGOOS perspective and recommendations for improvement. Front Mar Sci 7:129. https://doi.org/10.3389/fmars.2020.00129

Cavaleri L, Abdalla S, Benetazzo A, Bertotti L, Bidlot J-R, Breivik Ø, Carniel S, Jensen RE, Portilla-Yandun J, Rogers WE, Roland A, Sanchez-Arcilla A, Smith JM, Staneva J, Toledo Y, van Vledder GPh, van der Westhuysen AJ (2018) Wave modeling in coastal and inner seas. Progr Oceanogr 167: 164–233. https://doi.org/10.1016/j.pocean.2018.03.010

Cavaleri L, Alves J-HGM, Ardhuin F, Babanin A, Banner M, Belibassakis K, Benoit M, Donelan M, Groeneweg J, Herbers THC, Hwang P, Janssen PAEM, Janssen T, Lavrenov IV, Magne R, Monbaliu J, Onorato M, Polnikov V, Resio D, Rogers WE, Sheremet A, McKee Smith J, Tolman HL, van Vledder G, Wolf J, Young I (2007) Wave modeling—the state of the art. Prog Oceanogr 75(4):603–674. https://doi.org/10.1016/j.pocean.2007.05.005

Chai F, Dugdale R, Peng T-H, Wilkerson F, Barber R (2002) One-dimensional ecosystem model of the equatorial Pacific upwelling system. Part I: model development and silicon and nitrogen cycle. Deep Sea Res II: Topical Stud Oceanogr 49(13–14):2713–2745. https://doi.org/10.1016/s0967-0645(02)00055-3

Chai F, Johnson K, Claustre H, Xing X, Wang Y, Boss E, Riser S, Fennel K, Schofield O, Sutton A (2020) Monitoring ocean biogeochemistry with autonomous platforms. Nat Rev Earth Environ 1:315–326. https://doi.org/10.1038/s43017-020-0053-y

Chant R (2010) Estuarine secondary circulation. In: Valle-Levinson A (ed) Contemporary issues in Estuarine physics. Cambridge University Press, Cambridge, pp 100–124. https://doi.org/10.1017/CBO9780511676567.006

Davis R, Sherman J, Dufour J (2001) Profiling ALACES and other advances in autonomous subsurface floats. J Atmos Oceanic Tech 18(6):982–993

Fennel K, Gehlen M, Brasseur P, Brown CW, Ciavatta S, Cossarini G, Crise A, Edwards CA, Ford D, Friedrichs MAM, Grégoire M, Jones E, Kim H-C, Lamouroux J, Murtugudde R, Perruche C, The GODAE OceanView Marine Ecosystem Analysis and Prediction Task Team (2019) Advancing marine biogeochemical and ecosystem reanalyses and forecasts as tools for monitoring and managing ecosystem health. Front Mar Sci 6: 89. https://doi.org/10.3389/fmars.2019.00089

Funkquist L (2001) HIROMB: an operational eddy-resolving model for the Baltic Sea. Bull Mar Inst Gdansk 28:7–16

Gehlen M, Barciela R, Bertino L, Brasseur P, Butenschön M, Chai F, Crise A, Drillet Y, Ford D, Lavoie D, Lehodey P, Perruche C, Samuelsen A, Simon E (2015) Building the capacity for forecasting marine biogeochemistry and ecosystems: recent advances and future developments. J Oper Oceanogr 8(sup1):s168–s187. https://doi.org/10.1080/1755876X.2015.1022350

Grashorn S, Stanev E (2016) Kármán vortex and turbulent wake generation by wind park piles. Ocean Dyn 66(12):1543–1557

Grayek S, Stanev E, Schulz-Stellenfleth J (2015) Assessment of the Black Sea observing system. A focus on 2005–2012 Argo campaigns. Ocean Dyn 65:1665–1684

Grayek S, Staneva J, Schulz-Stellenfleth J, Petersen W, Stanev E (2011) Use of FerryBox surface temperature and salinity measurements to improve model-based state estimates for the German Bight. J Mar Sys 88(1):45–59

Gutow L, Ricker M, Holstein JM, Dannheim J, Stanev EV, Wolff J-O (2018). Distribution and trajectories of floating and benthic marine macrolitter in the south-eastern North Sea. Marine Pollut Bull 131(Part A): 763–772

Haidvogel DB, Arango H, Budgell WP, Cornuelle BD, Curchitser E, Di Lorenzo E, Fennel K, Geyer WR, Hermann AJ, Lanerolle L, Levin J, McWilliams JC, Miller AJ, Moore A. M, Powell TM, Shchepetkin AF, Sherwood CR, Signell RP, Warner JC, and Wilkin J (2008) Journal of Computational Physics, 227:3595–3624.

Haidvogel DB, Arango HG, Hedström KS, Beckmann A, Malanotte-Rizzoli P, Shchepetkin A.F (2000) Model evaluation experiments in the North Atlantic Basin simulations in nonlinear terrain-following coordinates, Dynamics of Atmospheres and Oceans 32:239–281.

Haid V, Stanev E, Pein J, Staneva J, Chen W (2020) Secondary circulation in shallow ocean straits: observations and numerical modeling of the Danish Straits. Ocean Model 148. https://doi.org/10.1016/j.ocemod.2020.101585

Hasselmann K, Barnett TP, Bouws E, Carlson H, Cartwright DE, Enke K, Ewing JA, Gienapp H, Hasselmann DE, Kruseman P, Meerburg A, Mller P, Olbers DJ, Richter K, Sell W, Walden H (1973) Measurements of wind-wave growth and swell decay during the Joint North Sea Wave Project (JONSWAP). Ergnzungsheft zur Deutschen Hydrographischen Zeitschrift Reihe A(8) (Nr. 12): 95

Jacob B, Stanev E (2021) Understanding the impact of bathymetric changes in the German Bight on coastal hydrodynamics: one step toward realistic morphodynamic modeling. Front Mar Sci 8: 640214

Jacob, B., Stanev, EV., Zhang, YJ (2016) Local and Remote Response of the North Sea Dynamics to Morphodynamic Changes in the Wadden Sea. Ocean Dynamics, 66, 5, 671-690, https://doi.org/10.1007/s10236-016-0949-8

Jarosz E, Teague WJ, Book JW, Beşiktepe Ş (2011) On flow variability in the Bosporus Strait. J Geophys Res 116:C08038. https://doi.org/10.1029/2010JC006861

Johnson KS, Claustre H (2016) Bringing biogeochemistry into the Argo age. Eos 97. https://doi.org/10.1029/2016EO062427

Kitaigorodskii SA (1962) Applications of the theory of similarity to the analysis of wind-generated wave motion as a stochastic process, Bull.Acad.Sci. USSR, Geophys. Ser, Engl. Transl, N1, 105–117

Korotaev G, Oguz T, Riser S (2006) Intermediate and deep currents of the Black Sea obtained from autonomous profiling floats. Deep Sea Res Pt II 53:1901–1910

Kundak S, Baypınar M (2011) The crazy project—Canal Istanbul, Tema. J Land Use Mobility Environ 4(3):53–63

Latif MA, Ozsoy E, Oguz T, Unluata U (1991) Observations of the Mediterranean inflow into the Black Sea. Deep-Sea Res 38(Suppl 2):5711–5723

Le Traon PY (2013) From satellite altimetry to Argo and operational oceanography: three revolutions in oceanography. Ocean Sci 9:901–915

Meier HEM (2005) Modeling the age of Baltic Sea water masses: quantification and steady state sensitivity experiments. J Geophys Res 110:C02006. https://doi.org/10.1029/2004JC002607

Meier HEM (2007) Modeling the pathways and ages of inflowing salt- and freshwater in the Baltic Sea. Estuar Coast Shelf Sci 74(4):610–627

Murray JW, Top Z, Ozsoy E (1991) Hydrographic properties and ventilation of the Black Sea. Deep-Sea Res 38:S663–S689

Oke P, Balmaseda M, Benkiran M, Cummings JA, Dombrowsky E, Fujii Y, Guinehut S, Larnicol G, Le Traon P-Y, Martin MJ (2009) Observing system evaluations using GODAE systems. Oceanography 22:144–153

Pein J, Grayek S, Schulz-Stelleneth J, Stanev EV (2016) On the impact of salinity observations on state estimates in Ems Estuary. Ocean Dyn 66:243–262

Pein J, Stanev E, Zhang Y (2014) The tidal asymmetries and residual flows in Ems Estuary. Ocean Dyn 64:1719–1741

Peixoto JP, Oort AH (1992) Physics of climate. American Institute of Physics, New York, NY, p 520

Pierson WJ Jr, Moskowitz LA (1964) Proposed spectral form for fully developed wind seas based on the similarity theory of S. A. Kitaigorodskii. J Geophys Res 69:5181–5190

Ricker M, Stanev E (2020) Circulation of the European northwest shelf: a lagrangian perspective. Ocean Sci 16(3):637–655

Roemmich D, Boebel O, Desaubies Y et al (2001) Argo: the global array of profiling floats. In: Koblinsky CJ, Smith NR (eds) Observing the Oceans in the 21st century. https://archimer.ifremer.fr/doc/00090/20097/

Saçu Ş, Erdik T, Stanev E, Şen O, Duricic Erdik J, Öztürk I (2020) Hydrodynamics of Canal Istanbul and its impact in the northern Sea of Marmara under extreme conditions. Ocean Dyn 70:745–758

Schulz-Stellenfleth J, Stanev E (2010) Statistical assessment of ocean observing networks—a study of water level measurements in the German Bight. Ocean Model 33: 270–282

Schuttelaars HM, de Jonge VN, Chernetsky A (2013) Improving the predictive power when modeling physical effects of human interventions in estuarine systems. Ocean Coast Manag 79:70–82

Şen O, Saçu Ş, Erdik T, Özturk I, Stanev EV (2022) Assessing the potential impacts of the Canal Istanbul on the physical oceanography of the Turkish Straits system. Cont Shelf Res 240: 104723. ISSN 0278-4343

Sloyan BM, Wilkin J, Hill KL, Chidichimo MP, Cronin MF, Johannessen JA, Karstensen J, Krug M, Lee T, Oka E, Palmer MD (2019) Evolving the physical global ocean observing system for research and application services through international coordination. Front Mar Sci 6:449. https://doi.org/10.3389/fmars.2019.00168

Sotillo MG, Wilkin J (2022) Definition of ocean forecasting systems: temporal and spatial scales solved by marine modeling systems. In: Fanjul EAS, Ciliberti S, Bahurel P (eds) Implementing operational ocean monitoring and forecasting systems; IOC-UNESCO, GOOS-275, pp 8–29

Stanev EV (2005) Understanding Black Sea dynamics: overview of recent numerical modeling. Oceanography 18:56–75

Stanev EV, AlNadhairi R, Staneva J, Schulz-Stellenfleth J, Valle-Levinson A (2014a) Tidal wave transformations in the German Bight. Ocean Dyn 64:951–968

Stanev EV, Al-Nadhairi R, Valle-Levinson A (2015a) The role of density gradients on tidal asymmetries in the German Bight. Ocean Dyn 65:77–92

Stanev E, Badewien T, Freund H, et al (2019a) Extreme westward surface drift in the North Sea: Public reports of stranded drifters and Lagrangian tracking. Cont Shelf Res 177: 24–32

Stanev EV, Chtirkova B (2021) Interannual change in mode waters: case of the Black Sea. J Geophys Res—Oceans 126:e2020JC016429. https://doi.org/10.1029/2020JC016429

Stanev EV, Chtirkova B, Peneva E (2021) Geothermal convection and double diffusion based on profiling floats in the Black Sea. Geophys Res Lett 48:e2020GL091788

Stanev EV, Grashorn S, Zhang YJ (2017a) Cascading Ocean basins: numerical simulations of the circulation and interbasin exchange in the Azov-Black-Marmara-Mediterranean seas system. Ocean Dyn 67:1003–1025

Stanev E, Grayek S, Claustre H, Schmechtig C, Poteau A (2017b) Water intrusions and particle signatures in the Black Sea: a Biogeochemical-Argo float investigation. Ocean Dynamics, 67: 1119–1136

Stanev EV, He Y, Grayek S, Boetius A (2013) Oxygen dynamics in the Black Sea as seen by Argo profiling floats. Geophys Res Lett 40(3085–3090)

Stanev EV, He Y, Staneva J, Yakushev E (2014b) Mixing in the Black Sea detected from the temporal and spatial variability of oxygen and sulfide—Argo float observations and numerical modeling. Biogeosciences 11:5707–5732. https://doi.org/10.5194/bg-11-5707-2014

Stanev EV, Le Traon PY, Peneva EL (2000) Sea level variations and their dependency on meteorological and hydrological forcing: analysis of altimeter and surface data for the Black Sea. J Geophys Res—Oceans 105(C7):17203–17216

Stanev E, Pein J, Grashorn S, Zhang, Y., Schrum, C (2018a) Dynamics of the Baltic Sea straits via numerical simulation of exchange flows. Ocean Model 131:40–58

Stanev E, Peneva E, Chtirkova B (2019b) Climate change and regional ocean water mass disappearance: case of the Black Sea. J Geophys Res—Oceans 124. https://doi.org/10.1029/2019JC015076

Stanev E, Poulain PM, Grayek S, Johnson K, Claustre H, Murray J (2018b) Understanding the dynamics of the oxic-anoxic interface in the Black Sea. Geophys Res Lett 45:864–871

Stanev E, Ricker M (2019) The fate of marine litter in semi-enclosed seas: a case study of the Black Sea. Front Mar Sci 6:660

Stanev E, Schulz-Stellenfleth J, Staneva J, Grayek S, Grashorn S, Behrens A, Koch W, Pein J (2016) Ocean forecasting for the German Bight: from regional to coastal scales. Ocean Sci 12:1105–1136

Stanev EV, Schulz-Stellenfleth J, Staneva J, Grayek S, Seemann J, Petersen W (2011) Coastal observing and forecasting system for the German Bight—estimates of hydrophysical states. Ocean Sci 7:569–583

Stanev EV, Wahle K, Staneva J (2022) The synergy of data from profiling floats, machine learning and numerical modeling: case of the Black Sea euphotic zone. J Geophys Res—Oceans 127:e2021JC018012. https://doi.org/10.1029/2021JC018012

Stanev E, Xi L, Grashorn S (2015b) Physical processes in the transition zone between North Sea and Baltic Sea. Numerical simulations and observations. Ocean Model 93:56–74

Staneva JV, Dietrich D, Stanev EV, Bowman M (2001) Rim current and coastal eddy mechanisms in an eddy-resolving Black Sea general circulation model. J Mar Sys 3:137–157

Staneva J, Stanev E, Wolff J, Badewien T, Reuter R, Flemming B, Bartholomä A, Bolding K (2009) Hydrodynamics and sediment dynamics in the German Bight. A focus on observations and numerical modeling in the East Frisian Wadden Sea. Cont Shelf Res 29(1):302–319. https://doi.org/10.1016/j.csr.2008.01.006

Thewes D, Stanev EV, Zielinski O (2020) Sensitivity of a 3D shelf sea ecosystem model to parameterizations of the underwater light field. Front Mar Sci 6: 816

Thewes D, Stanev EV, Zielinski O (2021) The North Sea light climate: analysis of observations and numerical simulations. J Geophys Res—Oceans 126:e2021JC017697

Vanhellemont Q, Ruddick K (2014) Turbid wakes associated with offshore wind turbines observed with Landsat 8. Remote Sens Environ 145:105–115

Waller J (2004) Fabulous science: fact and fiction in the history of scientific discovery. Oxford University Press, UK, p 368

WAMDI Group (1988) The WAM model—a third generation ocean wave prediction model. J Phys Oceanogr 18: 1775–1810. https://doi.org/10.1175/1520-0485(1988)018<1775:TWMTGO>2.0.CO;2

Wang B, Fennel K, Yu L, Gordon C (2020) Assessing the value of biogeochemical Argo profiles versus ocean color observations for biogeochemical model optimization in the Gulf of Mexico. Biogeosciences 17(15):4059–4074. https://doi.org/10.5194/bg-17-4059-2020

Wigner EP (1960) The unreasonable effectiveness of mathematics in the natural sciences. Richard Courant lecture in mathematical sciences delivered at New York University, May 11, 1959. Commun Pure Appl Math 13:1–14

Xi L, Soomere T, Stanev EV, Murawski J (2012) Identification of the environmentally safe fairway in the South-Western Baltic Sea and Kattegat. Ocean Dyn 62:815–829

Xiu P, Chai F (2011) Modeled biogeochemical responses to mesoscale eddies in the South China Sea. J Geophys Res 116:C10006. https://doi.org/10.1029/2010JC006800

Yakushev EV, Pollehne F, Jost G, Kuznetsov I, Schneider B, Umlauf L (2007) Analysis of the water column oxic/anoxic interface in the Black and Baltic seas with a numerical model. Mar Chem 107:388–410

Zhang Y, Baptista AM (2008) SELFE: a semi-implicit Eulerian Lagrangian finite element model for cross-scale ocean circulation. Ocean Model 21(3–4):71–96

Zhang YJ, Fei Y, Stanev EV, Grashorn S (2016a) Seamless cross-scale modeling with SCHISM. Ocean Model 102:64–81

Zhang YJ, Stanev EV, Grashorn S (2016b) Unstructured-grid model for the North Sea and Baltic Sea: validation against observations. Ocean Model 97:91–108

Zhang Z, Stanev E, Grayek S (2020) Reconstruction of the basin-wide sea-level variability in the North Sea using coastal data and generative adversarial networks. J Geophys Res—Oceans 125:e2020JC016402

Operational Oceanography and Future Scientific and Societal Challenges

By integrating *observations, theory,* and *numerical modeling*, operational oceanography provides routine oceanographic information needed for decision-making purposes. These three foundations on which operational oceanography rests have been discussed in previous chapters. Now, it will be analyzed how monitoring and forecasting activities are organized in Europe and worldwide. First, a description of how marine observations are structured in Europe, in particular EuroGOOS, the European component of the Global Ocean Observing System (GOOS), is presented. The development of the concepts and organization of the European Ocean Observing System (EOOS), which coordinates and integrates European communities and organizations operating, supporting, and maintaining ocean observing infrastructures and activities, is also presented. A large part of this chapter focuses on the evolution of European operational monitoring and forecasting, starting from the MyOcean project and reaching the present-day status of the Copernicus Marine Environment Monitoring Service (CMEMS). Here, the author presents, among others, his personal recollection of numerous events and concrete steps in the field of R&D and operational practice. The evolution of operational systems is also described, which includes provision of wind wave products, enhancements in the fields of biogeochemistry and marine biology, high-resolution monitoring and forecasting, and machine learning. Among these, developing and implementing concepts of how to merge national and European coastal monitoring and forecasting systems with regional and open sea systems has a high priority. Similar worldwide activities are presented on the example of the Global Ocean Data Assimilation Experiment (GODAE) and its continuation in the frame of GODAE OceanView (GOV) and the OceanPredict. The latter initiative will give weight to operational oceanography during the UN Decade of Ocean Science for Sustainable Development. It will contribute to the growing public awareness of climate change threats and provide support for sustainable development.

E. Stanev, *Trajectories in Oceanography*, Springer Oceanography,
https://doi.org/10.1007/978-3-031-33720-8_5

1 Evolution of Science and Operational Oceanography

Four paradigms define the development of science (Hey et al. 2009; Bell et al. 2009; Gray and Szalay 2007). A thousand years ago, science empirically described natural phenomena. In this early period of development, science was *experimental*. With Kepler's and Newton's laws and Maxwell's equations, the development of theoretical science began. Thus, in the seventeenth century, scientists recognized experimental and *theoretical science* as the two main paradigms for understanding nature. During the past century, and in particular in its second half, many problems represented by theoretical models became too complex to be solved using the analytical tools of mathematics. For this reason, scientists began to simulate processes in nature (and in many areas of human activity) with the help of computers. These simulations led to a new leap in the development of science, i.e., the third paradigm, namely that *science is computational*, was established. In the last few decades, computational science-simulating complex phenomena, together with the new technologies on which modern measurements are based, have contributed to a high growth in data volume. Today, scientists do not look directly through telescopes and microscopes, but use interfaces connecting them to the raw or processed data received in data centers, and only then, using sophisticated methods, study the information on their computers. Hence, the techniques and technologies for processing and understanding scientific data are not just a computational process but something new, giving rise to a fourth paradigm, namely *data-intensive science*. The *eScience* is defined as a synthesis of information technology (IT) and science that enables scientific challenges to be solved on a scale previously unimagined. This new paradigm brings together theory, experiment, and simulation and is much more than their simple sum.

Today, we are witnessing a "data deluge". The volumes of data generated in some areas of science just a decade ago were tens of thousands times (or even more) smaller than those coming from modern Earth-observing satellites, telescopes, or accelerators. In many fields of science, experiments or computer simulations generate petabytes (1 petabyte $= 10^{15}$ bytes) of data per year. Since the turn of the century, we have also seen a change in the modes of access to data, many of which have become freely available.

Operational oceanography is a good example that demonstrates the fourth paradigm.[1] Together with the integration of theory, experiment, and simulation, it explores data for predicting environmental processes, thus contributing to the conservation of environment and the improvement of the quality of life. In this part, I will talk about those who created and established this scientific and practical field, this scientific field itself, its European specificity, and, above all, the projection of trends in the development of operational oceanography at a global and regional level onto scientific and practical activities in coastal ocean areas. In contrast to the previous text, where I was exploring the scientific developments in oceanography, this section

[1] I heard this idea from N. Pinardi during her talk.

will discuss science management in depth, as operational oceanography is not just science, but consists of multiple activities involving scientists, routine practitioners, users, managers, decision-makers, and policymakers.

Davidson et al. (2019) define operational oceanography as the provision of routine oceanographic information needed for decision-making purposes.[2] Operational oceanography is dependent upon sustained R&D through the end-to-end framework of an operational service, from observation collection to delivery mechanisms. Of utmost importance is the rapid interpretation of routine measurements in the seas, oceans, and atmosphere using numerical models. The results of operational oceanography are important products of (1) the current state of the sea, including its living resources, (2) predictions of its future state with the longest possible predictability, and (3) analyses of past ocean states assembled from long-term observations to pinpoint trends and changes. Examples of the products of operational oceanography include warnings of coastal flooding, ice and storm damage, harmful algal blooms and pollutants, electronic charts, optimal ship routes, predictions of seasonal or annual primary productivity of the biological system, ocean currents, changes in ocean climate, etc. Final products should be disseminated rapidly to interested users, government agencies, and regulatory authorities. These activities were carried out on a smaller scale prior to 2000 by some national offices, agencies, and institutes. In the last twenty years, there is a tendency to harmonize different operational activities and put them on a more solid scientific footing. Similar process started earlier in other fields of physics. Nicholas Metropolis states: "…the fact is that quite some time ago the tables were turned between theory and applications in the physical sciences. Since World War II the discoveries that have changed the world were not made so much in lofty halls of theoretical physics as in the less-noticed labs of engineering and experimental physics. The roles of pure and applied science have been reversed; they are no longer what they were in the golden age of physics, in the age of Einstein, Schrödinger, Fermi and Dirac (Metropolis 1992)". The situation in oceanography is similar, and fundamental ocean science was increasingly giving way to applied research. The recent consolidation of research and operational activities in oceanography was additionally enhanced thanks to the growing role of European institutions in the governance of science.

2 Marine Observations in Europe

2.1 EuroGOOS

So far, I have covered many aspects of ocean observations and numerical modeling, but I have not yet explained what the coordination between the measurement communities and the ocean forecasters is. Coordination within and between these separate

[2] See also (Schiller et al. 2018).

communities is necessary to avoid repetition, to apply best practices, and to advance observational and forecast science. Whereas in the twentieth century, ocean observations and forecasting, carried out by different institutes and countries, were relatively loosely interconnected, the early twenty-first century marks a new trend, and let us call it the globalization of oceanographic practice.

Observations are known to be the main instrument to increase human knowledge and understanding of nature. Henry Stommel, one of the fathers of the theories of global ocean circulation, said: "The chief source of ideas in oceanography comes, I think, from new observations" (Stommel 1989). As of early 2021, eighty-six countries are directly involved in marine data collection, using some 8,933 different observing platforms, continuously monitoring the World Ocean and atmosphere (GOOS 2021). Nowadays, observations are not only instruments of generating ideas, but also become an attribute of models used for predictive purposes.

The Global Ocean Observing System (GOOS) is a platform for international collaboration on sustainable ocean observations and the creation of oceanographic products and services. It stimulates interaction between scientific, operational, and user communities and serves ocean researchers, coastal managers, parties to international conventions, national meteorological and oceanographic agencies, hydrographic services, marine and coastal industry, policymakers, and the interested general public.

The Globalization of Ocean Observations

The Global Earth Observing System of Systems (GEOSS) is an initiative of the Intergovernmental Group on Earth Observations (GEO[3]), aimed at achieving a closer cooperation in observing the planet. GEOSS has more than 400 million open data resources from over 150 national and regional providers such as NASA and ESA, as well as international organizations such as the WMO. GEO is a partnership of more than 100 national governments and over 100 participating organizations, which coordinates comprehensive and sustained observations of the Earth, to better integrate observing systems and to share and make available data by linking existing infrastructures using common standards. The GMES (Copernicus) program was also developed in line with these global trends.

The Global Ocean Observing System (GOOS) is the oceanographic component of GEOSS. GOOS is administered by the IOC, and together with the Global Climate Observing System (GCOS) and the Global Terrestrial Observing System (GTOS), it constitutes one of the three main building blocks of GEOSS. GOOS provides oversight and coordination to support the global unification of the measurement networks. The strategy of GOOS is to facilitate effective links between those providing data and their users at national and regional scales.

[3] https://www.earthobservations.org/geo_community.php. Last accessed on 03.02.2023.

GOOS Regional Alliances coordinate work in the different areas of the World Ocean. The European Global Ocean Observing System (EuroGOOS) is one such alliance. It was founded in 1994, and the British scientist John Woods from Imperial College London, was its first chairman. EuroGOOS is an association of national government agencies, research organizations, and private oceanographic companies. Its strategy (Woods et al. 1996) sets priorities, strengthens cooperation, and aims to maximize the use of operational oceanographic products and services by users in need of marine information. The number of EuroGOOS members is gradually growing, having reached 44 from 18 European countries by 2022.

EuroGOOS coordinates five European Regional Operational Oceanographic Systems (ROOS): Arctic ROOS, Baltic Operational Oceanographic System (BOOS), North West Shelf Operational Oceanographic System (NOOS), Ireland-Biscay-Iberia Regional Operational Oceanographic System (IBI-ROOS) as well as the Mediterranean Operational Network for the Global Ocean Observing System (MONGOOS). The Black Sea is not included, but separately there is a Black Sea Alliance (Black Sea GOOS), which is a regional association established by the Black Sea countries to promote operational oceanography in the region and establish links with other regional and global organizations with similar objectives. The regional approach is appropriate because the problems with observations and marine forecasts are not identical from region to region and regional cooperation is much more active and direct than the cooperation between different ROOS-es. Working with the regional operational alliances forms the basis of EuroGOOS' work as overall coordinator of ROOS-es.

After several years of GKSS involvement in BOOS and NOOS, F. Colijn and H. von Storch made a request for EuroGOOS membership. At the 17th regular annual meeting in 2010, GKSS became a member of EuroGOOS, and since then, I am the official representative of GKSS (HZG/Hereon) in this alliance.

The Change of the EuroGOOS Status

During the annual meeting at the end of 2011 in Rome, the chairman of EuroGOOS Peter Ehlers[4] raised the issue of the future of EuroGOOS. Functioning as a voluntary association without being a legal entity was causing problems because EuroGOOS was an informal association. The alliance could only act toward third parties through one of its members. In most cases, this was the Swedish Meteorological and Hydrological Institute (SMHI), where the secretariat has been located for the last 10 years or more.

It was decided that EuroGOOS should become a legal entity in order to be able to assume certain responsibilities. Brussels was accepted as the most appropriate seat, not least because of its central position in Europe and its proximity to the European institutions. EuroGOOS became officially known as EuroGOOS AISBL (Association Internationale Sans But Lucratif—International Non-Profit Association), and in 2013, it was established as an international organization. The EuroGOOS Director Hans Dahlin, from the SMHI, retired and Kostas Nittis from the HCMR was elected in his place as a secretary general. It is a great pity that Kostas, whom I knew from his Ph.D. years, passed away too young, during the course of his mandate. What remains is the medal in his name that EuroGOOS awards to young scientists every year.

During the chairmanship of P. Ehlers, the interaction and coordination of the Copernicus and EuroGOOS maritime operational activities were further clarified. In Copernicus, there was a move toward harmonization of standards for operational oceanography, but this was not one of the strengths of EuroGOOS, where regional approaches (e.g., use of different models in different seas) predominated. However, not all the regional work of the national services within EuroGOOS could be fed into Copernicus. This was particularly true for coastal forecasting systems. To summarize, the relative weight of numerical modeling and forecasting in EuroGOOS has decreased, but this activity has not disappeared, and there is still a place for it in EuroGOOS. The other important change during the initial years of the EuroGOOS AISBL was that the foundations were laid for the association to be actively involved in projects in the field of operational oceanography and their governance and strategic direction. In carrying out these new trends, P. Ehlers found strong support from the DMI representative E. Buch. In 2013, E. Buch replaced P. Ehlers as EuroGOOS Chairman.[5]

2.2 *Structuring Marine Observations in Europe*

The greater interest in atmospheric science in the past might explain, at least in part, why oceanography has lagged behind in some of its developments. Trying to identify problems and find solutions to accelerate ocean observing practices, the Policy Brief "Sustaining in situ Ocean Observations in the Age of the Digital Ocean"[6] compares the provision of continuous observations in the atmosphere and ocean. In the atmosphere, the primary source of data is the weather service. The measured parameters of main interest relate to weather, climate, atmospheric composition, air quality, and space weather. Most of the data are derived from continuous meteorological observations and are mainly used by meteorological services. Long-term measurements are largely (68%) financially supported. There is strong coordination between the main meteorological organizations involved in observations, as well as common (operational) standards and clearly defined national and international mandates (e.g., WMO Convention).

In the ocean, the situation is very different. The origin of oceanographic data is mainly from scientific research and only loosely related to operational practice. Parameters of primary interest are measured for climate, biogeochemistry, ecosystems, and the seabed. There is no established infrastructure for obtaining data, with the exception of those supporting the characterization of meteorological conditions, and some measurements in the *exclusive economic zones* (*EEZs*) of countries. The

[4] From 1989 to 2008 P. Ehlers served as the President of the BSH.

[5] The next chairs were George Petiakis from the HCMR and Henning Vehde from the Institute of Marine Research in Norway.

[6] https://www.marineboard.eu/publications/sustaining-situ-ocean-observations-age-digital-ocean. Last accessed on 03.02.2023.

measurements are carried out by various organizations as (scientific) projects, and long-term funding is only 28% secured. Furthermore, in many cases, marine research is inherently costly and relies on research facilities that are not always accessible to all researchers. The share of international mandates is low. The need for more co-ordination between different observational activities is evident, particularly due to the realization that the oceans and seas are under enormous pressure from human activity and climate change. The vulnerability of coastal areas and the continued deterioration of the marine environment require that more attention should be given to the coordination of research and monitoring of the seas and oceans.

Many of the above challenges are cross-cutting and involve different fields of knowledge and practice. There is a perception that research programs focused on a specific narrow field of knowledge or on a particular topic alone are suboptimal. On the contrary, an integrated and coherent approach at European level highlights new perspectives for optimizing current and developing new practices. The alignment of objectives, available financial resources, and capacities is expected to bring a new quality of work and to stimulate the translation of scientific information and knowledge into research applications. In pursuit of the above ideas, the Joint Programming Initiative (JPI) "Healthy and Productive Seas and Oceans", known as JPI Oceans, was established in 2011 as a pan-European intergovernmental platform aiming to increase efficiency and impact of research and innovation, especially in the long term.

JPI Oceans

JPI Oceans is as an international non-profit association (AISBL) under Belgian law and works in close coordination with the European Commission. JPI Oceans provides its member countries with a strategic agenda and action plan related to ocean research that cannot be addressed entirely at national level. In this way, it increases the value of national investment in research and innovation by aligning national priorities and implementing joint activities.

This is achieved by[7]:

- planning and launching joint calls for transnational research and innovation projects;
- sharing research infrastructures and resources;
- strengthening cooperation between science and policy and involving stakeholders in order to make the voice of science better heard in policy decisions;
- initiating new forms of collaboration between projects and scientists;
- building strategic communities and disseminating research results to support their exploitation and facilitate mutual learning.

[7] https://climate-adapt.eea.europa.eu/metadata/organisations/jpi-oceans. Last accessed on 03.02.2023.

Unlike JPI Oceans, which is an intergovernmental platform, the European Marine Board (EMB) is an independent non-governmental advisory body representing a leading European think tank in the field of marine science. It provides a strategic forum to develop marine research foresight, initiate analyses of the state of the art, and translate these into clear policy recommendations for European institutions as well as national governments. The EMB facilitates the exchange of knowledge between the scientific community and decision-makers, promoting European leadership in marine research and technology.

European Marine Board (EMB[8])

The EMB's strategy aims to identify scientific challenges and opportunities and, through foresight activities, analyses, and studies, to provide high-level recommendations. As a forum, it brings together marine research stakeholders and expresses their collective vision of European marine research priorities. The EMB promotes the idea that existing and future data infrastructures should be supported as a public good even if they are small, for example, if they consist of measurement networks with smaller elements. EMB Council members are calling for infrastructure commitments to be linked to international agreements related to ocean action, for example, as part of mandatory national contributions. Support for sustainable monitoring is being considered through funding models that are appropriate for infrastructure (rather than short-term projects).

There is now increasing interest in quantifying the benefits of ocean observations. Although the economic value is difficult to estimate, there are already some attempts at economic analyses. For example, the cost of maintaining the global weather observing network (including the surface ocean) is US $5 billion per year, and the benefit–cost ratio[9] is 25 (European Marine Board 2021). The JERICO-NEXT project found a high benefit–cost ratio (of about five) for the European coastal component of the observing system (Gaughan et al. 2019). Sustainable offshore investments could provide at least five times greater returns than costs (Konar and Ding 2020). The need to establish an ongoing process for reviewing the cost and effectiveness of the measurement system and determining its economic and environmental benefits is apparent. The philosophy "collect once and use many times" promoted by EMODnet will benefit all marine data users, including scientists, policymakers, private industry, and the public. It is expected that such an integrated marine data policy will open up new opportunities for innovation and growth.

---

[8] https://www.marineboard.eu/about-european-marine-board. Last accessed on 03.02.2023.

[9] The benefit–cost ratio (BCR) is an indicator showing the relative costs and benefits of a proposed action expressed in monetary terms. BCR > 1.0 means that the action is expected to generate a profit for the company and its investors.

In the spirit of the above needs, the European Ocean Observing System (EOOS[10]) framework has been established and designed to coordinate and integrate Europe's ocean observing capacity, to promote a systematic and collaborative approach to the collection of information on the state and variability of the seas and to support the sustainable management of the marine environment and its resources. EOOS aims to provide a flexible system to help manage and improve existing observing efforts, making them more efficient and effective at different geographical scales and for different users. Three expert bodies contribute to specific aspects of EOOS' development: an Advisory Committee, which brings together a broader stakeholder base; a Resources Forum,[11] which represents European ministries and funding agencies responsible for ocean observing; and an Operations Committee, which represents the diversity of the ocean observing implementers at national, regional, and pan-European levels. The pathways, through which marine data are used and transformed into practical information, creating systematized value chains, are expected to be at the forefront of EOOS interest and activities in the years to come (Jolly et al. 2021).

3 From MyOcean to CMEMS

3.1 MyOcean

Europe has been one of the pioneers in the use of space technology for Earth observation, investing heavily in various space missions. GMES (Copernicus), which is beginning to play a key role in managing the effects of natural disasters and climate change, is largely based on the use of satellite data. Research funded by the European Commission supports policies in the fields of agriculture, environment, fisheries, transport, and telecommunications. The Seventh Framework Programme (FP7) of the EU[12] explicitly identified the space subprogram (SPACE) for development of a European Space Policy and funding cooperation projects in this field. One of the three areas of research activity under SPACE was defined as "Space-based application at the service of European society (developing satellite observation systems and the GMES services for the management of the environment, security, agriculture, forestry and meteorology, civil protection and risk management)[13]". The space program periodically launches calls for proposals for funding projects on a competitive basis, setting out guidelines and financial frames. In response to one such call, MyOcean was born.

[10] https://www.eoos-ocean.eu/publications/eoos-strategy-2023-2027/. Last accessed on 28.02.2023.

[11] Since 2021, the author has acted as Germany's representative to the Resources Forum.

[12] Seventh Framework Programme for Research and Technological Development brought between 2007 and 2013 the major research-related EU initiatives together.

[13] https://cordis.europa.eu/programme/id/FP7-SPACE. Last accessed on 03.02.2023.

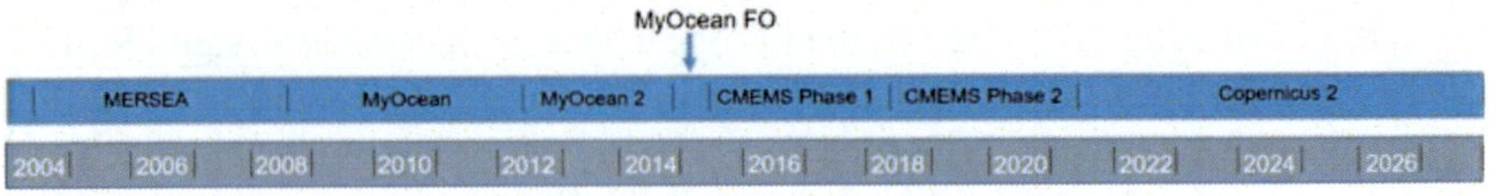

Fig. 1 Approximate timeline of the European (pre-)operational projects

MyOcean

MyOcean is a major project funded by the European Commission aimed at creating European infrastructures and resources for the preparation and operational deployment of the first basic marine services. It responds to the GMES objectives to develop and consolidate operational oceanography. It started on January 1, 2009, ended on March 31, 2012 (Fig. 1), and was led by P. Bahurel of the French ocean monitoring and forecasting center Mercator Ocean. The project consortium consisted of 67 partners and brought together operational staff as well as leading scientists, independent experts, and key users of marine information. The validation and promotion of the services that the project provided free of charge was one of the main achievements of this wide range of partnership. Among the main tasks of the project was the definition of an initial set of operational marine core services and the development of European operational practices that were to be harmonized and standardized for Member States. The development of research in support of the project objectives was also included.

MyOcean benefited from the experience gained in the MERSEA project, as well as from the cooperation with national metrological and oceanographic services, EuroGOOS, the European Environment Agency (EEA), and the European Maritime Safety Agency (EMSA). The main information about the ocean at global and European regional scales that MyOcean makes publicly available comes from satellite and in situ observations, as well as from numerical models that assimilate data in real time (Le Traon et al. 2019). The so-called baseline data include currents, temperature, salinity as a function of the three spatial coordinates and time, sea level height, sea ice, sea surface wind, and data on the BGC properties of water. They are catalogued, and an easy-access interface leads to them, allowing users to discover, visualize, and download some 240 basic products.[14] For the sake of clarity, I will recall that some of the data provided to users are similar to those produced by weather forecast centers. However, they are not easily understood by the citizens. Therefore, there are companies that "reformat" the data so that it can be understood by the general public. Such secondary data, for example, maps or graphs, are shown on television to help presenting weather forecast. In MyOcean terminology, companies that produce new data based on primary data are performing downstream services (Fig. 2).

[14] https://cordis.europa.eu/project/id/218812/reporting/de. Last accessed on 03.02.2023.

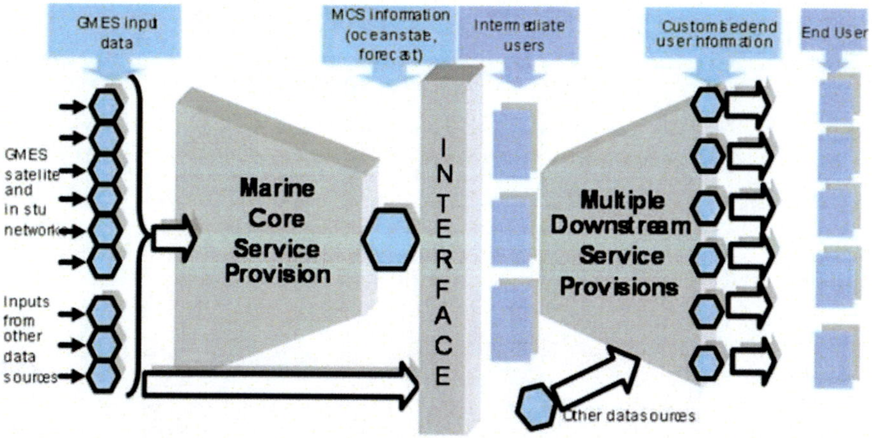

Fig. 2 Marine services: from core products to intermediate and final users. The diagram gives an overview of the activities in MyOcean within the GMES activities and downstream service lines. From Bahurel et al. (2010)

MyOcean consists of 13 main production centers (Fig. 3), of them 6 Thematic Assembly Centres (TACs) providing observational products, and 7 Monitoring and Forecasting Centres (MFCs) dealing with modeling and data assimilation. Each production center is responsible for its own area of activity or geographic region, thus avoiding duplication. The regions (Fig. 4) and the lead organizations for each were defined as Arctic (Nansen Environmental and Remote Sensing Center (NERSC), Norway), Baltic (DMI, Denmark), Atlantic-North West Shelf (Met Office[15], UK), Global (Mercator Ocean, France), Atlantic-Ireland-Biscay-Iberia (Mercator Ocean), Mediterranean (Istituto nazionale di geofisica e vulcanologia (INGV), Italy), and Black Sea (Marine Hydrophysical Institute (MHI), Ukraine). The production centers are integrated into the overall MyOcean system through a central information system managing the data flows. One original idea that has been put forward prioritizes user feedback in the design and operation of monitoring and forecasting systems. In short, MyOcean moved toward creating a user-driven system, i.e., one that gives high priority to user interactions at every level.

At the time of MyOcean's start, the MHI in Sevastopol had the greatest overall potential to conduct operational work in the Black Sea. The Black Sea MFC (BS MFC) was therefore located in Ukraine and led by G. Korotaev (Fig. 5).

In early 2009, shortly after MyOcean was launched, I spoke with P. Bahurel and P.-Y. LeTraon, the latter had become the Scientific Director of Mercator Ocean, about the possible involvement of GKSS in the next phase of the project. "We are supportive", they both said. "This has to go through the board. Talk to BSH.[16] If they support you, you have no problems", P. Bahurel added. I shared this information with

[15] There are some considerations on how to reorganise this MFC after the Brexit.

[16] BSH was at that time the main German partner in MyOcean.

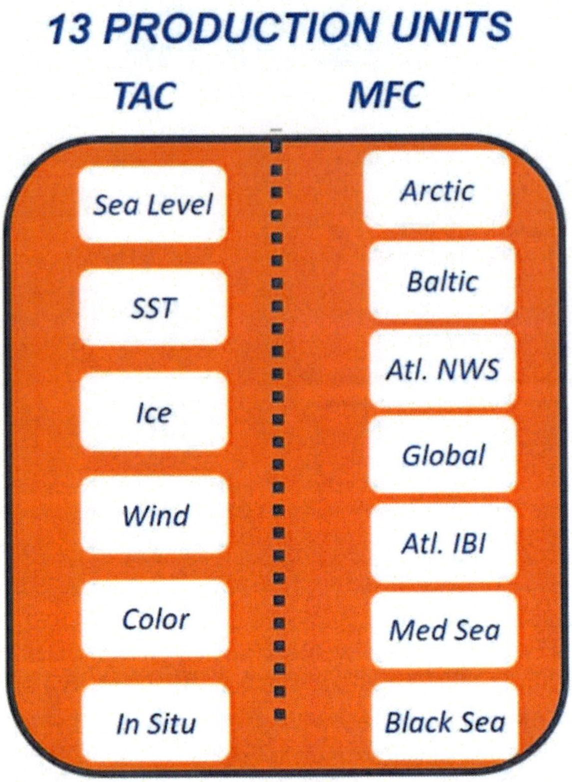

Fig. 3 Production units, which include Thematic Assembly Centres (TAC) and Monitoring and Forecasting Centres (MFC). Other abbreviations: NWS—North West European Shelf; IBI—Iberia-Biscay-Irish zone. From Bahurel et al. (2010)

H. von Storch. "Write a letter expressing our interest to join MyOcean, and I'll talk to Bernd Brugge,", he said. B. Brugge had become the head of BSH's Oceanography Department the previous year. The Board meeting at BSH took place in the fall of that year. H. von Storch and I were both present, and our request was granted. Hans suggested that I represent the institute further in the project.

3.2 *MyOcean2 and MyOcean-FO*

I first became a member of the MyOcean Science Advisory Committee (SAC). This was an auxiliary body to advise the head of the scientific work package, P. Brasseur, on strategic directions for the development of research in operational oceanography. Pierre Brasseur, an old friend of mine from my stays in Liège, had in the meantime moved to Grenoble, France. Along with the overall management of the scientific work

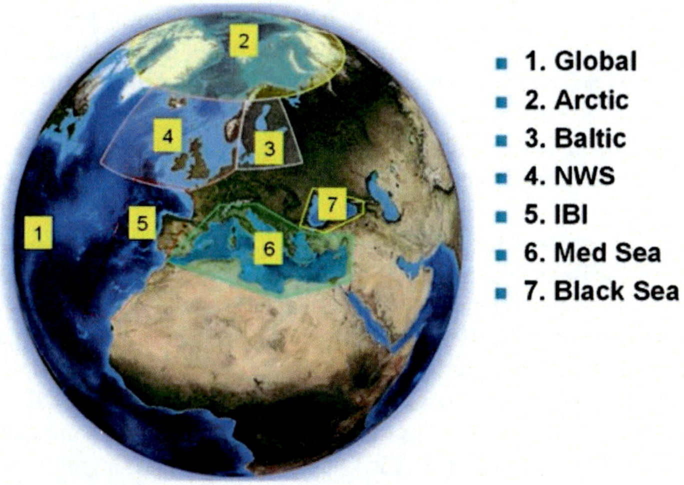

Fig. 4 Geographical areas of activity in MyOcean. Credit to Mercator Ocean International

Fig. 5 Author (on the left), G. Korotaev (middle) and Hans Dahlin (the then director of EuroGOOS) in front of the Church of the Resurrection of Christ, a popular tourist attraction in the vicinity of Yalta in Crimea, best known for its picturesque location

at MyOcean, he was also the main organizer of the first European colloquium on the scientific challenges facing the GMES marine service, entitled MyOcean Science Days. This colloquium was held on December 1 and 2, 2010, in Toulouse, France, at the International Conference Centre of the French Meteorological Service (Météo France). The MyOcean Science Days were concluded with a meeting of the SAC. During one of the breaks, I was approached by P. Bahurel and P.-Y. Le Traon. "We need to start preparing the next phase of the project and we think you can be of great help. Pierre Brasseur worked quite hard during the first phase, would you like to take over the scientific leadership of the second?" I had not expected such an invitation, nor did I have any idea how much work it would be, but I was sure it would not be small. I expressed reticence due to the fact that many of the organizational problems in this business were unfamiliar to me. "Pierre Brasseur will not "get off the train" completely", P.-Y. Le Traon continued, "He will become your deputy and introduce you to the work". I trusted P. Brasseur, which made me more agreeable.

"There is something else in our proposal", P. Bahurel continued. "You know that our goal is to use one model for all MFCs". At that time, the Arctic, Baltic, and Black Sea MFCs did not use NEMO. "Korotaev promised to switch to NEMO. It is more difficult with the Baltic community and they are not very willing to change the model. We think it is good to test how well NEMO can work in the Baltic Sea. You have used this model, you have worked on modeling the Baltic Sea and you can probably lead the work on the application of NEMO in this sea". My Ph.D. student L. Xi did work successfully on modeling currents in the Baltic Sea, however, not with NEMO, but with the HBM that was used by most of the Baltic operational community. With R. Kandilarov and my postdoc student in Oldenburg Sebastian Grayek, we had already published a paper on the use of NEMO (Grayek et al. 2010), but in the Black Sea. I knew that the extent to which GKSS (meanwhile renamed HZG) would be included in the next phase would depend on my positive response on whether I agree to lead the scientific development of MyOcean. I agreed, thereby taking a step toward my greatest involvement in international scientific management work. I had a year of active work ahead of me in developing the scientific part of MyOcean2.

The first framework that I had to take into account was that the relative financial weight of the scientific work to the whole remains the same as in MyOcean. This number was only eight percent.[17] The percentage seemed very small to me, but I had to learn to distinguish between routine and scientific work, to comply with the framework imposed, and to understand that MyOcean was not a scientific project, like many of those in which I had been involved up to that point, but a practical initiative to lay the foundations for a consolidated European operational oceanography underpinned by science.

There was also a delicate problem. The Baltic Sea modeling community did not want to hear about using an alternative model. Their model, first developed in BSH, was subsequently updated and adapted by colleagues at DMI to work with modern software and hardware platforms. They had put a lot of work into this model, including calibrating it to give good predictions. In addition, the model had a module that provided high resolution in the straits region,[18] making it suitable to be used for both the North Sea and the Baltic Sea. And all these developments, which they were proud of and which worked successfully, had to give way to a model that had not even been tested for that area by then. In this delicate situation, my Finnish colleague Tipani Stipa committed the Finnish Meteorological Institute to take over

[17] The European Commission subsequently approved a total budget of around €41 million with around €30 million in EU funding.

[18] In the version of HBM which was used by Stanev et al. (2015), the resolution in the straits was 900 m. This was superior to the other structured grid models at that time, but still inferior in comparison with what was developed in Zhang et al. (2016), see also Sect. 7.1 in chapter "New Observing and Modeling Systems and New Scientific Applications ".

the pre-operational implementation of NEMO in the Baltic Sea. Shortly afterward, the SMHI also started running NEMO.

Old and New Numerical Models

Modern modeling of environmental processes requires great computational professionalism and continuous development so that the model satisfies the latest scientific and applied trends. Associations of developers, leading scientists in the field of numerical mathematics, which are not very numerous, and associations of users are usually involved in the process of model development. Users may be leading scientists or routine users, but are not necessarily competent in all software problems of the model. Typically, there are many more users of numerical models than those who develop them. Users can develop new modules extending the capabilities of the models for their own scientific needs, for example. The introduction of these enhancements to the basic numerical model is subject to strict scientific rules and discipline and is usually done by the main developers of the models. The small number of users does not give enough feedback to the development specialists so that they can improve one or the other. There is a tendency for some old models without massive scientific and software support to lose users and gradually disappear.

MyOcean2 started on April 1, 2012 (Fig. 1), and with it, the European ocean monitoring and forecasting activities entered their pre-operational phase. After the completion of this phase, the "pre" was to be dropped and the ocean monitoring and forecasting was to become fully operational. The consortium numbered 59 partner institutes from 28 countries and included more than 350 experts and scientists from the marine field. The overall objective of MyOcean2 was to ensure a controlled continuation and expansion of the services and systems already implemented in MyOcean. From 2014, the transition to full operations was to be prepared. In other words, the prototype tools and the necessary management and coordination environment had to be developed in order to ensure that users had continuous access to GMES products and services.

Sebastian Grayek, who was currently working in Oldenburg on the WIMO project, accepted to move to HZG and take over the work on NEMO. Joanna Staneva took the modeling work on the ocean–atmosphere interaction and the resulting effects on sea surface currents and waves. Johannes Schulz-Stellenfleth, with whom we had already done some pioneering work on assimilating HF radar data into numerical models, undertook similar work in MyOcean2, which was to clarify the usefulness of such model complications in pre-operational practice (von Storch et al. 2015). Here, we were disseminating the experience gained in COSYNA toward European monitoring and forecasting systems.

Opening MyOcean2 to the Scientific Communities

Although MyOcean2 was a continuation of MyOcean, there was one significant difference between them in terms of R&D. It can be defined as an increase in the transparency of the work and opening it up to individuals and organizations not directly involved in MyOcean2. The aim was to attract scientific leaders and intermediate users who are interested in the development of research directions and downstream products. The implementation of this strategy has been supported by the European Commission, which has committed to launch calls for research projects in specific areas of interest for MyOcean2 and, more generally, for GMES. This has created a construct (Fig. 6) which, as a halo around the central project (MyOcean2), develops specific scientific or applied topics. They are aimed to support scientific development of MyOcean2. It was expected that these projects would either produce a result directly applicable to MyOcean2 or help, through collaborative work, raise the scientific level of the personal delivering operational services. As it was expected, a 100% rate of transfer of scientific developments to operations could not be achieved, but practice has shown that some of these projects have opened new horizons for the marine part of GMES.

From November 19 to 21, 2012, the "MyOcean2 Science Days" were organized in Geesthacht, which were attended by almost all MyOcean2 scientists, as well as by those from other active GMES-related projects and experts in research of interest to operational oceanography. Now, re-reading the recommendations of the workshop, I am reminded that it articulated some trends that are still relevant today. One was about the need to further develop operational oceanography in the coastal domain and link it to existing regional and global forecast systems. The interest of scientists was now also turning to combining satellite and in situdata, with a major focus on assessing the impact of upcoming Sentinel missions (see Fig. 37 of chapter "Further into the Ocean"). It was proposed that research activities on the ocean–atmosphere coupling should focus not only on a global scale, but also on a regional scale.

In 2014, GMES was renamed Copernicus. At the same time, MyOcean2 had to be extended from October 1, 2014, to May 31, 2015, under the name MyOcean Follow On (MyOcean-FO), in order to continue supporting pre-operational maritime activities and to avoid a break in the critical handover phase between pre-operational and fully operational services. At the end of the MyOcean-FO project, more than 5000 users were registered on the service and more than 140 operational users were using it every day to provide their own service streams to end users. A readiness was ensured to move from a prototype of operational oceanography in May 2015 to full operation, which was to happen in the frame of the Copernicus Marine Environment Monitoring Service (CMEMS[19]) in the period 2015–2021. The Blue Book

[19] See for more details Le Traon et al. (2017) and (2019).

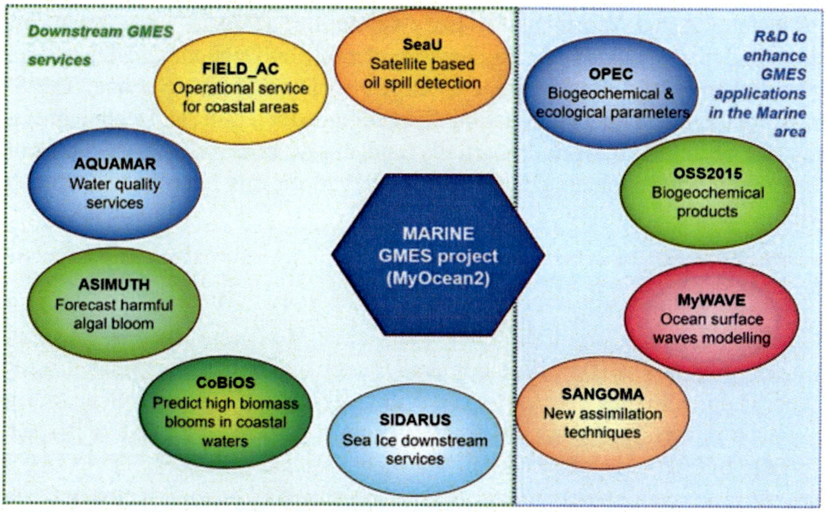

Fig. 6 GMES network concept. In the middle are the main projects (e.g., MyOcean2), which are surrounded by projects developing scientific prototypes for downstream services (left) and scientific projects reinforcing the application of GMES activities at sea (right)

"Copernicus for a Sustainable Ocean"[20] published in 2019 promotes and tells the story of the Copernicus Marine Service, which has been implemented and operated since 2015.

3.3 Scientific Evolution in CMEMS

We skip forward five or six years in time after the end of MyOcean-FO. The first operational phase of Copernicus ended in 2021, and registered users of CMEMS services had reached 32,000 by then. The web portal was accessed 280,000 times in 2021. About 53% of downloads were from business organizations and 41% from the public sector. Mercator Ocean became Mercator Ocean International, a non-profit organization providing sound science-based services focused on the conservation and sustainable use of the ocean and marine resources. This company is owned by a dozen European institutions working in the field of operational oceanography. It has received a mandate from the European Commission to organize the operation of CMEMS for a new six-year period.[21]

[20] https://www.mercator-ocean.eu/en/ocean-science/blue-book-copernicus-sustainable-ocean/. Last accessed on 03.02.2023.

[21] The book of Alvarez Fanjul et al. (2022) provides an extensive overview of the entire value chain of an Operational Ocean Monitoring and Forecasting System, thus promoting the development of marine forecasting systems around the globe.

A Scientific and Technical Advisory Committee (STAC[22]) was set up within Mercator Ocean to advise the company on scientific and technological issues related to the provision and improvement of the Copernicus Marine Service. The STAC consists of 12 independent internationally recognized scientific and technical experts. They are required to work impartially and apply strict rules to avoid conflicts of interest. Some, including myself, have served on this board since the days of MyOcean's SAC.

STAC and the Evolution of Copernicus Marine Service

The primary function of the STAC is to assist Mercator Ocean International in defining the CMEMS development strategy and to organize related activities. These activities help to prepare future versions of the Copernicus Marine Service aimed at meeting existing and future user requirements and needs, taking full account of scientific and technological advances. The STAC is working to define and regularly update the CMEMS evolution strategy. This strategy identifies the key R&D activities required to evolve the service. Particular attention is given to the evolution of the medium- and long-term R&D activities that are funded internally by CMEMS, with STAC issuing public calls for tenders for the development of the services. The longer-term evolution of CMEMS scientific and technological development is also ensured through external research projects funded, for example, by the European Horizon 2020 program or the new Horizon Europe program for the period 2021–2027.

STAC is led by Neville Smith from Australia. Prior to his retirement, Neville was the lead scientist and the head of the Australian Bureau of Meteorology Research Centre. He played a key role in the development of ocean and climate forecast systems. Together with other leaders in the international scientific community, he helped develop ocean and climate observing systems, initiated and led the Global Ocean Data Assimilation Experiment (GODAE). Neville Smith participates in world bodies at the highest level such as the IOC, IPCC, and World Climate Research Programme (WCRP).

What is my current involvement with STAC? CMEMS is in the process of opening up to coastal operational practice, and as a scientist with a background in coastal oceanography, I am involved in developing future strategies and practical steps in this area. I have also tried to initiate the Hereon joining Mercator Ocean International. The new director of the Institute for System Analysis and Modeling,[23] Prof. Corinna Schrum, has supported my initiative and taken over the activities in this direction. If it works out, I will be glad I did my job.

---

[22] https://marine.copernicus.eu/about/stac. Last accessed on 03.02.2023.

[23] Hans von Storch had retired in 2015 and the sector of the institute he led was transformed into an institute.

4 Whose Are the Waves—The Ocean's or the Atmosphere's?

4.1 Marine Meteorology and Storm Gorbush

Wind waves are a fundamental element in the process of air–sea interaction. Hence, they do not belong only to the ocean or only to the atmosphere. As for their belonging to the different branches of science, the matter is a little more complicated. Marine meteorology is a subfield of meteorology that deals with weather and climate and the associated conditions in the ocean and coasts. At its core lies the interaction between the ocean and the atmosphere. The importance of marine meteorology is explained by the need for information to support marine and coastal activities including shipping, fishing, tourism, offshore oil drilling and production, oil spill response, offshore wind turbines, search and rescue, and other marine operations. Forecasts of sea swell have long been made by weather centers. However, in their forecasts, the influence of ocean processes on swell is usually accounted for in a rather incomplete form.

A meteorological-oceanographic case has been recorded in recent political history which is worth recounting here. On December 2 and 3, 1989, just a few weeks after the fall of the Berlin Wall, a summit was held in Malta between US President George Bush Sr. and CPSU[24] General Secretary Mikhail Gorbachev. During the meeting, Bush and Gorbachev declared the end of the Cold War. Many believe it was the most important meeting since 1945, when British Prime Minister Winston Churchill, Soviet leader Joseph Stalin, and US President Franklin D. Roosevelt reached an agreement in Yalta on a post-war plan for Europe. The Soviet and American delegations spent the night on the missile cruisers Slava and Belknap, moored opposite a small fishing village on Malta's southeast coast. Rough seas led to the cancelation or rescheduling of some of the meetings. They were eventually held aboard the Soviet cruise ship Maxim Gorky. As Condoleezza Rice, the 66th US Secretary of State, wrote[25]: "... it took a long time to get it arranged, finding a place, a place that would not be ceremonial, a place where you didn't have to do a lot of other bilateral. And fortunately—or unfortunately—they chose Malta, which turned out to be a really horrible place in December. Although the Maltese were wonderful, the weather was really bad".[26]

[24] Communist Party of the Soviet Union.

[25] https://nsarchive2.gwu.edu/coldwar/interviews/episode-24/rice1.html. Last accessed on 03.02.2023.

[26] This was not the first such case. W. Bjerkness mentioned in his inaugural address: At last, the storm of November 14, 1854, over the Black Sea caused so much damage to the fleets of the combined Western Powers that it gave the needed impetus toward a speedier development of the science. The eminent astronomer Leverrier received an official order to study conditions of that storm with a view to the possibility of foreseeing such occurrences and the establishing of a storm-warning service. Upon this foundation rose the present organized international telegraphic weather service (Bjerknes 1914).

Not enough, I might add, the organizers cared what the weather would be like. And even if they had cared, it would not have helped them much, because at that time, marine meteorology was using coarse atmospheric models that did not adequately describe the position of the areas with maximum winds (Dell'Ossio et al. 1992). This is essential because local deviations of wind direction can cause significant errors in the prediction of swell. As shown by the evaluations of the authors of the above publication, who used the WAM, the problem in the forecasts was largely one of adequate representation of wind waves. Because of the global resonance, these authors named the storm "Gorbush" after the names of the main participants at the summit.

4.2 MyWave

It is not correct to consider wind waves and currents separately, especially in coastal areas. There, under sea storm conditions, the interactions between waves and currents become strong. Such interactions in coastal zones were not studied enough, and the FIELD_AC (see Sect. 5.2.2 in chapter "New Observing and Modeling Systems and New Scientific Applications") was one important contribution of the coastal wave community in this field. This project was initiated as a response to the call on "Stimulating the development of downstream GMES services" (Fig. 6-left). In 2011, FP7-SPACE launched a new call, this time on "Research and development to improve future applications of GMES in marine and atmospheric areas". Six projects in total in this field were accepted for funding (Fig. 6-right). Among them was the proposal "MyWave (A pan-European concerted and integrated approach to operational wave modeling and forecasting—a complement to GMES MyOcean services)" led by Øyvind Saetra from the Norwegian Meteorological Institute.

The Main Objectives and Tasks of MyWave

MyWave aimed to lay the foundation for a future marine GMES service that incorporates ocean waves. The major tasks were to (1) increase the use of terrestrial observations and (2) to improve data processing algorithms and data assimilation systems. Solving the second task necessitated improving the description of the physical processes in the wave models. All this was intended to provide a framework for the development of common model systems (atmosphere–waves–ocean). The third task (3) was to establish a new standard for probabilistic predictions, and the fourth (4) was to develop standard protocols for product validation. These research developments were intended to enable the operational oceanographic community to develop a new generation of products exceeding in quality those previously provided by marine meteorology.

The work at HZG in the frame of MyWave was led by Arno Berenz, who was committed to maintaining and distributing versions of the WAM via GitHub.[27] Heinz Günther worked on plugging new physical processes into the model and on writing a new FORTRAN95 code for WAM. Joanna Staneva was testing the model in real conditions in the German Bight and was investigating the interaction between wind waves and currents in extreme situations.

In the meantime, work was progressing in MyOcean2 and MyOcean-FO on coupled ocean–atmosphere modeling, which is being conducted globally by the UK Met Office. On the HZG side, J. Staneva and her Ph.D. student were advancing ocean–atmosphere modeling on a regional scale. Interesting results were coming out, confirming that the influence of wind waves can propagate high in the atmosphere, and that even the atmosphere over land feels this influence (Wiese et al. 2019). There was still much work to be done, but the path for bringing together the operational wave forecast and available to the MFCs forecasting tools was being traced. Wave forecasting became a part of Copernicus Marine Service during its operational phase, which started on June 1, 2015.

MyOcean2 and MyWave

MyOcean2 and MyWave were developing in parallel; therefore, efforts were made to coordinate the work of the two consortia. Representatives from MyWave were invited to Oslo on January 29 and 30, 2014, where the MyOcean2 Executive Committee and Board meetings were held. At that meeting, Ø. Saetra presented MyWave. Together with Luigi Cavaleri from the Institute of Marine Sciences, Venice, they also presented the "roadmap" that their consortium had developed, which contained recommendations on how waves could become part of a future Copernicus service. The emerging view was that the integration of operational wave modeling services should be a gradual process. P. Bahurel informed MyWave colleagues that the MyOcean2 consortium had a positive view on the integration of waves into a future Copernicus service and pointed out the need for a joint R&D strategy. In addition to addressing a number of scientific and practical issues, data policy issues needed to be clarified.

[27] A network version control service for software development projects. It is named after the Git version control system that tracks changes to a file or set of files over time so that it is possible to recall specific versions later.

5 Black Sea Operational Oceanography

The events in Crimea in February and March 2014 changed the situation of operational oceanography in the Black Sea. It was in Sevastopol, where the BS MFC was headquartered. Colleagues continued to do their pre-operational work; G. Korotaev regularly attended the Executive Committee meetings and declared that they would continue their participation. At some point, it was realized that the marine activities in the Black Sea within Copernicus could not be conducted by an institute that was in a territory that was not recognized by the European Union as part of the Russian Federation. Earlier I had repeatedly criticized the fact that it was not right that only one institute does all the forecasting activities (this was the real situation in the BS MFC), but no change was made. There were objective conditions for this. There was not enough operational capacity in the European Union Member States bordering the Black Sea to issue marine forecasts. On the other hand, there was not much affinity from Crimean colleagues to include MyOcean partners from the Black Sea countries in practical activities. After the new political situation, it appeared that there was no one to take over the work.

The question was to what extent the future BS MFC could benefit from the experience of the other centers established within MyOcean2 and MyOcean-FO. Nadia Pinardi took the initiative to provide outside logistical and personnel support to the Black Sea Centre that was to be established. The Italians offered to provide the Black Sea countries with their know-how, which they had developed within the Mediterranean MFC (Coppini et al. 2021). However, they had no experience in the field of Black Sea modeling, which has many specific features compared to Mediterranean modeling. This gap had to be filled by E. Peneva, E. Özsoy, J. Staneva, M. Grégoire, and myself. Numerous meetings and contacts followed. The MyOcean2 board decided that the BS MFC would have to catch up with the backlog caused by MHI's exit from the consortium.

On the Operational Personnel in the Field of Numerical Modeling in the Black Sea Countries

The situation in the Black Sea countries (excluding Russia) with numerical modeling of the Black Sea was roughly the following. Romania had only some experience in numerical modeling of wind waves. Bulgaria had experience in modeling currents (University of Sofia, E. Peneva), waves (NIMH, Anna Korcheva), and ecological processes (Institute of Biodiversity and Ecosystem Research, Georgi Daskalov). Unfortunately, in each of the institutes, the respective scientific groups were very small or represented by only one or two people. In Turkey, the scientific and staff support was approximately the same as in Bulgaria. In Ukraine, the group around Vladimir Maderich in Kiev had experience in particular areas of numerical modeling of the dynamics of the sea.

In short, there were some important conditions which predetermined further developments. The Bulgarians (including those who worked outside Bulgaria) had the most experience in modeling, and the Italians had the necessary infrastructure. A leader from a Black Sea country was sought. The choice fell on the IO-BAS in the person of the then director A. Palazov as a coordinator. An uneasy period of specifying the work and its organization began. There were technical problems related to the specific features of the Black Sea, as well as organizational problems. In the initial phases, the option of Bulgaria and Italy jointly doing the forecasts of the physical field and Belgium taking over the ecosystem modeling was discussed. I insisted that Sofia and Varna work as a team so that in the near future, they would take over all the work of issuing the operational forecasts. Atanas Palazov preferred to leave this work with the Italians. After all, the Italians now provide the currents and other elements of the Black Sea circulation, whereas J. Staneva took over the wind wave forecasting in Germany, and M. Gregoire took over ecosystem forecasting in Belgium. The actual work remained outside the Black Sea region. This was contrary to my ideas about the evolution of Black Sea cooperation. To my deepest regret, no structure has been established anywhere in Bulgaria, to this day (2023), that can take over marine operational forecasting.

6 Some Global Trends

6.1 From GODAE to OceanPredict

Data assimilation in operational oceanography in Europe has developed in parallel with developments in this field by other actors in the global data assimilation community. Here, I will briefly present the global framework without repeating the scientific issues and advancements already discussed. In 1996 and 1997, N. Smith and the French scientist Michel Lefebvre proposed the idea of a GODAE (Bell et al. 2009), that is, to test the feasibility and usefulness of real-time global ocean forecasting over a ten-year period. The main objectives of GODAE required the application of state of the art ocean models and data assimilation methods, the assessment of the predictability of ocean processes, and the improvement of the design and performance of the GOOS. The implementation of GODAE was supervised by an International Steering Team involving selected experts from operational and research institutes in Australia, Canada, China, France, Norway, Japan, the UK, and the USA. Neville Smith was the first chair of GODAE.

The Progress in GODAE

The main phases of GODAE were 1998–2000 conceptual development, 2000–2003 prototype development, and 2004–2008 major demonstration and consolidation phase. The end of GODAE was marked by a final symposium held in Nice, France, in November 2008, where over 300 participants critically evaluated the achievements of the last 10 years and discussed the future of operational ocean forecasting. The feasibility and usefulness of global ocean observing and forecasting were successfully demonstrated. In the early 2000s, there were eight different global ocean systems with data assimilation capabilities covering the World Ocean. Three of them were eddy-resolving, three were eddy-permitting, and the other two were of lower resolution (Dombrowsky et al. 2009; Drévillon et al. 2008). Several groups have also implemented regional systems covering individual ocean basins. The productive scientific collaboration during GODAE has given a considerable impetus to the establishment and improvement of operational ocean forecast systems in a number of countries. In more practical terms, GODAE has expanded the capability to reliably collect and process measurements in real time, as well as to generate and disseminate analyses and forecasts.

Consultations between members of the scientific community and stakeholders in observing systems and operational oceanography have shown that a continuing initiative is needed to further mature the science that underpins operational oceanography. The next phase named GODAE OceanView (GOV) has been programmed for a period of 10 years, starting in 2009 (Bell et al. 2015; Schiller et al. 2015). This was a period, in which the global modeling and data assimilation systems were refined, put into practice and compared with each other.

Working groups, GOV Task Teams (or TTs), have been established to coordinate scientific developments, focusing on a range of different topics and involving the wider ocean science communities. The Coastal Ocean and Shelf Seas Task Team (COSS-TT) that was initiated in 2012 addressed scientific issues in support of multidisciplinary analysis and forecasting in the transition zone between the coast and the open ocean, as well as exchange of water and substance between the shelf and the open ocean. The first COSS-TT workshop held in Miami during the same year provided an opportunity for direct exchange of information between scientists working on specific coastal observing and forecasting systems. COSYNA was one of the leading ones, and our experience with running this operational system was met with interest.

COSS-TT

COSS-TT is an international team of scientists from all continents. This team meets approximately every 18 months. The meetings aim to increase the convergence of the participating scientific communities in the field of regional and coastal ocean forecasting and to consolidate international coastal observing programs, e.g., satellite observations, coastal observatories, etc. There was ample scope for presenting European experience gained in ECOOP, Jerico, MyOcean, etc. Part of the activities of COSS-TT was materialized in two collections of papers (De Mey et al. 2017; Cirano et al. 2021).

In 2019, GOV was renamed OceanPredict, to emphasize ocean prediction within the general operational oceanographic context. By 2021, forecasting systems matured. The CMEMS global system operated by Mercator Ocean International has been set up with a 1/12° resolution, providing 3D global ocean predictions for up to 10 days, which were updated daily. Over the next decade, OceanPredict aims to improve science and operational activities in partnership with international organizations. This will give weight to operational oceanography during the UN Decade of Ocean Science for Sustainable Development. The rapidly growing public awareness of the threats posed by climate change provides an opportunity for OceanPredict to be in the forefront of actions in support of sustainable development.

6.2 The Decade of Ocean Science for Sustainable Development

The UN proclaimed the period 2021–2030 as "Decade of Ocean Science for Sustainable Development" (Ryabinin et al. 2019) with the aim to mobilize the scientific community as well as policymakers, companies, and civil society around a common agenda for research and technological innovation. The resolution entitled "Transforming our World: the 2030 Agenda for Sustainable Development",[28] which was adopted by all UN Member States in 2015, provides a common blueprint for peace and prosperity for people and planet now and in the future. At its core are the 17 Sustainable Development Goals (SDGs), which represent an urgent call to action by all countries, developed and developing, in a global partnership. SDGs recognize that ending poverty and other deprivations must go hand in hand with strategies that improve health and education, reduce inequality, and boost economic growth—all while tackling climate change and working to protect oceans and forests.

SDG 14 is worded like this: "Conserve and sustainably use the oceans, seas and marine resources for sustainable development". Key to achieving this goal will be

[28] https://undocs.org/Home/Mobile?FinalSymbol=A%2FRES%2F70%2F1&Language=E&DeviceType=Desktop&LangRequested=False. Last accessed on 03.02.2023.

improving knowledge of all parts of the marine system, from coastal seas to the deepest parts of the World Ocean. Target 14a states: "Increase scientific knowledge, develop research capacity and transfer marine technology, taking into account the Intergovernmental Oceanographic Commission Criteria and Guidelines on the Transfer of Marine Technology, in order to improve ocean health and to enhance the contribution of marine biodiversity to the development of developing countries, in particular small island developing States and least developed countries".

A key objective of the Decade of Ocean Science for Sustainable Development is to support efforts to reverse the cycle of ocean health decline and to bring together stakeholders from all over the world around a common framework to ensure that ocean science can fully support countries in creating better conditions for sustainable ocean development. The UN supports the view that scientific understanding of the ocean response to external pressures and the development of necessary management actions are fundamental to sustainable development. This view suggests that ocean observations and research are essential to predict the effects of change, design mitigation measures, and guide adaptation. As mandated by the UN General Assembly, UNESCO's IOC is coordinating the Decade's preparatory process, inviting the global ocean community to plan the next ten years in ocean science and technology to secure "the science *we need* for the *ocean we want*". With the narrative in this part, I hope to have detailed the background to the Decade related to the development of operational oceanography in European coastal and regional seas. The latter has a lot more to do in the coming years.

References

Alvarez Fanjul E, Ciliberti S, Bahurel P (eds) (2022) Implementing operational ocean monitoring and forecasting systems. Paris, France, IOC-UNESCO, p 376 & Annexes, (GOOS-275). https://doi.org/10.25607/OBP-1774

Bahurel P, Adragna F, Bell M, Jacq F, Johannessen J, Le Traon P-Y, Pinardi N, She J (2010) Ocean monitoring and forecasting core services, the European MyOcean example. In: Hall J, Harrison DE, Stammer D (eds) Proceedings of OceanObs'09: sustained ocean observations and information for society, vol 1, Venice, Italy, 21–25 September 2009, ESA Publication WPP-306. https://doi.org/10.5270/OceanObs09.pp.02

Bell G, Hey T, Szalay A (2009) Beyond the data deluge. Science 323(5919):1297–1298

Bell M, Schiller A, Traon P-YL, Smith N, Dombrowsky E, WilmerBecker K (2015) An introduction to GODAE OceanView. J Oper Oceanogr 8:s2–s11. https://doi.org/10.1080/1755876X.2015.1022041

Bjerknes V (1914) Meteorology as an exact science. Mon Weather Rev 42(1):11–14

Cirano M, Charria G, De Mey-Frémaux P, Kourafalou VH, Stanev E (2021) Coastal Ocean forecasting science supported by GODAE OceanView coastal oceans and shelf seas task team (COSS-TT)—part II. Ocean Dyn 71:779–783

Coppini G, Clementi E, Cossarini G, Korres G, Drudi M et al (2021) The copernicus marine service ocean forecasting system for the Mediterranean Sea. In: 9th EuroGOOS international conference, Shom; Ifremer; EuroGOOS AISBL, May 2021, Brest, France, pp 272–279 (hal-03334358v2)

Davidson F, Alvera-Azcárate A, Barth A, Brassington GB, Chassignet EP, Clementi E, De Mey-Frémaux P, Divakaran P, Harris C, Hernandez F, Hogan P, Hole LR, Holt J, Liu G, Lu Y,

Lorente P, Maksymczuk J, Martin M, Mehra A, Melsom A, Mo H, Moore A, Oddo P, Pascual A, Pequignet A-C, Kourafalou V, Ryan A, Siddorn J, Smith G, Spindler D, Spindler T, Stanev EV, Staneva J, Storto A, Tanajura C, Vinayachandran PN, Wan L, Wang H, Zhang Y, Zhu X, Zu Z (2019) Synergies in operational oceanography: the intrinsic need for sustained ocean observations. Front Mar Sci 6:450. https://doi.org/10.3389/fmars.2019.00450

Dell'Ossio L, Bertotti L, Cavaleri L (1992) The Gorbush storm in the Mediterranean Sea: atmospheric and wave simulation. Month Weather Rev. 120:77

De Mey P, Stanev E, Kourafalou V (2017) Science in support of coastal ocean forecasting—part 1. Ocean Dyn 67:665–668

Dombrowsky E, Bertino L, Brassington GB, Chassignet EP, Davidson F, Hurlburt HE, Kamachi M, Lee T, Martin MJ, Mei S, Tonani M (2009) GODAE systems in operation. Oceanography 22(3):80–95. https://doi.org/10.5670/oceanog.2009.68

Drévillon M, Bourdallé-Badie R, Derval C, Lellouche JM, Rémy E, Tranchant B, Benkiran M, Greiner E, Guinehut S, Verbrugge N, Garric G, Testut CE, Laborie M, Nouel L, Bahurel P, Bricaud C, Crosnier L, Dombrowsky E, Durand E, Ferry N, Hernandez F, Le Galloudec O, Messal F, Parent L (2008) The GODAE/Mercator-Ocean global ocean forecasting system: results, applications and prospects. J Oper Oceanogr 1(1):51–57. https://doi.org/10.1080/1755876X.2008.11020095

European Marine Board (2021) Sustaining in situ ocean observations in the age of the digital ocean. EMB Policy Brief No. 9, June 2021. ISSN: 0778-3590 ISBN: 9789464206081. https://doi.org/10.5281/zenodo.4836060

Gaughan P, Hallinan D, Reilly K (2019) Using economic cost benefit analysis methodologies to underpin the sustainability and strategic planning of Coastal Ocean research infrastructures in Europe. OCEANS 2019–Marseille, 2019, pp 1–8. https://doi.org/10.1109/OCEANSE.2019.8867276

GOOS (2021) A 5-year strategic plan for OceanOPS 2021–2025, GOOS observations coordination group, united nations educational, scientific and cultural organization, intergovernmental oceanographic commission. https://www.goosocean.org/index.php?option=com_oe&task=viewDocumentRecord&docID=28415

Gray J, Szalay A (2007) eScience—a transformed scientific method. Presentation to the Computer Science and Technology Board of the National Research Council, Mountain View, CA, 11 January 2007. http://itre.cis.upenn.edu/myl/JimGrayOnE-Science.pdf

Grayek S, Stanev EV, Kandilarov R (2010) On the response of Black Sea level external forcing: altimeter data and numerical modeling. Ocean Dyn 60:123–140

Hey T, Tansley S, Tolle K (eds) (2009) The fourth paradigm: data-intensive scientific discovery. Microsoft Research, Redmond. http://fourthparadigm.org

Jolly C, Jolliffe J, Postlethwaite C, Heslop E (2021) Value chains in public marine data: a UK case study. In: OECD science, technology and industry working papers, No. 2021/11, OECD Publishing, Paris. https://doi.org/10.1787/d8bbdcfa-en

Konar M, Ding HA (2020) Sustainable Ocean Economy for 2050: approximating its benefits and costs. World Resources Institute, Washington. Available online at: https://oceanpanel.org/publication/a-sustainable-ocean-economy-for-2050-approximating-its-benefits-and-costs/

Le Traon P-Y, Ali A, Alvarez Fanjul E et al (2017) The Copernicus marine environmental monitoring service: main scientific achievements and future prospects. Special Issue Mercator Océan International #56. Available at: https://marine.copernicus.eu/it/node/594

Le Traon P-Y, Reppucci A, Alvarez Fanjul E et al (2019) From observation to information and users: the copernicus marine service perspective. Front Mar Sci 6:234. https://doi.org/10.3389/fmars.2019.00234

Metropolis N (1992) The age of computing: a personal memoir. Daedalus 121:120

Ryabinin V, Barbière J, Haugan P, Kullenberg G, Smith N, McLean C, Troisi A, Fischer A, Aricò S, Aarup T, Pissierssens P, Visbeck M, Enevoldsen HO, Rigaud J (2019) The UN decade of ocean science for sustainable development. Front Mar Sci 6:470. https://doi.org/10.3389/fmars.2019.00470

Schiller A, Bell M, Brassington G, Brasseur P, Barciela R, De Mey P, Dombrowsky E, Gehlen M, Hernandez F, Kourafalou V, Larnicol G, Le Traon P-Y, Martin M, Oke P, Smith GC, Smith N, Tolman H, Wilmer-Becker K (2015) Synthesis of new scientific challenges for GODAE OceanView. J Oper Oceanogr 8(sup2):s259–s271. https://doi.org/10.1080/1755876X.2015.1049901

Schiller A, Mourre B, Drillet Y, Brassington G (2018) An overview of operational oceanography. In Chassignet E, Pascual A, Tintoré J, Verron J (eds) New frontiers in operational oceanography, GODAE OceanView, pp 1–26. https://doi.org/10.17125/gov2018.ch01

Stanev E, Xi L, Grashorn S (2015). Physical processes in the transition zone between North Sea and Baltic Sea. Numerical simulations and observations. Ocean Modeling, 93:56–74

Stommel H (1989) Why we are oceanographers. Oceanography 2(2):48–54

von Storch H, Emeis HK, Meinke I, Kannen A, Matthias V, Ratter BW, Stanev E, Weisse R, Wirtz K (2015) Making coastal research useful. Oceanologia 57(1):3–16. https://doi.org/10.1016/j.oceano.2014.09.001

Wiese A, Stanev E, Koch W, Behrens A, Geyer B, Staneva J (2019) The impact of the two-way coupling between wind wave and atmospheric models on the lower atmosphere over the North Sea. Atmosphere 10(7):386. https://doi.org/10.3390/atmos10070386

Woods JD, Dahlin H, Droppert L, Glass M, Vallerga S, Flemming NC (1996) The strategy for EuroGOOS EuroGOOS Publication No. 1, Southampton Oceanography Centre, Southampton. ISBN 0-904175-22-7

Zhang YJ, Stanev EV, Grashorn S (2016) Unstructured-grid model for the North Sea and Baltic Sea: validation against observations. Ocean Model 97:91–108

Afterword

The reflections on science that I present in the afterword focus on recent developments in science and services for a better understanding of planet Earth and its optimal use by humans. A number of concerns about the degradation of the planet are addressed, as well as some questions about how to better manage the living environment. Among the tools needed to build strategies to avoid climate catastrophes, the digitalization of science and scientific applications is placed at the forefront as a must in the drive to make the ocean predictable, clean, healthy, and resilient to impacts. The new role that science and technology will play in the future is also discussed.

1 Reflections on Science

Jeremy Baumberg, the author of "The Secret Life of Science" (Baumberg, 2018), has synthesized some thoughts that largely coincide with what I have also arrived at. At a first glance, there are two main trends: the digitalization of knowledge and the growing number of scientists. Figuratively speaking, J. Baumberg presents science as an ecosystem in which competition is growing. Digitalization supports the development of universal standards, launches best practices, makes it easy to extract scientific output from publicly available data, should reduce duplication, and, most importantly, facilitates connectivity between scientists. One wonders: does the increase in the number of scientists mean that more science is being produced? Let us look at what is happening. Universities are struggling to climb the research rankings, which can be roughly accounted for by their *h*-index.[1] Scientific journals compete to attract the strongest authors to offer them the most impactful articles. New journals are springing up which, on the one hand, weaken the large influx of

[1] The *h*-index (Hirsch index) is a key figure for the worldwide perception of a scientist in professional circles. A scientist has an index h if h of his N articles have been cited at least h times each. It can also be applied to the productivity and impact of a group of scientists, such as a department or university.

E. Stanev, *Trajectories in Oceanography*, Springer Oceanography,
https://doi.org/10.1007/978-3-031-33720-8_6

articles to established journals, but on the other hand, redirect valuable manuscripts to themselves, often luring authors with a more expeditious peer-review process and free publication. Not to mention the emerging electronic journals, where the process from manuscript submission to publication is much faster.

The competition to attract the most distinguished speakers does not pass conferences by, and there are so many of them, often happening at the same time, that one cannot attend them all even if one wanted to. Institutes and journalists flood the public with information about the most important discoveries. I have already spoken about the role of funders in connection with the example of funding within the European Commission, which directs research toward the development of science-based services for improved understanding of the world and innovative solutions to environmental problems, for example.

Science is subject to constant verification, and those on the inside must develop the fortitude to discover and admit that they are wrong about something. This is not a prime characteristic of political and religious leaders who insist that they have a monopoly on truth. In his book "It's About Time", Nathaniel David Mermin says: "This process of discovering that one's former beliefs are wrong, and the painstaking search to identify the old errors, enabling one to construct better founded beliefs to replace them, is what makes the pursuit of science so engrossing. The world would be a far better place for all of us if this joy in exposing one's own misconceptions were more common in other areas of human endeavor" (Mermin 2009). I believe that what also distinguishes scientists from much of society are the motivational characteristics, the thirst for knowledge as opposed to the thirst for material values. The reward for the scientist is to lead the struggle for the new, for the perfect, to touch the unknown. There are numerous examples in history showing that for the elevated men of science and culture, material gratifications and compensations are surmountable temptations on the road to perfection.

Unfortunately, things are neither so ideal nor so simple. In the distant past, science was the province of the elite. In the not so distant past, the small number of funders made the distribution of research funds seem more transparent. Today, in addition to fame (the prestige of universities) and wealth (of universities and of funding corporations), the factor of lobbying intervenes. The latter plays an important role when it comes to determining which areas of research deserve the most funding. The dependence of scientific policies in some countries and their modification as a function of larger policy frameworks of the respective government is well known.

It often happens that the best scientists are not always the best at attracting the funding needed for their research. A moderately capable person can do good scientific work by making use of their computer literacy and freely available data. This work can be done by many other colleagues of similar abilities if given the same task.[2] However,

[2] This is in line with the requirement of many journals that papers be written so that other scientists can reproduce the work. Many journals now also require that the data used be made available. This is very good because it subjects the research to critical scrutiny.

it is routine. What is more exciting is that every now and then someone comes along with a new idea. Unfortunately, originality and different (one might say alternative) thinking sometimes goes unnoticed in the huge stream of new publications. I direct my collaborators, who are very happy that something works well, to the question "Under what conditions will it not work" or "how do you go beyond what you know". And here again I remember with what curiosity I opened, disassembled, and reassembled my father's pocket watch as a child until it finally stopped working, just curiosity leading me to a "new quality".

Digitalization and communications are changing the way we think and act. The new metrics of science do, too. The *h*-index has become a kind of currency. A scientist's competitiveness is based on the citations his or her work receives in the scientific press. In some places, this metric is a determinant of salaries, but it is also a guiding metric when the question of subsidizing their research is decided. This creates pressure on them. J. Baumberg says: "What is known is that the population of scientists is increasing at a faster rate than the human population as a whole. More scientists do not mean more discoveries. What they can cause, through intensifying competition within the ecosystem, is *h*-index inflation, just as printing more money can cause price inflation".

As always, the coin has a flip side. With the current massification of science, competition is needed to stimulate scientists to work, i.e., some pressure through the use of the toolkit of scientific metrics. My observations show that if the drive for curiosity is absent, the scientist needs periodic evaluation of success to force him to give more of himself. Sometimes I feel sorry for these persons, who remind me of unbelieving clergymen suffering deprivation just to find a cozy and quiet place to spend their lives. I assume that the education these colleagues received is at least partially responsible for their condition. "An education that does not teach to ask questions is bad", claims the French poet, writer, and literary critic Paul Valéry. Austrian physicist and theoretician of science Erwin Schrödinger, who is considered one of the founders of quantum mechanics, is much more clear about the science education: "...scientific education is fabulously neglected ... This is an evil that is inherited, passed on from generation to generation. The majority of educated persons are not interested in science, and are not aware that scientific knowledge forms part of the idealistic background of human life. Many believe—in their complete ignorance of what science really is—that it has mainly the ancillary task of inventing new machinery, or helping to invent it, for improving our conditions of life. They are prepared to leave this task to the specialists, as they leave the repairing of their pipes to the plumber. If persons with this outlook decide upon the curriculum of our children, the result is necessarily such as I have just described it".[3]

[3] Opening remarks of the second of four public lectures for the Dublin Institute for Advanced Studies at University College, Dublin (Feb 1950), "The Practical Achievements of Science Tending to Obliterate its True Import", collected in *Science and Humanism: Physics in Our Time* (1951). Reprinted in "*Nature and the Greeks" and "Science and Humanism"* (1996), 113.

And again, we come to the main driving force of scientific knowledge: scientific curiosity and ability to be fascinated. Although this can be stimulated in the process of education, it is perhaps also an individual given.

At the interface between science and the outside world (funding, for instance), there can be an erosion of scientific goals and the quality of the scientific product. Even more challenging is the coexistence between science and policy. It opens up a wide field for opportunism on the part of scientists, for fitting into the political agenda. Here, a major contradiction arises. At one point, politicians need the objective opinion of scientists, say, on how to tackle climate change, and it turns out that some scientists have already been "educated" to fit the politicians' opinions. Perhaps, some inadequacies of the response of politicians and political scientists to the global pandemic of COVID-19 are one illustration of the result of the blurred line between politics and science. Isn't it better for everyone to know their place: for scientists to work solely to uphold scientific truth and for politicians to be responsible for policies based on scientific evidence, not influenced by outside interference and distortion?

2 Quo Vadis?

The field of action for climate change deniers is steadily narrowing. The inexorable degradation of the planet is evident to anyone unbiased. According to the International Union for Conservation of Nature (IUCN), 467 animal species have become extinct in the last 10 years alone. Nearly 10% of marine species are found to be at risk of extinction. When asked how many years it would take nature to create 300 new species of mammals—as many as have gone extinct in the time humans have been on Earth, the answer is 3–7 million years. Will mankind outlive that time? We are only talking about mammals here. According to the United Nations, about one million species are at risk of extinction. We are currently experiencing the largest mass extinction since the end of the dinosaurs. However, human population is growing. In 1950, it was 2.5 billion people. Today, we are about 8 billion. Such a population increase has never happened before. Naturally, this cannot happen without increasing food production (Costello et al., 2020), without urbanizing the countryside, without absorbing air and sea space. It has come to the point where the atmosphere and the ocean have become like highways on which thousands of aircrafts and high-tonnage ships cruise. Do we care how this affects the environment? Living standards are rising across the board, but at what cost? The British clergyman Thomas Malthus taught in his 1798 Essay on the Laws of Population that unless it is curbed, humankind is heading for demographic catastrophe, because the population is increasing geometrically, while the means of feeding it are increasing arithmetically. It cannot be argued that such a late eighteenth-century theory is universally applicable, but the issue of the finiteness of the Earth's resources cannot be sidestepped.[4]

The time has come to ask ourselves how to better manage the planet. We may even be too late in asking this question. Humans have become so powerful that they are significantly influencing nature; unfortunately, the changes are not always in the right direction. We may have the resources to reverse the trends, but is there a global consensus on what needs to be done and how urgently? For almost 20 years, United Nations Member States have been negotiating to agree on a legal framework to protect parts of the ocean beyond national borders, known as the *High Seas Treaty*. This is an instrument of the UNCLOS, adopted by an intergovernmental conference at the UN on 4 March 2023. The treaty will provide a legal framework for establishing marine protected areas (MPAs) over 30 percent of the world's oceans to protect against the loss of wildlife and share out the genetic resources. I wonder how long it will take to implement the *High Seas Treaty*.

The enormous increase in energy consumption somehow went unnoticed by the public. Today, people use more than five times as much energy as they did in 1950, with a threefold increase in population over that period. To me, the most striking here is that what has been created over millions of years, humanity can consume over several generations. For the first time in Earth's history, a single species (Homo sapiens) is using colossal amounts of energy, leading to drastic changes in the atmosphere, oceans, and solid ground. It looks strange that some changes brought about by Homo sapiens (literally meaning wise man) proceed in an unwise way. Many people do not understand the magnitude and significance of what is happening. We do not worry about what is happening to nature until we notice certain symptoms that trigger, for example, migration waves and subsequent reactions in politics and the mass media. It seems that the question of humanity's future is far from meaningless.

According to a WMO report, over the past 50 years, between 1970 and 2019, more than 11,000 weather, climate, and water-related disasters have been recorded, or almost one a day. The death toll is two million, or 115 per day. The number of disasters has increased fivefold in the last 50 years, and the associated economic costs have risen sharply. This trend is not expected to change. However, there is something positive, and that is, that the number of victims has decreased almost three times. Better forecasting and more coordinated disaster management planning have contributed to this.

In his March 22, 2022, appeal on the World Meteorology Day, UN Secretary-General, António Guterres, states: "We must invest equally in adaptation and

---

[4] A rough estimate for the twentieth century shows a population increase of about 5 billion in a century, while for the whole history of mankind up to 1800, the population of the Earth could not reach 1 billion. Such explosive changes in physical systems are often associated with enormous perturbations, instabilities, or catastrophic developments. Hopefully, "softer feedbacks" are at work in socio-economic and ecological systems that prevent catastrophic developments.

resilience. That includes the information that allows us to anticipate storms, heatwaves, floods and droughts".[5] His Special Representative for Disaster Risk Reduction, Japan's Mami Mizutori, recalls[6] that in 1736, the famous statesman and inventor Benjamin Franklin advised the fire-threatened people of Philadelphia that "An ounce of prevention is worth a pound of cure". She continues: "These famous words are as relevant today as they were hundreds of years ago. They speak to the power of early warning and early action…" Unfortunately, one-third of the world's population, mainly in the least developed countries and small island developing states, is still not covered by early warning systems. "This is unacceptable, particularly with climate impacts sure to get even worse", claims António Guterres. "Early warnings and action save lives. To that end, today I announce the United Nations will spearhead new action to ensure every person on Earth is protected by early warning systems within five years".

What I presented in this book provides only a glimpse of some of the tools needed to build strategies to avoid climate catastrophe. Their judicious use, together with the many other tools available, for example, from the field of geoengineering, can save us from falling into a completely chaotic and unpredictable future.

Stopping the burning of fossil fuels is only a first, albeit very difficult step in combating climate change. In addition, how will food production be maintained or increased, given that the fertility of nearly 75% of the plants we eat will suffer a decline if bees disappear? On the other hand, without using chemicals (many of them bee-killing), production would be reduced. The contradictions are enormous. How to reconcile increasing consumption with the Earth's limited resources? Should population growth be controlled and limited? Are limiting consumption and increasing production compatible? Was not the growth of production one of the main conditions for the development of capitalism? Moreover, what constraints does environmental and climate protection impose on future economic and socio-political systems?

I would remind you again of the motto of the Decade of Ocean Science for Sustainable Development: "The ocean we need for the future we want". A consensus is emerging that the ocean we want must be clean, healthy, and resilient to impacts, predictable, and safe. Making the necessary changes will be difficult because they will harm many interests. In addition to science, technology, and vision, we need a transformation in the way people think. We need global awareness and enlightenment. The challenge is enormous.

The pandemic COVID-19 has given yet another signal of the global problems and the difficulties of humanity to cope with them. Many scientists are adamant that the deforestation and the destruction of many wildlife habitats are helping to spread diseases from animals to humans. Disturbed biodiversity may also be increasing the possibility of reaching risky situations. Intensive air travel can carry viruses around the world in just a few hours. Some believe that the pandemic is one illustration of the breakdown of the human–animal barrier. Everybody wants a return to "normal

---

5 https://media.un.org/en/asset/k1n/k1neyrj398. Last accessed on 03.02.2023.

6 https://www.undrr.org/news/message-mami-mizutori-world-meteorological-day-2022. Last accessed on 03.02.2023.

life", but is it not time to consider what exactly it is that we should return to? Here, the active intervention of science, I suggest, will not be superfluous. It should be remembered again that public confidence is highest in medical doctors, followed by scientists, with politicians far behind. I believe that the latter would improve their ratings if they listened more to what scientists have to say. While concerning the climate, politicians have long slowed the "ball"; the pandemic required them to react more quickly. It is not only about how or whether politicians will improve their ratings. It is about something much deeper, namely about fundamental changes to some paradigms that have been in place for most of human history. Wendell Berry, an American novelist, poet, and environmental activist, declared: "Whether we and our politicians know it or not, Nature is party to all our deals and decisions, and she has more votes, a longer memory, and a sterner sense of justice than we do".[7]

3 The Information Society—Will Numbers Drown Man?

We live in an information society without a good understanding of where it is leading. However, did anyone, at the time know where Gutenberg's discovery of printing, which today is thought to have catalyzed the transition from feudalism to capitalism, was leading? What will be the socio-political consequences of the information society? More than 50 years ago, when I turned to the physics of the Earth, I had not the slightest idea how vital this science would become for humanity. When I chose to pursue numerical modeling, no one was talking about digitalization. Without having planned for it, it turned out that my work took place in one of the most information-saturated fields. One of the pillars of the development of the Decade of Ocean Science for Sustainable Development activities is the "predicted ocean". The European Commission has recently led the initiative to develop a digital, multidimensional, and near real-time high-resolution representation of the world's ocean, integrating data infrastructures and modeling and forecasting services and using IT. Now, the development of digital twins of the ocean[8] is under way. On these digital replica, we will be able to test, for example, the impact of climate and disaster scenarios on the climate system, as well as the efficacy of mitigation and adaptation plans. The conflict and union of man and numbers is another major challenge. I hope the numbers do not drown man in the incoming ocean of the New World[9]. I do not want the beautiful ocean we know to remain a distant memory, displaced by the emerging digital twins.

[7] The quote is part of an endorsement statement for the book by Little (1997).

[8] https://digitaltwinocean.mercator-ocean.eu/. Last accessed on 03.02.2023.

[9] In this book, I have addressed some issues related to the positive outcome of using machine learning in oceanography. However, I do not want to reflect on the broader issues of artificial intelligence in the afterword because this is a very wide and controvercial problem, which would need a more in-depth and extensive analysis.

In this book, I hope to have convinced you that numerical modeling and observational technologies are at the heart of this future development of science. More generally, I wonder whether numerical modeling does not lie at the heart of the information society. When we say that we are beginning to live in an information society, does it not follow that the weight of its various driving forces must shift in the direction of scientists and scholarship? Perhaps, the time has come for them to lead the world to better and safer "shores".

I will conclude by quoting I. Newton: "I do not know what I may appear to the world, but to myself I seem to have been only like a boy playing on the seashore, and diverting myself in now and then finding a smoother pebble or a prettier shell than ordinary, while the great ocean of truth lay all undiscovered before me".

References

Baumberg JJ (2018) The secret life of science. Princeton University Press, Princeton, p 248

Costello C, Cao L, Gelcich S et al (2020) The future of food from the sea. Nature 588:95–100. https://doi.org/10.1038/s41586-020-2616-y

Little CE (1997) The dying of the trees. Penguin Books, 288 pp. ISBN: 10: 0140158723/ISBN 13: 9780140158724

Mermin N D (2009) It's about time: understanding Einstein's relativity. Course Book ed. Princeton: Princeton University Press. Project MUSE, 208 p

Index

E. Stanev, *Trajectories in Oceanography*, Springer Oceanography,
https://doi.org/10.1007/978-3-031-33720-8